Recovering Bioactive Compounds from Agricultural Wastes

Recovering Bioactive Compounds from Agricultural Wastes

Edited by Van Tang Nguyen

University of Newcastle, Australia
Nha Trang University, Vietnam

WILEY

This edition first published 2017

Registered Offices
John Wiley & Sons, Inc., 111 River Street, Hoboken, NJ 07030, USA
John Wiley & Sons Ltd, The Atrium, Southern Gate, Chichester, West Sussex, PO19 8SQ, UK

Editorial Office
The Atrium, Southern Gate, Chichester, West Sussex, PO19 8SQ, UK

For details of our global editorial offices, customer services, and more information about Wiley products visit us at www.wiley.com.

Wiley also publishes its books in a variety of electronic formats and by print-on-demand. Some content that appears in standard print versions of this book may not be available in other formats.

Library of Congress Cataloging-in-Publication Data

Names: Nguyen, Van Tang, 1979- editor.
Title: Recovering bioactive compounds from agricultural wastes / edited by Van Tang Nguyen.
Description: Hoboken, NJ : John Wiley & Sons, 2017. | Includes bibliographical references and index. |
Identifiers: LCCN 2017009461 (print) | LCCN 2017018525 (ebook) | ISBN 9781119168836 (ePDF) | ISBN 9781119168843 (ePub) | ISBN 9781119168829 (cloth)
Subjects: LCSH: Bioactive compounds. | Agricultural wastes–Recycling.
Classification: LCC QP517.B44 (ebook) | LCC QP517.B44 R43 2017 (print) | DDC 363.72/88–dc23
LC record available at https://lccn.loc.gov/2017009461

Cover Design: Wiley
Cover Images: Courtesy of Van Tang Nguyen

Set in 10/12pt, WarnockPro by SPi Global, Chennai, India

10 9 8 7 6 5 4 3 2 1

Contents

This book is dedicated to my dad Van Tac Nguyen, my mom Thi Thuy Duong, my wife Thi Le Nguyen, my son Trong Nhan Nguyen and my daughter Dan Thanh Nguyen.

List of Contributors

Md. Ariful Alam
Department of Pharmaceutical Technology, Faculty of Pharmacy, International Islamic University Malaysia (IIUM), Pahang, Malaysia

Fiorella P. Cárdenas-Toro
Department of Engineering, Section of Industrial Engineering, Pontifical Catholic University of Peru, San Miguel, Lima, Peru

Sahena Ferdosh
Department of Plant Science, Faculty of Science, International Islamic University Malaysia (IIUM), Kuantan Campus, Pahang, Malaysia

Kashif Ghafoor
Department of Food Science and Nutrition, King Saud University, Riyadh, Saudi Arabia

Rukshana Akter Happy
Department of Biochemistry and Biotechnology, Faculty of Basic Medical and Pharmaceutical Sciences, University of Science and Technology Chittagong (USTC), Foy's Lake, Chittagong, Bangladesh

Thang Trung Khong
Nha Trang University, Nha Trang, Khanh Hoa, Vietnam

Van Tang Nguyen
School of Environmental and Life Sciences, Faculty of Science and Information Technology, University of Newcastle, Ourimbah, NSW, Australia

and

Department of Food Technology, Faculty of Food Technology, Nha Trang University, Nha Trang, Khanh Hoa, Vietnam

Hong Ngoc Thuy Pham
School of Environmental and Life Sciences, Faculty of Science and Information Technology, University of Newcastle, Ourimbah, NSW, Australia

and

Department of Postharvest Technology, Faculty of Food Technology, Nha Trang University, Nha Trang, Khanh Hoa, Vietnam

Md. Zaidul Islam Sarker
Department of Pharmaceutical Technology, Faculty of Pharmacy, International Islamic University Malaysia (IIUM), Pahang, Malaysia

Nguyen Thi Thao
Department of Quality Management, School of Biotechnology and Food Technology, Hanoi University of Science and Technology, Hanoi, Vietnam

Hoang Quoc Tuan
Department of Quality Management, School of Biotechnology and Food Technology, Hanoi University of Science and Technology, Hanoi, Vietnam

Giovani L. Zabot
Federal University of Santa Maria (UFSM), Cachoeira do Sul, RS, Brazil

About the Editor

Van Tang Nguyen was born in Hai Duong province, Vietnam. He obtained an Engineer degree in Food Technology from the Hanoi University of Science and Technology, Vietnam. He then received a Master's degree in Food Science from the National Taiwan Ocean University, Taiwan and a PhD in Food Science from the University of Newcastle, Australia. He has worked as a Demonstrator in Food Science and Human Nutrition at the School of Environmental and Life Sciences, Faculty of Science and Information Technology, University of Newcastle, Australia. He also works as a Lecturer in Food Technology at the Department of Food Technology, Faculty of Food Technology, Nha Trang University, Vietnam. His research has focused on natural bioactive compounds, pharmacological activity, value-added products and functional foods. His current expertise is in the extraction, isolation and quantification of bioactive compounds from natural materials and the determination of biological activity *in vitro* and *in vivo*.

Van Tang Nguyen has been interested in the use of agricultural residues/byproducts from the processing and production of food products for the production of value-added products for over a decade. He has published more than 25 research papers in peer-reviewed journals (*Food Science, Chemical Papers, Drying Technology, Chemistry and Biodiversity, Industrial Crops and Products, Food Processing and Preservation,* etc.) and has authored/edited five book chapters and five books in the field of Food Science and Technology. He has presented over 15 scientific reports at international conferences and served as the reviewer for many reputed journals (*Food Science, Food Biochemistry, Pharmaceutical Biology, Industrial Crops and Products, Current*

Pharmaceutical Research, and so on). He is currently Editor-in-Chief and Founder of Bioactive Research and a member of the Institute of Food Technologist (USA) and Pancreatic Cancer Research Group (Australia).

Van Tang Nguyen is married to Thi Le Nguyen. They have two children, Trong Nhan Nguyen and Dan Thanh Nguyen.

Preface

Agriculture is regarded as one of the most important fields of human industry, due to its role in ensuring global food security for over 7 billion people around the world and supporting other industries. Agricultural production creates a great amount of residues/byproducts, which are considered 'wastes'. Interestingly, agricultural wastes contain many valuable bioactive compounds, possessing a wide range of potential pharmacological properties, which have great contributions to make in related industries, such as nutraceuticals/functional foods, medicines, pharmaceuticals and cosmetics. However, they are still underutilised as abundant, inexpensive, renewable and sustainable sources of natural bioactive compounds.

In order to increase the value of agricultural production, reduce pollution risks and promote the development of related industries, we have prepared *Recovering Bioactive Compounds from Agricultural Wastes* to introduce the potential of agricultural wastes obtaining from the different sectors of agricultural production, such as tea, coffee, cacao, cashew, fruit and vegetable, wine, edible oil, starch and sugar, and to present, discuss and recommend various techniques for the extraction, isolation, purification and application of these bioactive compounds in different fields. We also discuss the economic and market analysis of agricultural wastes and bioactive compounds derived from these sources, based on a number of actual recovery processes to be established at pilot and industrial scales. Hopefully, this book will be a helpful reference for researchers, producers and traders in agricultural production and related industries.

Van Tang Nguyen, Editor
University of Newcastle, Australia
Nha Trang University, Vietnam

Acknowledgements

I would like to express my special thanks to John Wiley & Sons for their suggestions and effective cooperation in the preparation and completion of the manuscript of this book. I also sincerely acknowledge the reviewers for their comments on the proposal. I kindly thank all the authors for their valuable contribution to the chapters; their effort is highly appreciated and adequately recorded in this book. Last but not least, I would like to thank my colleagues, friends and family for their support, encouragement and special interest in the preparation and publication of this book.

1

Potential, Uses and Future Perspectives of Agricultural Wastes

Van Tang Nguyen

School of Environmental and Life Sciences, Faculty of Science and Information Technology, University of Newcastle, Ourimbah, NSW, Australia
Department of Food Technology, Faculty of Food Technology, Nha Trang University, Nha Trang, Khanh Hoa, Vietnam

1.1 Introduction

Agriculture has a developmental history going back thousands of years and is considered one of the most important fields of human knowledge because of its special role in ensuring global food security for over 7 billion people around the world. It also has an important role in supporting and promoting the development of other industries, such as nutraceuticals, medicines, pharmaceuticals and cosmetics. In particular, agriculture produces a large amount of wastes, containing a significant quantity of valuable bioactive compounds, such as polyphenols, phenolic acids, flavonols, flavanols, flavonoids, procyanidins, proanthocyanidins, anthocyanins, glycosides, carotenoids, saponins, tannins, alkaloids, sterols, steroids, triterpenes, quinones, peptides and carbohydrates, which have been proved to possess a variety of biological activities, including antioxidant, antibacterial, antifungal, antiviral, antimicrobial, antidiabetic, anticancer, antidiarrhoeal, antihypertensive, antimutagenic, anti-inflammatory, anticholesterol and anticardiovascular properties (Figure 1.1) (Balasundram et al., 2006; Santana-Méridas et al., 2012). However, the utilisation of agricultural wastes as an abundant, biorenewable and low-cost source for the production of high value-added products is still under investigation, with limited outcomes. Therefore, research is needed into the application of environmentally friendly traditional and advanced techniques with low production costs in the extraction, isolation and purification of phytochemical compounds from agricultural wastes in high yields and at maximal quality. This strategy will increase the value of agricultural wastes and reduce pollution risks for the environment in both the short and the long term, and will enable sustainable development, one of the most important goals of modern global agricultural production.

1.2 Potential of Agricultural Wastes

According to the Food and Agriculture Organization of the United Nations (FAOSTAT, 2015), the total harvested crop area worldwide in 2013 was about 4.36 billion ha, producing approximately 21.70 billion tonnes, with a total gross production value of US$24

Recovering Bioactive Compounds from Agricultural Wastes, First Edition. Edited by Van Tang Nguyen.

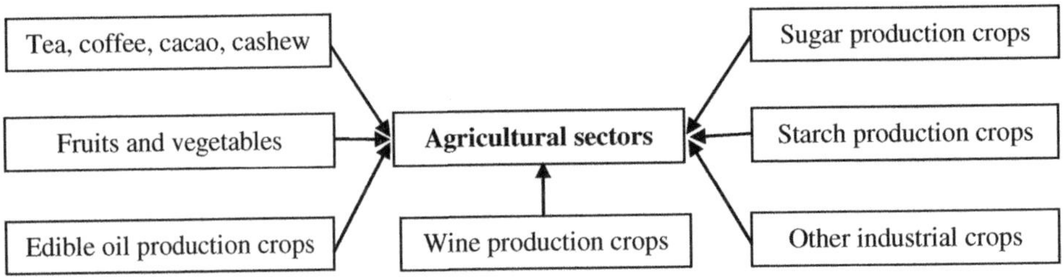

Figure 1.1 Major agricultural sectors in the recovery of bioactive compounds.

Table 1.1 Total harvested crop area, total production and total gross agricultural production value in 2013. *Source:* FAOSTAT (2015).

Location	Total harvested area (billion ha)	Total production (billion tonnes)	Total gross production value (billion US$)
Africa	0.71	1.90	1295.47
Americas	0.89	5.22	3503.55
Asia	1.82	9.94	15543.06
Europe	0.86	4.43	4350.89
Oceania	0.08	0.21	239.35
Total	4.36	21.70	24932.32

932.32 billion (Table 1.1). In Asia, the total harvested crops area was 1.82 billion ha (41.81% of total), producing 9.94 billion tonnes (45.80%), with a total gross production value of US$15 543.06 billion (62.34%).

Sugar cane had the highest production globally (1910 million tonnes), followed by maize, rice, wheat, potatoes, fresh vegetable, cassava, soybeans, palm fruit and sugar beet (1020, 741, 716, 376, 280, 277, 276, 266 and 247 million tonnes, respectively) (Table 1.2). All of these crops produced a large amount of relevant wastes, including leaves, tops and bagasse from sugar cane; straw, stalks, husk, bran and cobs from maize, rice and wheat; foliage, tops, peels and pulps from potatoes; leaves, stems, peels, skins and seeds from fresh vegetables; and peels, stalks and bagasse from cassava (FAOSTAT, 2015).

The wastes from crop-based residues, in terms of aerial biomass, roots, leaves, straw and stems, are rich sources of bioactive compounds, including polyphenols, flavonoids, sterols, anthocyanins and carbohydrates, which have direct links with potent pharmacological properties, such as antioxidant, antibacterial, antifungal, antimicrobial, anti-inflammatory and anticholesterol capacity (Table 1.3). Many valuable bioactive compounds, such as glycosides, procyanidins, proanthocyanidins, flavonols, flavanols, flavonoids, phenolic acids, carotenoids, saponins, tannins, alkaloids, steroids, triterpenes, quinones and peptides, can also be isolated from processing-based residues, such as from the fresh fruit, dry fruit, brewing, wine, cereal, oil, essential oil, sugarcane and tobacco industries (Table 1.4). The bioactive compounds isolated from these sources have been proved to possess a wide range of biological activities, including antioxidant, anticancer, antidiarrhoeal, antibacterial, antifungal, antiviral,

Table 1.2 Total harvested area, total production and total gross production value of major crops in 2013, as well as their main residues/wastes. *Source:* FAOSTAT (2015).

Crop	Total harvested area (million ha)	Total production (million tonnes)	Total gross production value (billion US$)	Main residues/wastes
Tea, coffee, cacao and cashew				
Tea	3.52	5.35	22.22	Fruit, flower, old leaves, dust, stalk and fibre
Coffee	10.14	8.92	16.37	Outer skin, pulp/mucilage, parchment/hull/husk, silver skin and spent coffee grounds
Cacao	10.01	4.59	6.95	Leaves, shell, husk, pulp/mucilage and hull
Cashew	5.46	4.44	2.82	cashew apple, outer shell, inner skin and nut shell
Fruits and vegetables				
Fruit fresh nes	5.01	33.52	13.39	Peels, pulps, seeds, stalks and skins
Vegetable fresh nes	19.79	280	135.42	Leaves, stems, peels, skins and seeds
Wine production crops				
Grapes	7.16	77.18	78.50	Pomace (seeds, skins), leaves, stalks, rachis and lees
Edible oil production crops				
Palm fruit	18.05	266	31.34	Shells, husks and fronts
Olives	10.31	20.40	19.47	Leaves and stalks
Coconuts	12.07	62.45	10.47	Shells, husks and fronts
Soybeans	112	276	131.26	Straw and pods
Sunflower seed	25.45	44.55	26.08	Foliage and stems
Rapeseed	36.50	72.70	53.13	Straw
Cotton seed	n/r	47.07	9.75	Stalks
Sugar production crops				
Beet	4.37	247	14.04	Roots, pulps and scums
Cane	26.94	1910	108.55	Leaves, tops and bagasse
Others	0.12	0.93	nr	Pulps and bagasse
Starch production crops				
Rice	165	741	429.27	Straw, husk and bran
Wheat	219	716	242.25	Straw
Maize	185	1020	382.34	Straw, stalks and cobs
Potatoes	19.34	376	149.51	Foliage, tops, peels and pulps
Cassava	20.39	277	47.31	Peels, stalks and bagasse
Barley	49.15	144	36.30	Straw

(Continued)

Table 1.2 (Continued)

Crop	Total harvested area (million ha)	Total production (million tonnes)	Total gross production value (billion US$)	Main residues/wastes
Other crops				
Beans, dry	29.05	22.81	16.06	Straw and pods
Beans, green	1.54	21.37	40.53	Straw and pods
Pepper	0.48	0.47	2.88	Leaves and stems
Seed cotton	32.17	73.05	82.62	Stalk

n/r, not reported

Table 1.3 High added-value products from crop-based residues. *Source:* Santana-Méridas et al. (2012). Reproduced with permission of Springer.

Activity	Species	Waste type	Bioactive compounds	Applications
Horticultural production				
Melon	*Cucumis melo*	Aerial biomass	Xanthan	Rheology modifier, food additive
Broccoli	*Brassica oleracea*	Aerial biomass	Glucosinolates, phenolic acids, flavonoids, vitamin C	Antioxidant
Carrot	*Daucus carota*	Roots	Hydroxycinnamic acid, anthocyanins	Antioxidant
Spinach	*Spinacea oleracea*	Leaves	Flavonoids	Antioxidant
Pepper	*Capsicum annuum*	Leaves, stems	Capsaicin, dihydrocapsaicin	Antioxidant, anti-inflammatory
Cucumber	*Cucumis sativus*	Leaves	Isovitexin, saponarin, vicenin-2, apigenin	Antioxidant
Tomato	*Lycopersicum sculentum*	Leaves	Solanesol	Antibacterial, anti-inflammatory
Cereal production				
Wheat	*Triticum* sp.	Straw	Xylose, polyphenols	Food ingredient
Others	sp.	Straw	Lignin	Value-added products
Tuber production				
Potato	*Solanum tuberosum*	Leaves	Solanesol	Antibacterial, anti-inflammatory

Table 1.3 (Continued)

Activity	Species	Waste type	Bioactive compounds	Applications
Fruit production				
Ginja cherry	*Prunus cerasus*	Stems, leaves	Polyphenols (catechin > 70%)	Antioxidant, antimicrobial
Pineapple	*Ananus comosus*	Straw (leaves)	Fibre	Polymer reinforcement
Grass production				
Ryegrass	*Lolium perenne*	Grass chaff	β-adenosine	Mushroom production
Miscanthus	*Miscanthus* × giganteus	Biomass	Lignin, phenols, sterols	Fuel, antioxidant, anticholesterol
Oil production				
Olive	*Olea europaea*	Leaves	Polyphenols	Antimicrobial, antioxidant
Medicinal and condimentary herbs				
Creosote bush	*Larrea tridentate*	Leaves	Poly and monomeric phenols	Antifungal
Saffron	*Crocus sativus*	Leaves	Kaempferol, orientin, vitexin	Antioxidant

Table 1.4 High added-value products from processing-based residues. *Source:* Santana-Méridas et al. (2012). Reproduced with permission of Springer.

Activity/crops	Species	Waste type	Bioactive compounds	Activity/applications
Fresh fruit industry				
Mango	*Mangifera indica*	Peels, pits/seeds	Tannins, vanillin, mangiferin	Antioxidant
Apple	*Malus domestica*	Pomace (peels, core, seeds, calyces, stems)	Pectin, catechins, hydroxycinnamates, phloretin glycosides, quercetin glycosides, procyanidins	Antioxidant
Watermelon	*Citrullus lanatus*	Rinds, flesh	Lycopene, citrulline, phenolic compounds	Antioxidant, food additives
Rambutan	*Nephelium lappaceum*	Peels	Ellagitannins	Antioxidant
Mangosteen	*Garcinia mangostana*	Pericarps	Proanthocyanidins	Antioxidant

(Continued)

Table 1.4 (Continued)

Activity/crops	Species	Waste type	Bioactive compounds	Activity/applications
Guajava	*Psidium guajava*	Bagasse	Epicatechin, quercetin, syringic acid, mirycetin	Antimicrobial
Banana	*Musa sapientum*	Dried leaves, pseudostems	Sugars	Fermentation
		Peels	α-amilasa, laccasa, citric acid	Enzyme production
Lemon	*Citrus limon*	Peels	Flavanoids, saponins, tannins, alkaloids, steroids, triterpenes	Antimicrobial
			Essential oil	Nematostatic activity
			Limonene	Insecticidal (larvicidal)
Orange	*Citrus sinensis*	Peels	Citric acid	Additive, detergent, cosmetic
			Essential oil	Nematostatic activity
Pineapple	*Ananus comosus*	Peels, core, crowns, stems	Bromelain	Food and textile industries, anti-inflammatory, anti-diarrhea, digestive
Pomegranate	*Punica granatum*	Husks	Poly- and monomeric phenols	Antifungal
Grapefruit	*Citrus paradisi*	Peels	Essential oil	Nematostatic activity
Mandarin	*Citrus reticulata*	Peels	Phenolic compounds	Antioxidant, antimicrobial
Papaya	*Carica papaya*	Peels, seeds	Phenolic compounds	Antioxidant, antimicrobial
Bergamot	*Citrus bergamia*	White tissues	Brutieridin, melitidin	Anticholesterolaemic
		Seeds	Limonoids	Antiviral
Satsuma mandarin	*Citrus unshiu*	Peels	Hesperidin, narirutin, quercetagetin	Antioxidant
Citrus fruits	*Citrus* sp.	Seeds, molasses	Limonoids	Anticancer
		Peels	Flavonoids (hesperidin, diosmin, narirutin, didymin, sinesetin), carotenoids (violaxanthin, β-crytoxanthin, β-carotene), vitamin C, essential oils (limonene), minerals	Antioxidant
Horticultural industry				
Artichoke	*Cynara scolymus*	Bracts, receptacles, stems, juice, heads	Neochlorogenic acid, chlorogenic acid, caffeoylquinic acids	Antioxidant

Table 1.4 (Continued)

Activity/crops	Species	Waste type	Bioactive compounds	Activity/applications
Beet	*Beta vulgaris*	Stalks	Azelaic acid	Antimicrobial
Onion	*Allium cepa*	Fresh peeling	Condensed tannins, flavonoids, quercetin aglycone	Antioxidant, textile dyes
		Skins, top–bottom wastes, scales, discarded onions	Flavonoids, fructans and alk(en)yl cystein sulphoxides, quercetin aglycone, minerals	Antioxidant, dietary fibre
Tomato	*Solanum lycopersicum*	Seeds, pulps, skins	Lycopene, β-carotene, sterols, tocopherols, terpenes, glycoalkaloids	Antioxidant, anticholesterol
Coffee industry				
Coffee	*Coffea Arabica*	Spent coffee grounds	Caffeine, chlorogenic acid	Allelopathy
Dry fruit industry				
Peanut	*Arachis hypogaea*	Skins, seed coats	Polyphenols oligomeric proanthocyanidins, indole alkaloids, phenolic acids	Antioxidant anticancer, Blood vessels protector, Antimicrobial
Almond	*Prunus dulcis*	Hulls	Triterpenes (olcanoic, ursolic, betulinic acids), daucosterol	Anticancer
Hazelnut	*Corylus avellana*	Skins, hard shells, leafy covers	Phenolic acids (gallic, caffeic, *p*-coumaric, ferulic, sinapic)	Antioxidant
Chestnut	*Castanea sativa*	Shells (outer, inner)	Tannins, polyphenols, tocopherols	Antioxidant
Walnut	*Juglans regia*	Shells	Holocellulose, α-cellulose, lignin	Panel manufacture
Pecan nut	*Carya illinoensis*	Shells (endocarp)	Poly and monomeric phenols	Antifungal
Pistachio	*Pistachia vera*	Hulls	Phenolic compounds	Antioxidant
Legume industry				
Pea	*Pisum sativum*	Husks	Growth factors (nitrogen and carbon)	Carrier for rhizobial inoculants
		Pods	Fibre, polyphenols	Antioxidant, texturing additive
Faba bean	*Vicia faba*	Pods	Fibre, polyphenols	Antioxidant, texturing additive
		Off-quality grains	Protein	Food ingredient

(Continued)

Table 1.4 (Continued)

Activity/crops	Species	Waste type	Bioactive compounds	Activity/applications
Chickpea	*Cicer arietinum*	Off-quality grains	Peptides	Metal chelating, antihypertensive, food ingredient
Common bean	*Phaseolus vulgaris*	Off-quality grains	Peptides	Antioxidant, metal chelating
Cereal industry				
Rice	*Oryza sativa*	Rice brand	Tocotrienols-tocopherols, γ-oryzanol, β-sitosterol	Anticholesterol
Wheat	*Triticum durum*	Wheat brand	Vitamin E, carotenoids, quinines	Nutrients, antioxidant
Wine industry				
Grape	*Vitis vinifera*	Stems	Betulinic acid, stilbenoid *trans*-resveratrol, *trans*-3-viniferin, sitosterol 6′-*O*-acylglucosides	Anticancer
		Pomace	Flavonols, flavonols glucosides, flavanols, gallate esters, anthocyanins, proanthocyanins	Antioxidant, antimicrobial
		Seeds	Epicatechin, caffeic and gallic acids	Antimicrobial
Brewing industry				
Barley	*Hordeum vulgare*	Spent grains	Xylitol, cellulose, hemicelluloses, lignin, xylose, glucose, arabinose, protein, ferulic and *p*-coumaric acids	Growth medium, lactic acid production, diabetes treatment (xylitol)
Oil industry				
Olive	*Olea europaea*	Olive mill wastewaters 'alpechin'	Hydroxytyrosol, gallic acid, oleuropein, ligstroside isomers and derivatives, squalene, tocopherols, triterpenes, soluble sugars, polyphenols	Antimicrobial, antioxidant, anti-inflammatory, textile dyes
		Pomace (solid wastes)	Hydroxytyrosol, tyrosol, caffeic protocatechuic, vanillic, p-coumaric and syringic acids, vanillin, oleuropein, apigenin	Antioxidant

Table 1.4 (Continued)

Activity/crops	Species	Waste type	Bioactive compounds	Activity/applications
		Flesh, stones, seeds	Polyphenols, tocopherol	Antioxidants
Rapeseed	*Brassica napus*	Meals	Gallic and syringic acids, kaempferol, naringenin	Antioxidant, anti-inflammatory
		Defatted cakes	Glucosinolates, peptides	Herbicide, food additive, foaming, emulsifying, anthitrombotic, antiviral
Turnip	*Brassica rapa*	Hulls, defatted cakes	Polyphenols	Antioxidant
Cotton	*Gossypium hirsutum*	Meals	Kaempferol, naringenin, rutin	Antioxidant, anti-inflammatory
Peanut	*Arachis hypogeal*	Defatted flours	Protein	Packaging biomaterials
Soybean	*Glycine max*	Meals	Caffeic acid, naringenin, daidzein	Antioxidant, anti-inflammatory
Sunflower	*Helianthus annuus*	Defatted cakes	Protein concentrates and isolates	Food ingredient
			Peptides	Antihypertensive, anticholesterol
Essential oil industry (residues after distillation)				
Lavandin	*Lavandula × intermedia*	Solid residues	Phenolic acids, flavonoids, hydroxycinnamoylquinic acid derivatives, glucosides of hydroxycinnamic acids	Antioxidant
Rosemary	*Rosmarinus officinalis*	Solid residues	Phenolic acids (rosmarinic, carnosic, caffeic, chlorogenic acid and *p*-coumaric acids)	Antioxidant
Sage	*Salvia officinalis, S. glutinosa*	Solid residues	Coumarins, hydrocarbons, monoterpenes, phenolic compounds, sesquiterpenes, diterpenes, triterpenes, fatty acid ester, hydroxycinnamic acid, luteolin	Antimicrobial, antioxidant, textile dyes
Thyme	*Thymus vulgaris*	Solid residues	Rosmarinic acid, hydroxycinnamic acid, luteolin	Antioxidant, textile dyes

(Continued)

Table 1.4 (Continued)

Activity/crops	Species	Waste type	Bioactive compounds	Activity/applications
Bay laurel	*Laurus nobilis*	Hydrolates	Terpinen-4-ol, α-terpineol, phenol	Antioxidant
Malt industry				
Malt		Malt sprouts	Growth factors (nitrogen and carbon)	Carrier for rhizobial inoculant
Tuber processing industry				
Potato	*Solanum tuberosum*	Peels	Proteins	Produce yeast biomass
		Peels	Glycoalkaloids	Pesticide, anticancer
Wood industry				
Pine	*Pinus* sp.	Vinegars	Aldehydes, ketones, acids, esters, phenols (cresols)	Repellent, insecticide, herbicide,
		Bark	Polyphenols	Antioxidant, anti-inflammatory
Other industrial crops				
Sisal	*Agave sisalana*	Waste liquids (juice)	Saponins, glycosides, terpenoids, tannins	Antimicrobial
Cork oak	*Quercus suber*	Black waxes	Friedelane triterpenes (friedelin, 3-hydroxyfriedel-3-en-2-*one*), β-sitosterol, campesterol, α-amyrin, sitost-4-en-3-*one*	Antiinflammatory, antibacterial, antifungal, antiviral, cytotoxic, insecticidal
Cassava	*Manihot esculenta*	Bagasse	Starch	Carbon source for microbial growth
Sugarcane industry				
Sugarcane	*Saccharum officinarum*	Bagasse	α-cellulose, pentosans	Enzyme production
Tobacco industry				
Tobacco	*Nicotiana tabacum*	Stems, leaves	Mono-caffeoylquinic acids	Antibacterial, antioxidant, anti-hyperglycemic antimutagenic

antimicrobial, antihypertensive, antimutagenic, anti-inflammatory and anticholesterol properties (Santana-Méridas et al., 2012).

Phenolic compounds, found ubiquitously in plants, are an essential part of the human diet. They are also of considerable interest due to their antioxidant properties. The phenolic compounds possess an aromatic ring bearing one or more hydroxyl groups and their structures range from that of a simple phenolic molecule to that of a complex high-molecular weight polymer. The chemical structures of hydroxybenzoic

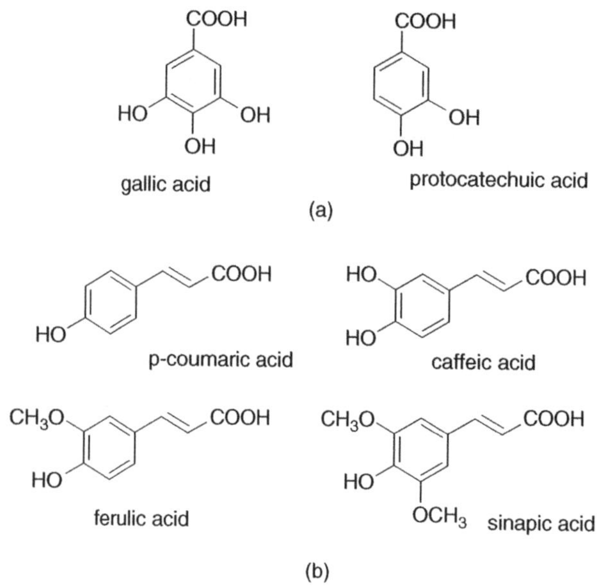

Figure 1.2 Chemical structures of (a) hydroxybenzoic and (b) hydroxycinnamic acids. *Source:* Balasundram (2006). Reproduced with permission of Elsevier.

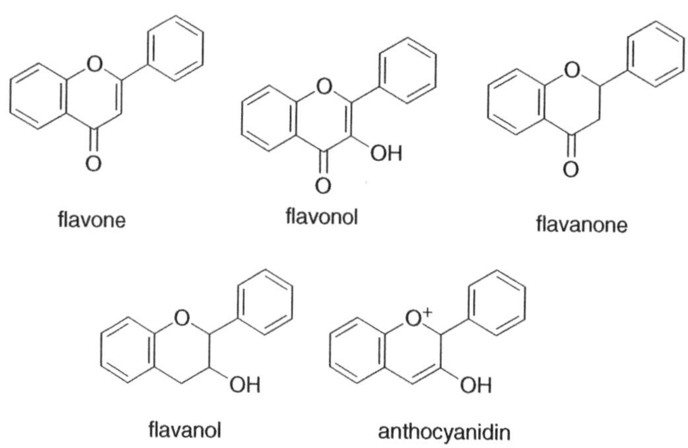

Figure 1.3 Chemical structures of major classes of flavonoids. *Source:* Balasundram (2006). Reproduced with permission of Elsevier.

and hydroxycinnamic acids (gallic acid, protocatechuic acid, ρ-coumaric acid, caffeic acid, ferulic acid and sinapic acid) are shown in Figure 1.2, while Figure 1.3 indicates the chemical structures of major classes of flavonoids (flavone, flavonol, flavanone, flavanol and anthocyanidin) (Balasundram et al., 2006).

Table 1.5 shows the phenolic compounds, such as phenolic acids, flavonols, flavanols, flavonoids and anthocyanins, obtained from agricultural byproducts, such as the hulls of rice, buckwheat, almond, Swedish oats and pistachio; the peels and seeds of citrus, lemons, oranges and grapefruit; the peels of apples, peaches, pears, flesh nectarines,

Table 1.5 Phenolic compounds from agricultural byproducts. *Source:* Balasundram (2006). Reproduced with permission of Elsevier.

Byproduct	Phenolic compounds	Levels[a)]
Almond (*Prunus dulcis* (Mill.) D.A. Webb) hulls	Chlorogenic acid	42.52 ± 4.50 mg/100 g fw
	4-*O*-caffeoylquinic acid	7.90 mg/100 g fw
	3-*O*-caffeoylquinic acid	3.04 mg/100 g fw
Apple peels	Flavonoids	2299 mg CE/100 g dw
	Anthocyanin	169 mg CGE/100 g dw
Artichoke blanching waters	Neochlorogenic acid	11.3 g phenolics/100 ml
	Cryptochlorogenic acid	
	Chlorogenic acid	
	Cynarin	
	Caffeic acid derivatives	
Buckwheat (*Fagopyrum esculentum* Möench) hulls	Protocatechuic acid	13.4 mg/100 mg dw
	3,4-dihydroxybenzaldehyde	6.1 mg/100 g dw
	Hyperin	5.0 mg/100 g dw
	Rutin	4.3 mg/100 g dw
	Quercetin	2.5 mg/100 g dw
Dried apple pomace	Flavonols	673 mg/kg dw
	Flavanols	318 mg/kg dw
	Dihydrochalcones	861 mg/kg dw
	Hydroxycinnamatcs	562 mg/kg dw
Dried coconut husk	4-hydroxybenzoic acid ferulic acid	13.0 mg phenolics/g dry weight

a) Expressed on fresh weight (fw) or dry weight (dw) basis.

pomegranate and bananas; apple pomace; olive mill wastewater (OMW) and leaves; grape seeds and skin; the seeds of mango, longan, avocado and jackfruit; the peels and seeds of tomatoes; and pineapple waste (residual pulp, peels and skin) (Balasundram et al., 2006).

Among the phenolic compounds derived from agricultural byproducts (Table 1.5), hydroxytyrosol, tyrosol, oleuropein and hydroxycinnamic acids are the major components of OMW, while oleuropein is the major component of olive leaves, followed by hydroxytyrosol, luteolin-7-glucoside, apigenin-7-glucoside and verbascoside. Apple peel is rich in flavonoids (approximately 23 mg catechin equivalents/g dry weight), while dried coconut husk contains a high amount of 4-hydroxybenzoic acid (13.0 mg phenolics/g dry weight).

In recent years, a number of studies have been conducted to recover the bioactive compounds from various agricultural wastes. Table 1.6 illustrates a wide range of traditional and emerging techniques that have been applied to the extraction and

Table 1.6 Molecule of interest, substrate from food wastes, extraction method and yield. *Source:* Baiano (2014), http://www.mdpi.com/1420-3049/19/9/14821/htm. Used under CC BY 3.0 https://creativecommons.org/licenses/by/3.0/.

Extractable biomolecule	Substrate	Extraction method	Yield
Pectin	Apple pomace, citrus peel, sugar beet, sunflower heads, wastes from tropical fruits	Solid–liquid extraction	10–15%, 20–30%
Flavanones	Citrus peels and residues from segments and seeds after pressing	Solid–liquid extraction	
Total and soluble dietary fibres	Apple pomace	Solid–liquid extraction	72 and 10%
Phenolic compounds	Apple pomace	Solid–liquid extraction	33%
Lycopene and β-carotene	Tomato pomace	Supercritical CO_2	50%
Anthocyanins	Grape skins	Heat treatment at 70 °C, Ultrasonics, high hydrostatic pressure, pulsed electric fields	Variable
Caffeine	Green tea leaves	Supercritical fluid extraction	97%
Essential oils (matricine, chamazulene and α-bisabolol	Chamomile	Supercritical fluid extraction	28.08, 0.05 and 2.68%, respectively
Capsaicinoids and colour components	Chilli pepper	Supercritical fluid extraction	66–86% and 26–34%, respectively
Oil	Rice bran	Supercritical fluid extraction	24.65%
γ-oryzanol	Rice bran	Solid–liquid extraction	1527–4164 mg/kg
β-glucans	Barley bran	Solid–liquid extraction	
Lignans	Flaxseeds	Solid–liquid extraction	
Phenolic acids	Wheat brans	Solid–liquid extraction, ultrasound-assisted extraction, microwave-assisted extraction	
Tocopherols, tocotrienols, sterols, and squalene	Palm fatty acid distillate	Liquid–liquid extraction	
Phenolic antioxidants	Aqueous byproducts from the palm oil extraction	Separation techniques through membranes	

(Continued)

Table 1.6 (Continued)

Extractable biomolecule	Substrate	Extraction method	Yield
Tocopherols and tocotrienols	Palm fatty acid distillate	Treatment with alkyl alcohol and sodium methoxide; distillation under reduced pressure; a cooling step; passage of the filtrate through an ion-exchange column with anionic exchange resin; removal of the solvent; molecular distillation	
Phenolic antioxidants	Aqueous byproducts from the extraction of palm oil	Without solvent; based on simple separation principles	

isolation of valuable bioactive compounds from food wastes, such as solid–liquid extraction (SLE), ultrasound-assisted extraction (UAE), microwave-assisted extraction (MAE), supercritical fluid extraction (SFE), high hydrostatic pressure (HHP), pulsed electric fields (PEF) and membrane and resin techniques. These techniques allow high yields of bioactive compounds to be achieved, such as pectin (10–30%), flavanones, soluble dietary fibres (10%), phenolic compounds (33%), lycopene and β-carotene (50%), anthocyanins, caffeine (97%), essential oils (matricine, chamazulene and α-bisabolol: 28.08, 0.05 and 2.68%, respectively), capsaicinoids and colour components (66–86% and 26–34%, respectively), γ-oryzanol (1527–4164 mg/kg), β-glucans, lignans, phenolic acids, tocopherols, tocotrienols, sterols and squalene (Baiano, 2014).

Table 1.7 describes the use of different solvents to obtain high yields of bioactive compounds and shows individual bioactive compounds in extracts from agro-industrial wastes, while Table 1.8 shows the antioxidant activity of extracts from agro-industrial residues (Moure et al., 2001). Perretti et al. (2003) extracted α-tocopherol and γ-oryzanol from rice byproducts (hulls, rice bran) using SFE and found that the γ-oryzanol content ranged from 8.2 to 18.0 mg/kg, while the α-tocopherol content was greatly affected by extraction conditions in terms of pressure and temperature (1176.9 and 1228.1 mg/kg at 5000 psi/40 °C and 10 000 psi/80 °C, respectively). Kim et al. (2006) reported that extracts from cereal brans possessed stronger antioxidant activity than free phenolic acids, with ferulic, vanillic and syringic acids found to be the major individual phenolic acids in wheat bran. Izydorczyk & Dexter (2008) found that β-glucans extracted from barley flour could improve lipid metabolism, reduce glycaemic index, lower plasma cholesterol and reduce risk of colon cancer, while Zanwar et al. (2011) reported biological activity of lignan concentrates from flaxseed as anticancer, antioxidant, antibacterial, antiviral and anti-inflammatory agents.

Bernini & Mincione (2007) described the recovery and reuse of phenolic compounds from OMW, such as tyrosol, hydroxytyrosol and oleuropein, which obtained at 0.6 g/l

Table 1.7 Extraction yields, total extractable polyphenols (TEP) and composition of crude extracts from agro-industrial wastes. *Source:* Moure et al. (2001). Reproduced with permission of Elsevier.

Residue	Solvent	Solubles yield (% dry weight) or TEP (as equivalents)	Identified compounds
Durum wheat bran	Ethanol	12.1 (% dry weight) 2.769 (HPLC)	PA, pBA, GA, CaA, VA, CA, SA, pCA, FA
Fraxinus ornus bark	Ethanol	14.5 (dry weight)	Hydroxycoumarin (Es, Est, Fx, Fxt)
Corn bran hemicelluloses	NaOH	3.259 (FA)	FA; pCA; dFA
Potato peel extract	Water (100 °C)	0.048 (HPLC)	CA, GA, PA, CCA
	Methanol (4 °C)	0.041 (HPLC)	CA, GA, PA, CCA
Shrimp shell waste	95% ethanol		1,2 diamino-1(o-hydroxyphenyl) propene
Red grape promace peels	Methanol water (1,1)	4,3 (tannic acid)	
	Acetone,water (7,3)	c. t., 27.0 (″)	
Apple pomace	70% acetone		EC, CA, 3- hPhlz, Phl-2-x, Phlz, Q-3-gal, Q-3-glu, Q-3-xyl, Q-3-ara, Q-3-rha
Buckwheat hulls	Ethanol	23.8 (HPLC)	PA, 3,4-dihydroxybcenzaldehyde, Hy, Ru, Q, Vi, iVi
Oat hulls	Methanol	0.035 (GC-MS)	FA, pCA, VA, pBA, V, 4-PhA, catechol, o-coumaric acid, SA, SaA
Lemon seeds	Methanol	0.2333[a]	CaA, pCA (*cis* and *trans*), FA, SA, Eri, Nar, Neh
Sweetorange seeds		0.0544	CaA, pCA (*cis* and *trans*), FA, SA, Nat, Nar, Hes
Sour orange peel		2.526	CaA, pCA (*cis* and *trans*), FA, SA, Ner, Nat, Nar, Hes
Bergamot peel		1.359	CaA, pCA (*cis* and *trans*), FA, SA, Ncr, Nar, Nch
Lentil seed coat (brown)	Water	22.3 (% dry weight)	L, Q, K, PC, PD, FA, PA, CaA
(green)		8.7 (% dry weight)	
Grape marc	Ethyl acetate	0.224 (HPLC)	K, Q, iRh3glu; K-3-glu; Q-3-glu; Q-3-gal; EC; C; VA; PA; GA
	Water	0.021 (HPLC)	M-3(6-a)glu; Po-3(6-a)glu; M-3-glu; Po-3-glu; Pt-3-glu; Cy-3-glu; D-3-glu; PC

(Continued)

Table 1.7 (Continued)

Residue	Solvent	Solubles yield (% dry weight) or TEP (as equivalents)	Identified compounds
Grape pomace	80% ethanol	42% (dry weight)	GA; GA 3-βglup ; GA 4-βglup; *trans*-CfA; *cis* and *trans* CtA; 2-h-5(2he) pβ-D-glp; C; EC; PCB1; Q-3glup; Q-3-glurcp; K-3-glup; K.3-galp; Ey; At; En
Lemon peel	Methanol-water	0.02% (dry weight)	coumarins (8-geranyloxypsolaren, 5-geranyloxypsolaren, 5 geranyloxy-7 methoxycoumarin)
Grape seeds	95% ethanol	–	pCA; GA; CAA; C,-EC, procyanidins (B1-B8)
Nonvolatile residue from orange essential oil	–	–	α-toc; 3, 3′, 4′, 5, 6, 7 hexamethoxyflavone; 3, 3′, 4′, 5, 6, 7, 8-heptamethoxyflavone; 4′, 5, 6, 7, 8 pentamethoxyflavone
Olive mill waste waters	Water ethanol	8.11 (% dry weight)	hT, T, EA, oleuropein derivatives, Lu-7-glu, Q, CnA derivatives, hT derivatives
Grape seed extract	Ethyl acetate 20% Ethanol	22.78 (% dry weight)	proanthocianidins, monomeric flavonols

a) glycosylated flavanone + phenolic acids content.
e.p., extractable polyphenols; n.e., non-extractable; UP; unidentified polyphenols; c.t., condensed tannins; GC-MS, gas chromatography mass spectroscopy; CA, chlorogenic; CaA, caffeic; Cf A, caftaric; CnA, cinamic; Ct A, Coutaric; EA, elenoic; FA, ferulic; dFA, diferulic; GA, gallic; KA, kiwic; QA, quinic; VA, vanillic; SA, syringic; CnA, cinnamic; PA, protocatechuic; pBA, p-hydroxybenzoic; GeA, gentisic; pCA, p-coumaric; VA, vanillic; At, astilibin; Ct, catechol; EC, Epicatechin; C, Catechin; Cy-3-glu, Cyanidin-3-glucoside; D-3-glu, Delphinidin-3-glucoside, En, Engeletin; Eri, eriocitrin; Es, Esculin; Est, Esculetin; Ey, eucryphinl; Fx, fraxin; Fxt, fraxetin; GA 3-βglup, gallic acid 3-β-glucopyranoside; GA 4-βglup, Gallic acid 4-β-glucopyranoside; 2-h-5(2he)p β-D-glp, 2-hydroxy-5-(2-hydroxyethyl(phenyl-β-D-glucopyranoside; Hes, hesperidin; Hy, hyperin; K, kaempferol; K-3-glu, kaempferol-3-glucoside; K-3-glup, kaempferol 3β-D-glucopyranoside; K-3-galp, kaempferol 3-β-D-galactopiranoside; Lu-7-glu, Lu-7-glucoside; M-3(6-a)glu, malvidin-3-(6-acetyl)-glucoside; M-3-glu, malvidin-3-glucoside; Nar, naringin; Nat, narirutin; Neh, neohesperidin; Ner, neoeriocitrin; PC, procyanidin; PD, prodelphinidin; 4-PhA, 4-hydroxyphenilacetic acid; 3,4-PhE, 3,4-hydroxyphenilethanol; 3,4-dB, 3,4-dihydroxybenzaldehyde; 3-hPhlz, 3-hydroxyphlorodzin; Phl-2-x, phloretin-2′-xyloglucoside; Phlz, phloridzin; Po-3(6-a)glu, peonidin-3-(6-acetyl)-glucoside; Pt-3-glu, petunidin-3-glucoside; Q, quercetin; Q-3-gal, quercetin-3-galactoside; Q-3-glu, quercetin-3-glucoside; Q-3-glucr, quercetin-3-glucuronide; Q-3-glucrp, quercetin 3-β-D glucuropyranoside; Q-3-glup, quercetin 3-β-D glucopyranoside; Q-3-xyl, quercetin-3-xyloside; Q-3-ara, quercetin-3-arabinoside; Q-3-rha, quercetin-3-rhamnoside; iRh3Glu, isoRhamnetin-3-glucoside; Ru, rutin; SaA, salicilyc acid; T, tyrosol; hT, hydroxytyrosol; V, vanillin Vi, vitexin; iVi, isovitexin.

Table 1.8 Antioxidant activity of extracts from agro-industrial residues. *Source:* Moure et al. (2001). Reproduced with permission of Elsevier.

Residue (solvent)	Antioxidant activity assay	Activity (conc. antioxidant)
Rape of olives (A + E)	Refined olive oil oxidation	PV (control), 52 meq/kg PV (1000 ppm), 18 meq/kg
	Refined olive oil oxidation	AV (control), 175 AV (1000 ppm), 60
Durum wheat bran (E)	Soy oil oxidation	PV (0.05%), 37.6–42.0 meq/kg PV (0.05% BHA–BHT), 22.0 meq/kg PV (control), 129.0 meq/kg
Fraxinus oxinus bark (E)	Triachylglycerols of lard and sunflower oil oxidation	F (0.05%), 3.6–4.8 F(0.1%), 4.0–6.1 ORR (0.05%), 0.6–0.28 ORR (0.1%), 0.5–0.28
Corn bran hemicelluloses fragments (FA sugar esters)	Lipid peroxidation of rat liver microsomes	$ABS_{510\ mm}$ (control), 0.23 $ABS_{510\ mm}$ (Toc 0.5 mM), 0.03 $ABS_{510\ mm}$ (0.5 mM), 0.065
Potato peel waste (W)	Sunflower crude oil oxidation	PV (200 ppm), 37.47 PV (200 ppm BHA), 37.47 PV (control), 49.15
Tamarind seed coats (E) (EA)	Linoleic acid oxidation	ILP (α-tocopherol 0.2 mg), 90% ILP (0.2 mg), 90% ILP (0.2 mg), 98%
Wild rice hulls (M)	—	TBARS (0.1%), 2.4 TBARS (0.2%), 0.9 TBARS (control), 3.8 TBARS (0.02%), 0.3
Peanut hulls (M)	DPPH radical scavenging	IP (extract 1.5 mg/ml), 89.3% IP (BHA 240 μM), 92.6% IP (catechin 8 μM), 89.3%
Peanut hulls (M)	Linoleic acid oxidation	AA (9.6 mg), 96.1–96.8%
Mung bean hulls (M)	Linoleic acid peroxidation	$ABS_{500\ nm}$ (100 ppm), 0.20 $ABS_{500\ nm}$ (BHA, 100 ppm), 0.23 $ABS_{500\ nm}$ (control), 0.95
	Soybean oil oxidation	PV (100 ppm), 70 PV (BHA, 100 ppm), 55 PV (control), 105
	Soybean oil oxidation	$ABS_{535\ nm}$ (100 ppm), 0.6 $ABS_{535\ nm}$ (BHA, 100 ppm), 0.75 $ABS_{535\ nm}$ (control), 1.4
Peanut hulls (M)	Soybean and peanut oil oxidation	O.S. (0.48%), 194 O.S. (0.01% BHA), 143 O.S. (control), 107
Grape seeds (EA – C pptn)	Lecithin liposome oxidation	ILLO (0.1 mM), 86% ILLO (0.1 mM BHT), 88.5% ILLO (+)-catechin, 40%
Rose hips (EA – C pptn)	Lecithin liposome oxidation	ILLO (0.1 mM), 68% ILLO (0.1 mM BHT), 88.5% ILLO (+)-catechin, 40%
Buckwheat hulls (E-separation by chromatography)	Methyl linoleate oxidation	PV (control), 3.5 mM
Oat groats (M)	Soybean oil oxidation	IO (0.3%), 98.4%
Oat hulls (M)	Soybean oil oxidation	IO (0.3%), 96.4%

(Continued)

Table 1.8 (Continued)

Residue (solvent)	Antioxidant activity assay	Activity (conc. antioxidant)
Lemon peel (M)	Citronellal oxidations	AOP, 0.16 l/g
Mandarin seeds (M)		AOP, 0.5 l/g
Sour orange peel (M)		AOP, 0.27 l/g
Sweet orange (M)		AOP, 0.2 l/g
Eucalyptus wood acid hyrdolysates (EA)	Linoleic acid and β-carotene oxidation	AAC (400 mg), 588 AAC (BHT 400 mg), 916
Grape pomace (enzyme treatment, A)	Cu-induced human LDL oxidation	NPIT (catechin 3 μM GAcq), 110.4 minutes NPIT (catechin 3 μM GAeq), 62.2 minutes
Grape seed extract (E)	Antinuclear activity (200 mg/kg)	Lesion length (control), 111 mm Lesion length (catechin), 88 mm Lesion length (extract), 4–20 mm
Grape pomace (M,W-A,W)	DPPH radical scavenging	IC_{50}, 0.2 g (4 mg extractable polyph) IC_{50} (D,L, α-tocopherol), 0.02 g
	Linoleic acid oxidation	IC_{50}, 0.7 g (14 mg extractable polyph) IC_{50} (D,L, α-tocopherol), 0.3 g
Olive mil waste water (W-E)	Cu-induced human LDL oxidation	LOOH (control), 0.35 mm
Olive mill waste water (EA)		LOOH (20 ppm), 0.1–0.2 mm
Shrimp shell waste (95% E)	β-carotene bleaching	34.09% BHA/BHT/citric acid, 11.36%
Lentil seed coat (MW)	50% inhibition in the reduction of the nitro blue tetrazolium (NBT)	IC_{50}, 4.52 μg/ml IC_{50} (ascorbic acid), 5.5 μg/ml IC_{50} (catechin), 1.9 μg/ml
Brown lentil husk (M-W) (EA) (B)	DPPH radical scavenging	IC_{50}, 12.58–14.83 μg/ml IC_{50}, 6.62 μg/ml IC_{50}, 4.04 μg/ml IC_{50} (catechin), 5.46 μg/ml IC_{50} (quercetin), 1.73 μg/ml IC_{50} (gallic acid), 0.63 μg/ml IC_{50} (ascorbic acid), 2.58 μg/ml
Nonvolatile residue from orange essential oil	Olive oil (70 °C, 48 hours)	Oxidation (%) (3000 ppm), 52–64% Oxidation (%) (BHA, 200 ppm), 72%
Red grape marc (EA)	Refined olive oil oxidation	IP (100 mg/kg), 22 hours IP (100 mg/kg BHA), 26 hours IP (100 mg/kg BHT), 24 hours
Grape seed (W)	Cu-mediated oxidation of rat plasma	Lag phase, 95 minutes (control) Lag phase 175 minutes (proanthocyanidin, incubation 30 minutes)
Lemon peel (M)	Haemoglobin-catalysed peroxidation of linoleic acid	IC_{50}, 122.0 ppm
Orange peel (M)		IC_{50}, 68.8 ppm

Table 1.8 (Continued)

Residue (solvent)	Antioxidant activity assay	Activity (conc. antioxidant)
Peanut hull		IC_{50}, 111 ppm IC_{50} (BHA), 0.65 ppm
Gevuina avellana hulls (E)	β-carotene bleaching	AAC (4000 mg/l), 767 AAC (230 mg BHA/l), 901 IO (1 g/l), 73.1%
*G. avellana*hulls (M)	Crude soybean oil oxidation	IO (0.01 g BHA/l), 12.8% IO (0.01 g BHT/l), 34.8%
Old tea leaves (M)	Rapeseed oil (60 °C, 20 days)	Control: AV, 80.7; PV, 290 meq/kg OTL 0.25%: AV, 9.5; PV, 30 meq/kg TBHQ 0.02%: AV, 2.7; PV, 5 meq/kg Rosemary 0.1%: AV, 4.8; PV, 25 meq/kg

A, acetone; C, chlorogorm; B, butanol; E, ethanol; EA, ethylacetate; M, methanol; W, water; IO, inhibition of soybean oil oxidation, after 20 days at 60 °C in the dark, $((PV_{control} - PV_{treatment})/PV_{control}) \times 100$; AA, antioxidant activity (thiocyanate method), calculated as percentage of inhibition of peroxidation of linoelic acid; AAC, antioxidant activity coefficient (absorbance of $extract_{120\,hours}$ − Absorbence of $control_{120\,hours}$)/(Absorbence of $control_{0\,hours}$ − Absorbence of $control_{120\,hours}$); AOP, reciprocal of the concentration required to double the half-life of citronellal (1/g dry matter peel or seed); $ABS_{500\,nm}$, measure of linoleic acid oxidation by the thiocyanate method after 12 days; $ABS_{510\,nm}$, measure of lipid peroxidation in rat liver microsome; $ABS_{535\,nm}$, formation of TBARS (thiobarbituric acid reactive substances) on soybean oil after accelerated oxidation; BHA, butylated hydroxoyaniosle; BHT, butylated hydroxytoluene; DPPH, *α* α-diphenyl-β-picrylhydrazil; F (protection factor), $IP_{inh}/IP_0 - IP_{inh}$ is the induction period in the presence of an inhibitor and IP_0 is the induction period of the non-inhibited system; ILLO, inhibition of lecithin liposome oxidation as increase in $ABS_{535\,nm}$; ILP, inhibition of linoleic acid peroxidation with respect to a control (thiocyanate method), I.P., induction period – the time required for the fluorescence spots on a silica gel TLC plate sprayed with a 3% solution of linoleic acid in hexane, which was considered the induction period for lipid oxidation; IC_{50}, inhibitory concentration for 50% inhibition in the reduction of oxidation – lag phase of CE-OOH accumulation in $CuSO_4$-induced oxidation of rat plasma (minutes); LOOH, lipid peroxide production; NPIT, net prolongation of induction time (minutes) for conjugated diene hydroperoxide formation; O.R.R., oxidative rate ratio $(W_{inh}/W_o) - W_{inh}$ is the initial oxidation rate in the presence of an inhibitor, W_o is the initial oxidation rate of the non-inhibited system; OS, oxidative stability (hours); AV, anisidine value; PV, peroxide value; TBHQ, tertiary butyl hydroquinone.

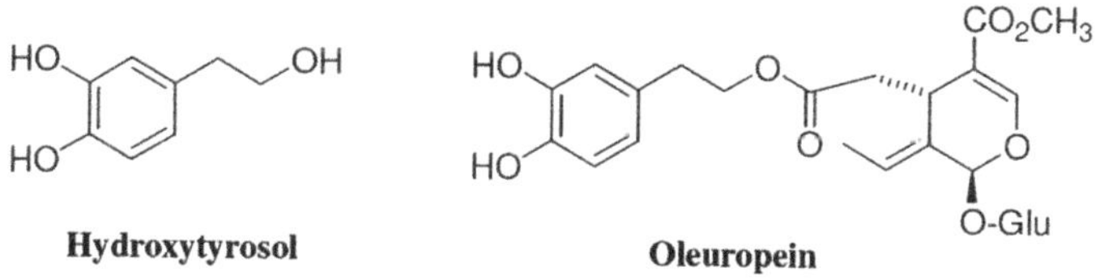

Figure 1.4 Chemical structures of some phenolic compounds from OMW. *Source:* Adapted from Bernini & Mincione (2007).

of tyrosol and 1 g/l of hydroxytyrosol with purity higher than 98% (Figures 1.4 and 1.5), while Martins et al. (2011) showed the potential for the successful conversion of inexpensive agro-industrial residues by the same bioprocess; for example, pomegranate wastes contain a significant amount of phenolic compounds, cranberry pomace is a good source of ellagic acid and other phenolic compounds, green coconut husk is a potential source of ferulic acid and agricultural refuses (including cereal and vegetable wastes) consist mainly of cellulose, hemicellulose and lignin, of which the lignin fraction

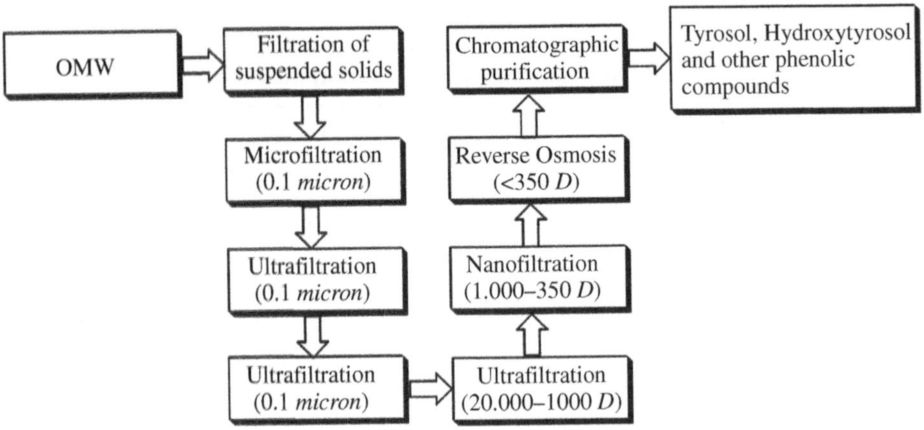

Figure 1.5 Recovery of phenolic compounds from OMW *Source:* Adapted from Bernini & Mincione (2007).

contains numerous phenolic components. Brunton & Wijngaard (2011) indicated the high potential of fruit and vegetable wastes as sources of bioactive compounds and found a pressurised liquid method for the extraction of antioxidants from apple pomace using 60% ethanol at 102 °C and an SLE method for the recovery of antioxidants from apple pomace and potato peel using 56% ethanol at 80 °C for 31 minutes and 75% ethanol at 80 °C for 22 minutes, respectively. Conde et al. (2013) reported the recovery and concentration of antioxidants in the form of phenolic compounds, peptides, proteins and oligosaccharides from industrial wastes and biomass, such as shells, husks,

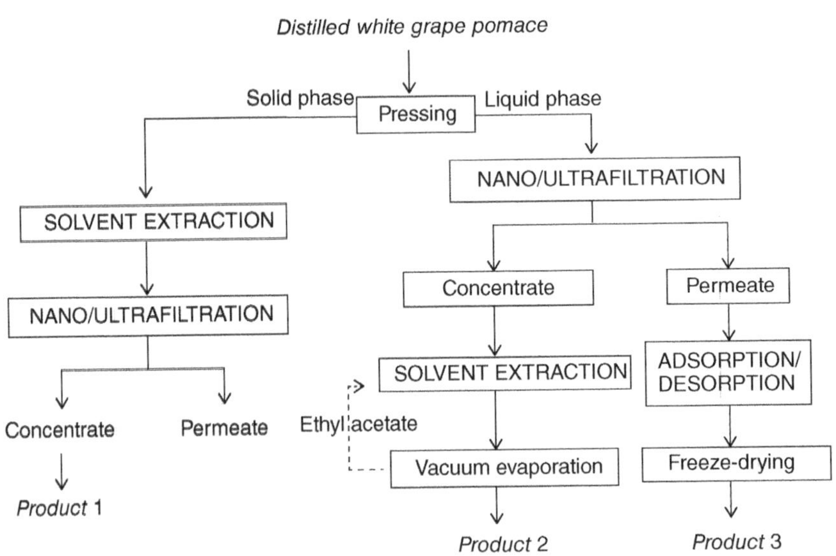

Figure 1.6 Recovery and concentration of phenolic components from wine byproducts. *Source:* Adapted from Conde (2013).

cob, burs, wood and pomace, using novel techniques such as membrane and renin technologies (Figure 1.6).

Baiano (2014) showed the recovery and reuse of high-value biomolecules, such as proteins, polysaccharides, fibres, flavour compounds and phytochemicals (flavanones, phenolic compounds, lycopene, β-carotene, anthocyanins, tocopherols, tocotrienols and sterols) from food wastes and co-products, which are produced from households (42%), food processing (38%) and whole chain (20%) (e.g. fruit and vegetable wastes, include trimmings, peelings, stems, seeds, shells and bran, and residues remaining after the extraction of oil, starch, sugar and juice using SLE, Soxhlet extraction, pressurised fluid extraction (PFE), SFE, UAE, MAE, PEF and enzyme-assisted extraction). Lozano-Sánchez et al. (2014) demonstrated the recovery of bioactive compounds from olive oil filter cake, such as phenolic alcohols (hydroxytyrosol and its oxidised products), secoiridoids (decarboxymethylated and hydroxylated forms of oleuropein and ligstroside aglycones), flavones (luteolin and apigenin) and elenolic acid derivatives, using pressurised liquid extraction (PLE) as an advanced technique.

Nguyen (2014) reported the extraction of pigments from cacao pod shell using 80% methanol as solvent and found that the optimal extraction temperature, extraction time and ratio of solvent to material in terms of pigment yield from cacao pod shell were 60 °C, 80 minutes and 10 : 1 (v/w), respectively. Nguyen & Scarlett (2016) studied the extraction of bioactive compounds in terms of phenolics and saponins from carrot peel, a major waste product of carrot processing, with a high proportion of 14.19% by fresh weight, using conventional extraction and various solvents, and evaluated the antioxidant capacity of the extracts obtained, finding that carrot peel is rich in saponins (approximately 250 mg escin equivalents/g dried sample) and that methanol is the most suitable solvent for the effective extraction of antioxidants from carrot peel.

The findings obtained from recent studies are a systematic and valuable information source indicating a great potential for agricultural wastes in terms of crop-based and processing-based residues as abundant, biorenewable, sustainable and inexpensive sources of bioactive compounds. Therefore, the principal aim of current and future research is to enhance the exploitation of bioactive compounds from agricultural wastes using various emerging techniques in order to obtain high added-value products with low production costs and low environmental risk, for potential application in the nutraceutical, medical, pharmaceutical and cosmetic industries. This strategy requires an effective cooperation between producers, researchers and traders.

1.3 Uses of Agricultural Wastes and Recovered Bioactive Compounds

In the past, agricultural wastes were mainly discarded to landfills, with only a small amount being used as a source of bioenergy, biomaterials, feeds for animals (include aquaculture), organic fertilisers (compost) and raw materials for the production of fertilisers. This dumping entailed a high risk for the environment and meant a great bioactive potential was wasted.

Regarding the benefits of bioactive compounds derived from agricultural wastes in food and food processing, xanthan obtained from melon wastes has been found to act as a suspending, stabilising, thickening and emulsifying agent that can be used for food and oil recovery as a rheology modifier or a food additive (López et al., 2004). Pectins extracted and purified from various fruit pomaces were applied as gelling agents in numerous food products, such as jams, fillings, sweets and so on, and fruit pomaces were also used to produce many other food additives, such as dietary fibres, lactic acid, pigments, vinegar, natural sweeteners and cellulose (Nawirska & Kwaśniewska, 2005). β-glucans have been used as a gelling agent in food products such as pasta, noodles, breakfast cereals and dairy products, with β-glucans-enriched pearled barley displaying a good cooking quality (Izydorczyk & Dexter, 2008), and lignan-enriched flaxseed or flaxseed meal found to be easy to add to various cereal-based formulations, such as bread, muffins and other bakery products (Baiano, 2014). Tarazona-Díaz et al. (2011) showed that lycopene, citrulline and phenolic compounds extracted from the rinds and flesh of watermelon are sources of food additives for drinks, juices and others products, while xylose and polyphenols isolated from wheat straw can be used as food ingredients (Akpinar et al., 2012). Durante et al. (2012) indicated that vitamin E, carotenoids and quinones separated from wheat bran are valuable nutrients and antioxidants, while glucosinolates and peptides obtained from defatted cakes of rapeseed can be used as foaming (Sánchez-Vioque et al., 2001) and emulsifying (Sánchez-Vioque et al., 2004) agents. Lycopene, β-carotene, sterols, tocopherols, terpenes and glycoalkaloids extracted from tomato byproducts (e.g. seeds, pulps and skins) possess strong antioxidant capacity, so they could be utilised as ingredients in the formulation of antioxidant rich functional foods (Kalogeropoulos et al., 2012). Thus, the bioactive compounds isolated from agricultural wastes can be used as food additives or supplements in order to enhance physicochemical, nutritious and sensory properties.

Numerous bioactive compounds derived from agricultural wastes display potent pharmacological properties, and can therefore be used by the medical and pharmaceutical industries as agents for the prevention and treatment of many chronic human diseases, such as hepatitis, diabetes, obesity, cancer and inflammatory and cardiovascular diseases. For instance, Sainvitu et al. (2012) indicated that lignin secoisolariciresinol isolated from flaxseed can be converted into enterolignans (Figure 1.7), which can bind to oestrogen receptors to act as an agent against breast, prostate and colon cancers. Santana-Méridas et al. (2012) described many valuable bioactive compounds

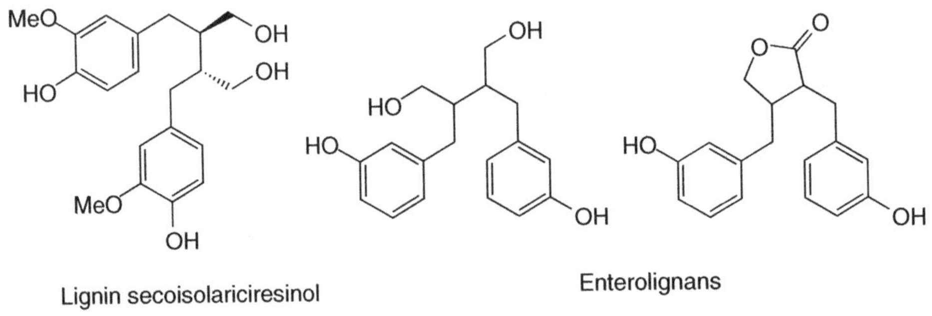

Figure 1.7 Chemical structure of lignin secoisolariciresinol from flaxseed and its conversion products. *Source:* Adapted from Sainvitu et al. (2012).

derived from crop- and processing-based residues (Table 1.4). For example, capsaicin and dihydrocapsaicin isolated from pepper leaves and stems possess antioxidant and anti-inflammatory activity; isovitexin, saponarin, vicenin-2 and apigenin extracted from cucumber leaves, kaempferol, orientin and vitexin separated from saffron leaves, tannins, vanillin and mangiferin isolated from mango peels, pits and seeds display antioxidant capacity; solanesol derived from tomato and potato leaves act against bacteria and inflammation; flavanoids, saponins, tannins, alkaloids, steroids and triterpenes extracted from lemon peels exhibit antimicrobial potential; brutieridin and melitidin isolated from bergamot white tissues display anticholesterolaemic activity; and limonoids derived from the seeds and molasses of citrus fruits, triterpenes (oleanoic, ursolic, betulinic acids) and daucosterol extracted from almond hulls, betulinic acid, stilbenoid transresveratrol, trans-3-viniferin and sitosterol 6′-O-acylglucosides isolated from grape stems and glycoalkaloids derived from potato peels possess anticancer capacity.

In the cosmetics industry, recent studies have found bioactive compounds in improving anti-aging, antiwrinkle, anticellulite and antidryness properties. For example, varieties of phytoconstituents, such as aloin, ginsenoside, curcumin, epicatechin, asiaticoside, ziyuglycoside I, magnolol, gallic acid, hydroxychavicol, hydroxycinnamic acids and hydroxybenzoic acids (Figure 1.8), which can scavenge free radicals from skin cells, prevent trans-epidermal water loss and contribute to protect skin from wrinkles (sun protection factor of 15 or higher), leading to glowing and healthy younger skin. In particular, different parts of medicinal plants have been found to possess anti-aging capacity via various mechanisms of action (Table 1.9). Herbal cosmetics are becoming popular across the world due to their significant effects on skin aging. This can take two forms: one is chronological aging, relating to the passage of time, and the other is premature aging or photoaging, relating to environmental influences (Mukherjee et al., 2011). Shaheda et al. (2014) have also suggested natural bioactive compounds as potential antiwrinkle agents. Anti-aging creams are predominantly moisturiser-based cosmeceutical skin-care products that promise to make the consumer look younger by reducing, masking or preventing signs of skin aging, including laxity (sagging), rhytids (wrinkles) and photoaging, such as erythema (redness), dyspigmentation (brown discolorations), solar elastosis (yellowing), keratosis (abnormal growths) and poor texture.

Other applications, such as the use of agricultural residues for the production of bioactive compounds with biopesticidal activities, are currently underutilised. Based on the insecticidal activity of selected residues, both crop-based and processing-based residues may be considered as cheap and abundant sources of insect pest-control agents (Santana-Méridas et al., 2012).

The use of agricultural wastes for the recovery of bioactive compounds is thus of great interest. This will not only increase the value of agricultural wastes, support and promote the development of related industries and contribute to protect human health, but also reduce pollution risks for the environment in both the short and the long term and lead towards sustainable development. However, before bioactive compounds derived from agricultural wastes can be utilised in the nutraceutical, medical, pharmaceutical and cosmetics industries, their safety must be carefully evaluated in order to increase their positive effects and reduce their negative ones (Gil-Chávez et al., 2013).

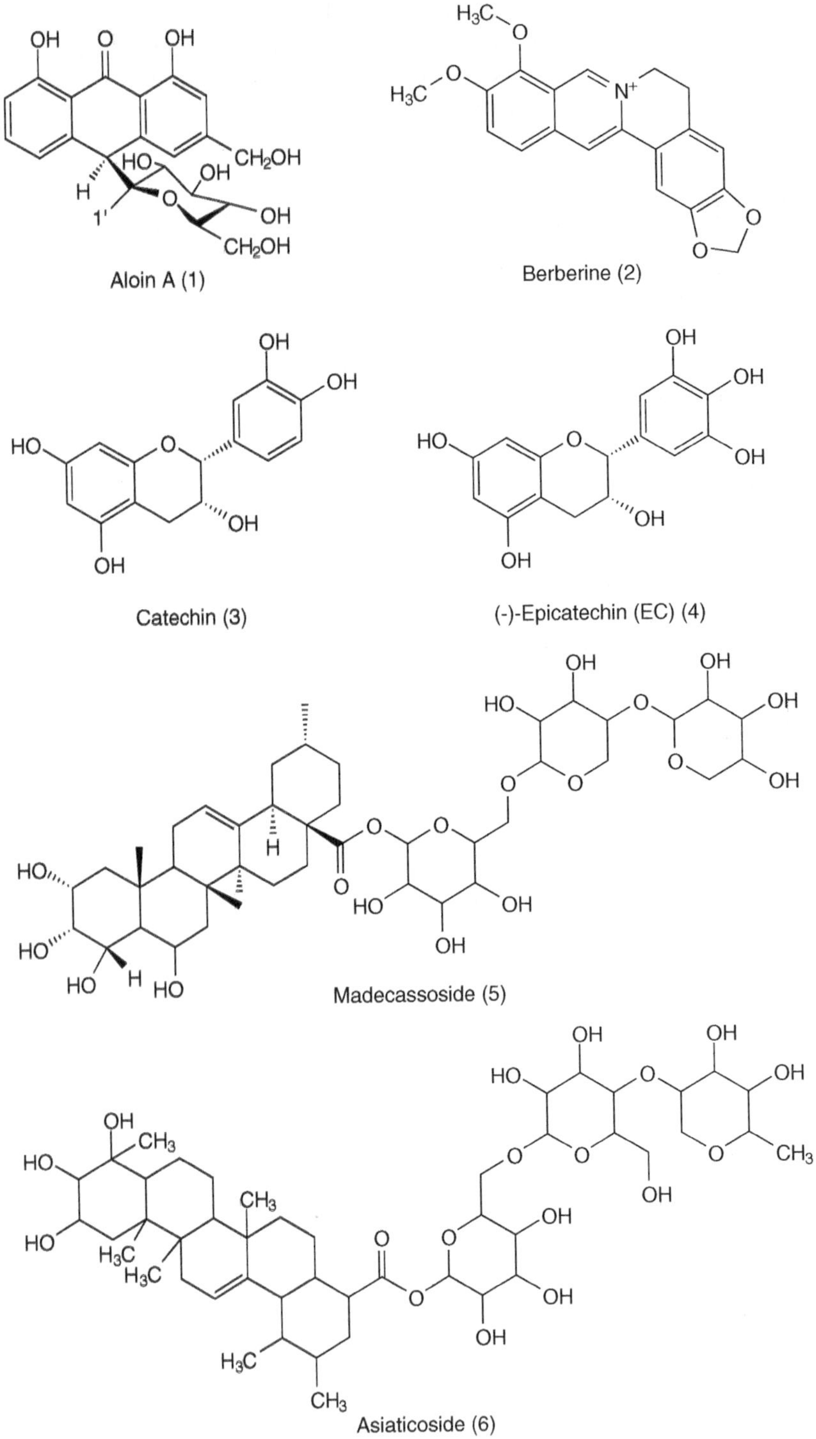

Figure 1.8 Chemical structures of phytochemicals with anti-aging potential. *Source:* Mukherjee et al. (2011). Reproduced with permission of Elsevier.

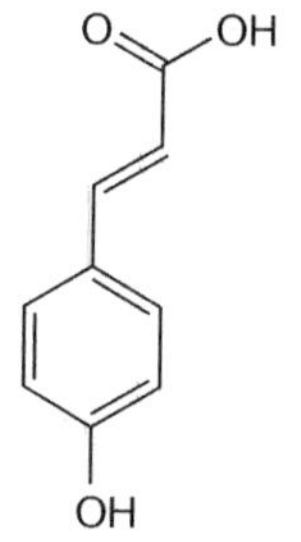

4-Hydroxycinnamic acids (7)

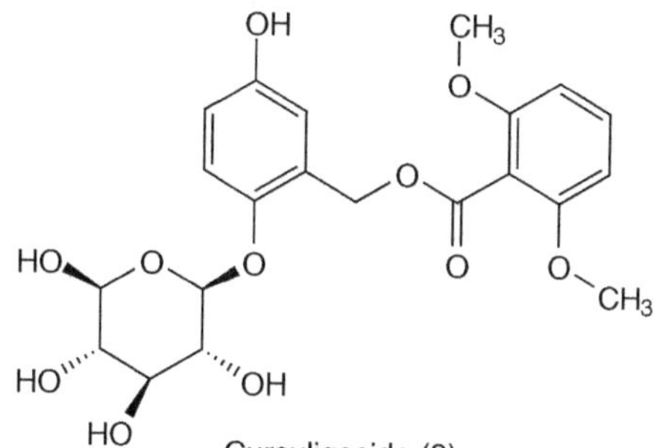

Curculigoside (8)

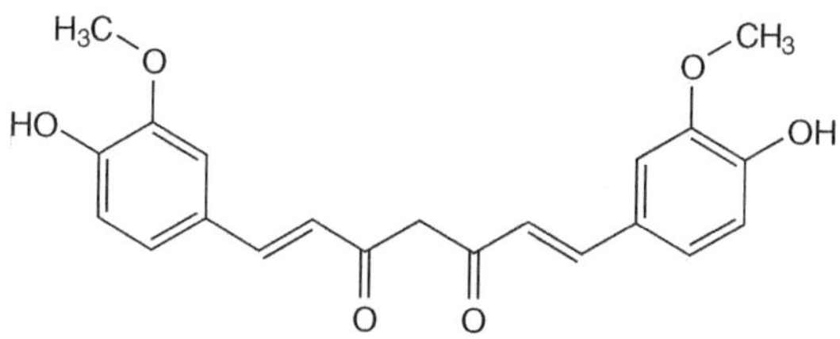

Curcumin (9)

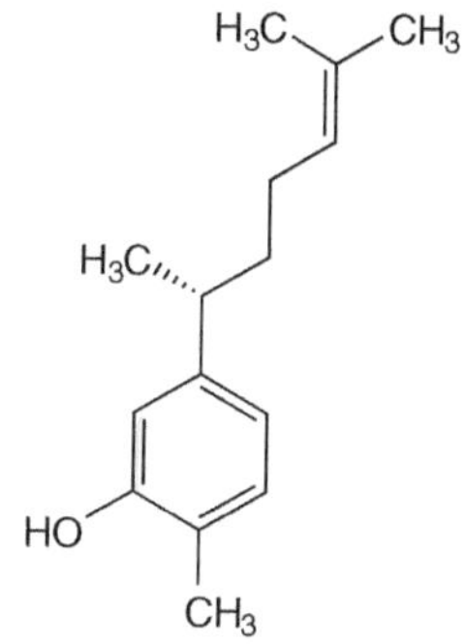

Xanthorrhizol (10)

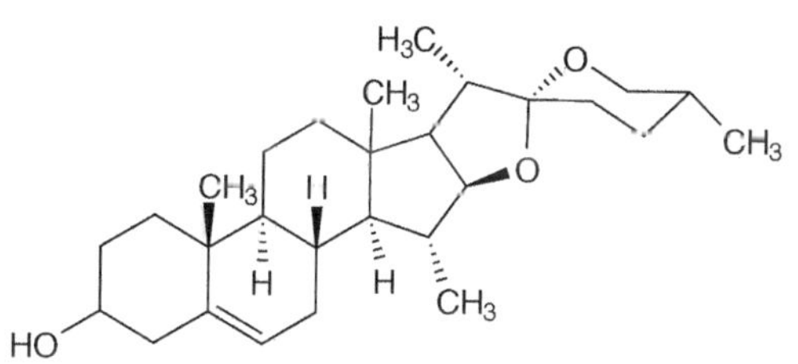

Diosgenin (11)

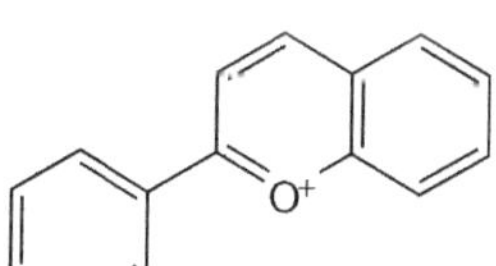

Anthocyanins (12)

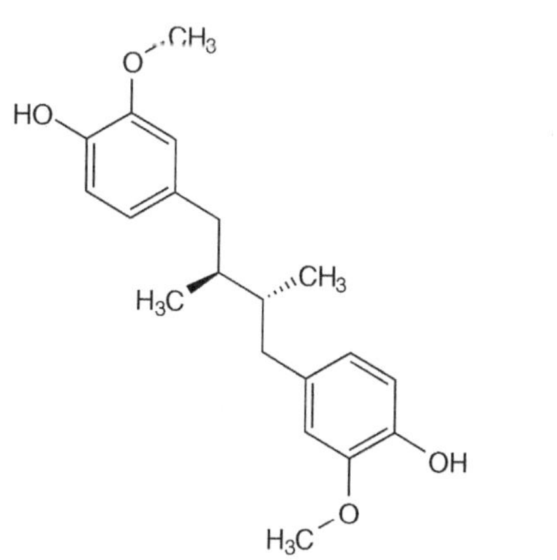

Meso-dihydroguaiaretic acid (14)

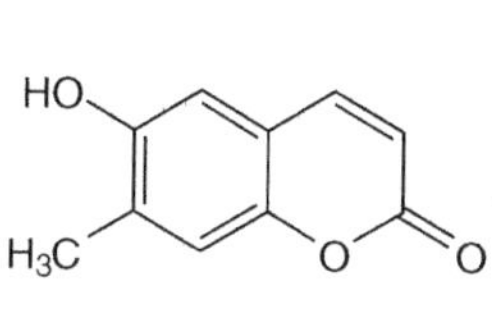

Esculetin (13)

Figure 1.8 *(Continued)*

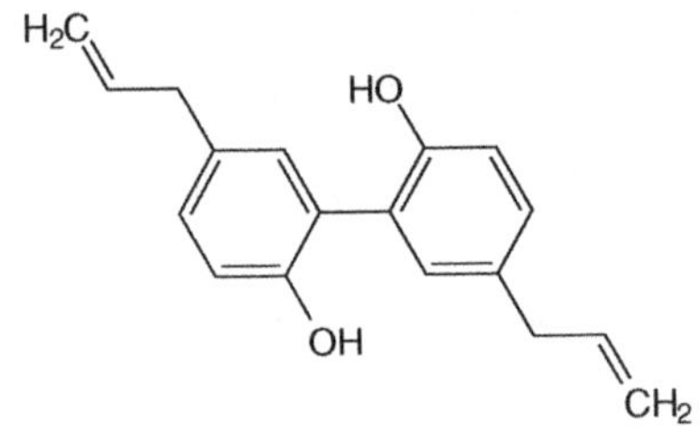

Magnolol (15)

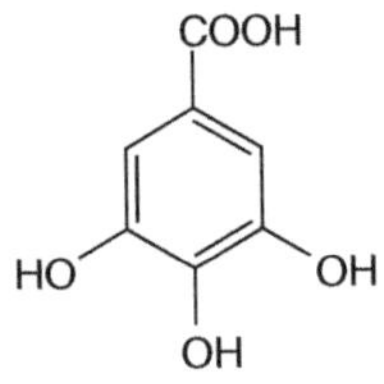

Gallic acid (16)

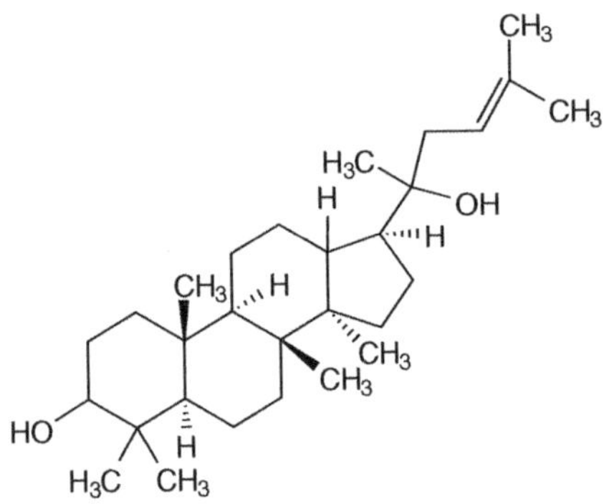

Ginsenoside (17)

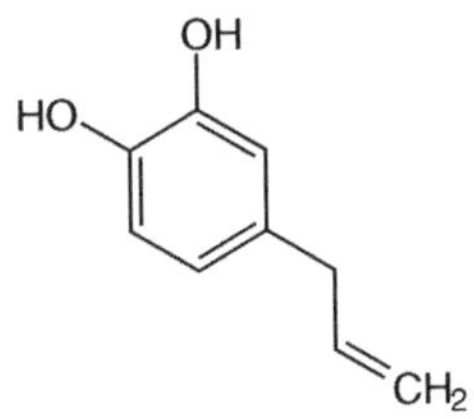

Allylpyrocatechol or hydroxychavicol (18)

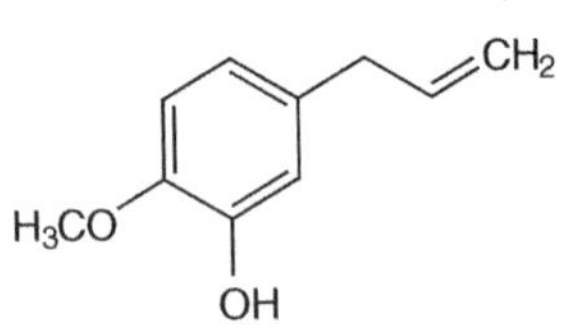

Chavibetol (19)

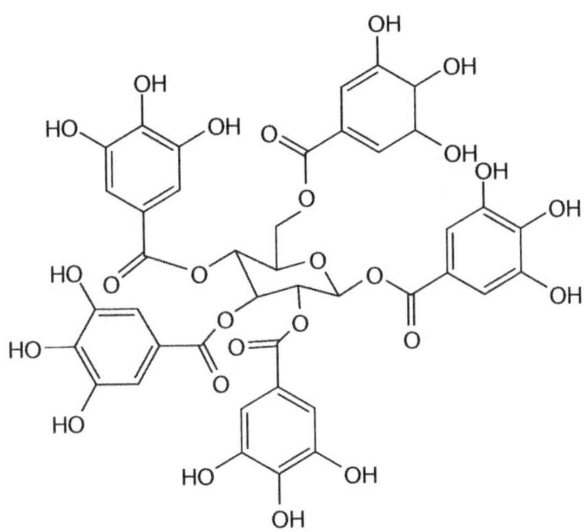

1, 2, 3, 4, 6-penta-O-galloyl-β-D-glucose (21)

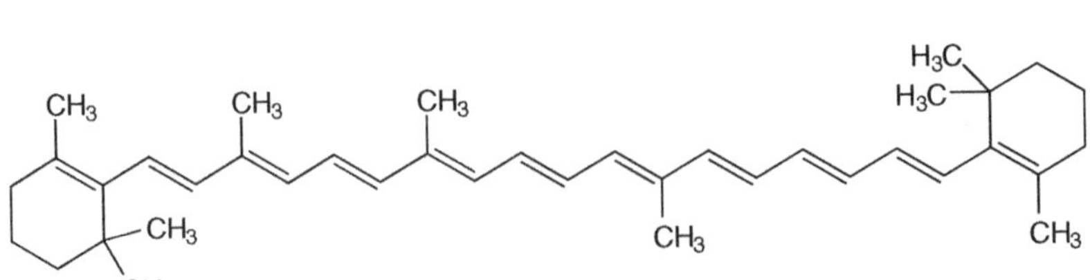

β-carotene (20)

Figure 1.8 *(Continued)*

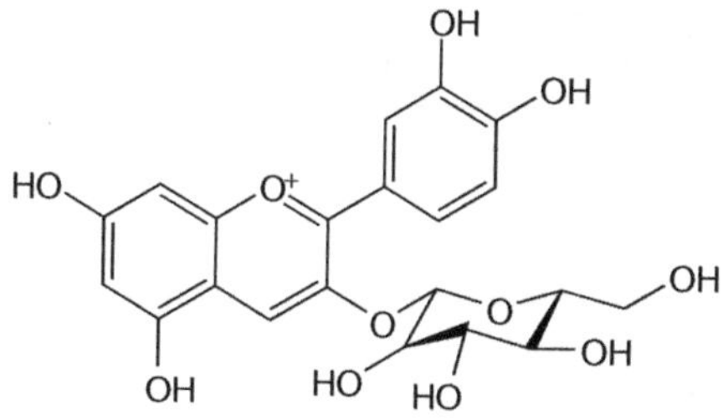

Cyanidin-3-glucoside (22)

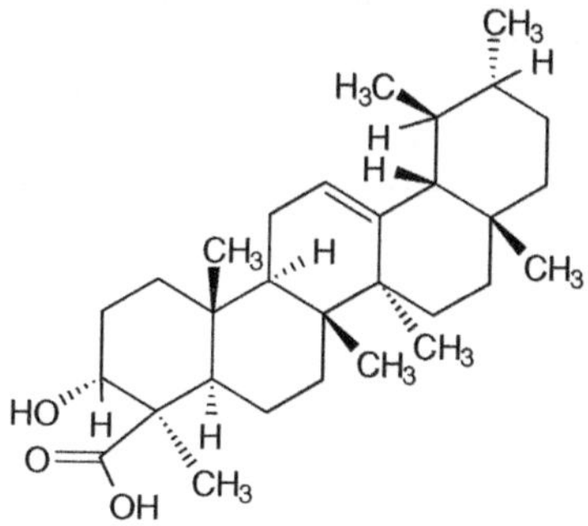

Boswellic acid (23)

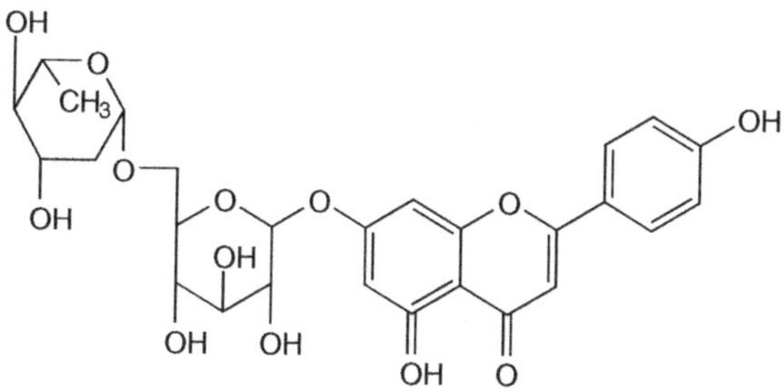

Linarin (24)

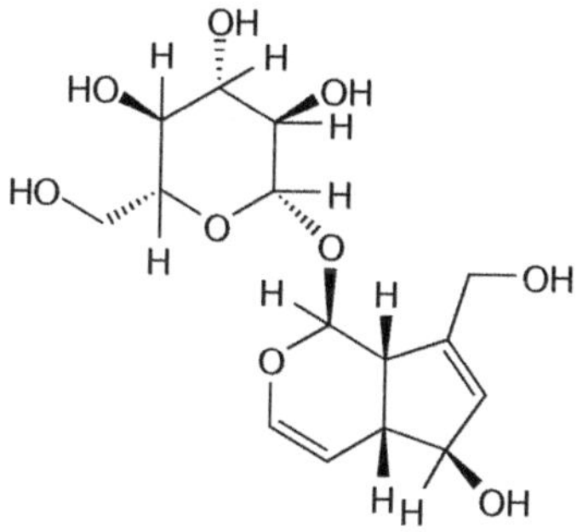

Aucubin (26)

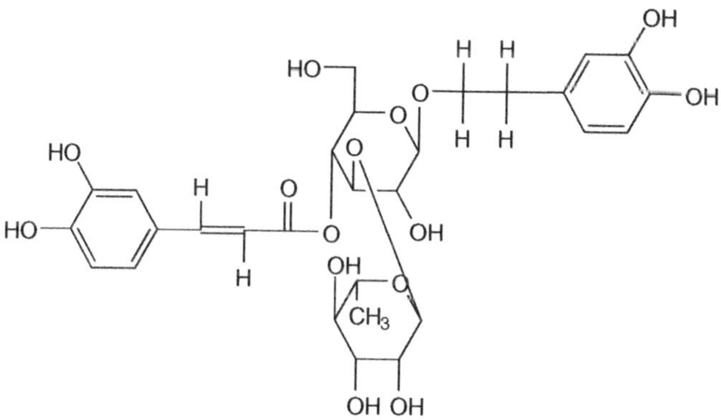

Verbascoside (25)

Figure 1.8 (*Continued*)

Table 1.9 Plants with anti-aging capacity and their mechanisms of action. *Source:* Mukherjee et al. (2011). Reproduced with permission of Elsevier.

S. no.	Plant and family names	Part used	Possible mechanism of action
1	*Aesculus hippocastanum* L. (Hippocastanaceae)	Leaves	Generation of contraction forces
2	*Aloe vera* (Liliaceae)	Gel	Inhibition of stimulated granulocyte MMPs Inhibition of tyrosine hydroxylase and 3,4-dihydroxyphenylalanine oxidase
3	*Astragalus membranaceus* (Fisch.) Bunge (Fabaceae)	Leaves	Increase in the content of hyaluronic acid Increase in the cellular expression of telomerase reverse transcriptase
4	*Berberis aristata* DC (Berberidaceae)	Berries	Inhibition of the expression of MMP-9 and suppressed TPA-induced IL-6 expression Increase in type-I procollagen expression
5	*Calendula officinalis* L. (Asteraceae)	Flower	Control of the activity/secretion of MMP-2 and MMP-9
6	*Camellia japonica* L. (Theaceae)	Oil	Induction of type-I procollagen synthesis and inhibition of MMP-1 activity
7	*Camellia sinesis* L. (Theaceae)	Leaves	Suppression of UV irradiation-induced cutaneous erythema, thickening of the epidermis, overexpression of CK5/6, CK16, MMP-2, MMP-9
8	*Centella asiatica* L. Urban. (Umbelliferae)	Whole plant	Improvement of the clinical score for deep and superficial wrinkles, suppleness, firmness, roughness and skin hydration Induction of type-I collagen synthesis
9	*Citrus sinesis* L. (Rutaceae)	Fruit	NF-κB and AP-1 translocation and procaspase-3 cleavage
10	*Curculigo orchioides* G. (Hypoxidaceae)	Rhizome	Inhibition of MMP-1 expression
11	*Curcuma longa* L. (Zingiberacease)	Rhizome	Inhibition of MMP-2 expression
12	*Curcuma xanthorrhiza* Roxb. (Zingiberaceae)	Rhizome	Inhibition of MMP-1 expression
13	*Dioscorea composita* or *Dioscorea villosa* L. (Dioscoreaceae)		Increased bromodeoxyuridine uptake and intracellular cAMP level in kerationocytes
14	*Emblica officinalis* L. (Euphobiaceae)	Fruit	Inhibition of type-I collagen collagenase Increase of TIMP-1 level
15	*Fraxinus chinesis* Roxb. (Oleaceae)		Decrease in MMP-1 mRNA expression
16	*Glycine max* L. Merr. (Fabaceae)	Seeds	Inhibition of melanosome phagocytosis and prevention of activation of the caspase-3 pathway

Table 1.9 (Continued)

S. no.	Plant and family names	Part used	Possible mechanism of action
17	*kaempferia pandurata* Roxb (Zingiberaceae)	Rhizomes	Inhibition of UV-induced phosphorylations of mitogen-activated protein kinases
18	*Labisia pumita* (Blume) (myrsinaceae)	Root	Inhibition of TNF-α, COX-2, MMP-1 and MMP-9 expression
19	*Machilus thunbergii* Sieb and Zucc (Lauraceae)	Stem bark	Strong inhibition of MMP-1
20	*Magnolia ovovata* Thunb. (Magnoliaceae)		Inhibition of NF-κB-mediated gene expression
21	*Melothria heterophylla* (Lour.) Cogn. (Cucurbitaceae)		Inhibition of MMP-1 activity
22	*Panax ginseng* L. (Araliaceae)	Roots	Type-I procollagen gene and protein expression, prevention of MMP-9 gene induction and elongation of fibrillin-1 fibre length Increase of expression of procolagen type I and decrease of MMP-1
23	*Piper betel* L. (Piperaceae)	Leaves	Protection of photosensitisation-mediated lipid peroxidation (LPO)
24	*Prunus dulcis* Mill. (Rosaceae)	Nut	Antioxidant and free radical scavenging potential
25	*Tagetes erecta* L. (Asteraceae)	Flowers	Photoprotection
26	*Terminalia chebula* Retz. (Combretaceae)	Fruits	Tyrosinase and MMP-2 inhibition Elastase, hyaluronidase, MMP-2 enzyme inhibition
27	*Theobroma cacao* L. (Sterculiaceae)	Bean	Downregulation of hydroxyproline and pepsin-resistant hydroxyproline content
28	*Vaccinium uliginosum* L. (Ericaceae)	Berries	Removal of reactive oxygen species, diminishment of UVB-augmented release of inflammatory interleukin (IL)-6 and IL-8
29	*Viola hondensis* W. Becker and H. Boissieu (Violaceae)	Whole plant	Inhibition of MMP-1 expression, at both the mRNA and the protein levels
30	*Vitis vinifera* L. (Vitaceae)	Shoot	Antioxidant and free radical scavenging capacity
31	*Zingiber officinale* L. (Zingiberaceae)	Rhizomes	Inhibition of fibroblast-derived elastase

1.4 Future Perspectives on the Use of Agricultural Wastes and Recovered Bioactive Compounds

With a growing global population and rapid industrialisation, climate change and environmental pollution, humans are facing many chronic diseases, such as hepatitis, diabetes, cardiovascular, obesity and cancer. Researchers are turning to natural bioactive compounds, which have been proven safe and nontoxic, in an attempt to prevent or cure these threats. The recovery of bioactive compounds from agricultural wastes is a sustainable and effective strategy. A few final points should be considered.

First, a wide range of different techniques can be used for the extraction, isolation and purification of bioactive compounds, including conventional and advanced extraction processes, enzymatic and/or chemical modifications, solid-state fermentation (Santana-Méridas et al., 2012) and metabolic engineering ('biotransformation' or 'bioconversion') (Gil-Chávez et al., 2013). The use of nontoxic or 'green' solvents not only increases extraction yield, reduces solvent and energy consumptions and reduces production costs but also protects human health and reduces environmental pollution in the short and long term.

Second, the reuse of components of the co-products of the food processing industry should be considered, due to the necessity of investing in research, new recovery technologies and/or new production lines. This may entail higher costs than the use of landfills and composting, and it may necessitate a greater collaboration between research and industry. The recovery of bioactive compounds from co-products may be facilitated by the adaptation and use of existing technologies and plants, but economies of scale must be applied to enable the treatment of large volumes of wastes/residues characterised by low specific economic values. The food industry will need to make an effort to reduce energy and water consumption and to increase the recovery of energy from wastes (Baiano, 2014).

Finally, in order to increase consumer acceptability of products containing recovered bioactive compounds, it will be necessary to reduce their price and increase their sensory quality (Baiano, 2014).

1.5 Conclusion

Agricultural wastes are an abundant, biorenewable, sustainable and inexpensive source of bioactive compounds with potential application in the nutraceutical, medical, pharmaceutical and cosmetic industries. Use of products from such sources will enable the application of a sustainable development society.

References

Akpinar, O., Sabanci, S., Levent, O. & Sayaslan, A. 2012. Evaluation of antioxidant activity of dilute acid hydrolysate of wheat straw during xylose production. *Industrial Crops and Products*, **40**, 39–44.

Baiano, A. 2014. Recovery of biomolecules from food wastes-a review. *Molecules*, **19**, 14821–14842.

Balasundram, N., Sundram, K. & Samman, S. 2006. Phenolic compounds in plants and agri-industrial by-products: Antioxidant activity, occurrence, and potential uses. *Food Chemistry*, **99**, 191–203.

Bernini, R. & Mincione, E. 2007. Agroindustrial wastes: a renewable source of materials for fine-chemicals and bioactive compounds by green chemistry methodologies. Available from www.infoil.gr (last accessed March 20, 2017).

Brunton, N. & Wijngaard, H. 2011. Bio-actives from by-products of food processing. *Technology Updates.*

Conde, E., Reinoso, B. D., González-Muñoz, M. J., Moure, A., Domínguez, H. & Parajó, J. C. 2013. Recovery and concentration of antioxidants from industrial effluents and from processing streams of underutilized vegetal biomass. *Food and Public Health*, **3**, 69–91.

Durante, M., Lenucci, M. S., Rescio, L., Mita, G. & Caretto, S. 2012. Durum wheat by-products as natural sources of valuable nutrients. *Phytochemistry Reviews*, **11**, 255–262.

FAOSTAT 2015. FAO Statistic Division. http://faostat3.fao.org/home/E (last accessed March 20, 2017).

Gil-Chávez, G. J., Villa, J. A., Ayala-Zavala, J. F., Heredia, J. B., Sepulveda, D., Yahia, E. M. & González-Aguilar, G. A. 2013. Technologies for extraction and production of bioactive compounds to be used as nutraceuticals and food ingredients: an overview. *Comprehensive Reviews in Food Science and Food Safety*, **12**, 1–19.

Izydorczyk, M. S. & Dexter, J. E. 2008. Barley β-glucans and arabinoxylans: molecular structure, physicochemical properties, and uses in food products – a review. *Food Research International*, **41**, 850–868.

Kalogeropoulos, N., Chiou, A., Pyriochou, V., Peristeraki, A. & Karathanos, V. T. 2012. Bioactive phytochemicals in industrial tomatoes and their processing byproducts. *LWT – Food Science and Technology*, **49**, 213–216.

Kim, K. H., Tsao, R., Yang, R. & Cui, S. W. 2006. Phenolic acid profiles and antioxidant activities of wheat bran extracts and the effect of hydrolysis conditions. *Food Chemistry*, **95**, 466–473.

López, M., Vargas-Garcia, M., Suarez-Estrella, F. & Moreno, J. 2004. Properties of xanthan obtained from agricultural wastes acid hydrolysates. *Journal of Food Engineering*, **63**, 111–115.

Lozano-Sánchez, J., Castro-Puyana, M., Mendiola, J. A., Segura-Carretero, A., Cifuentes, A. & Ibáñez, E. 2014. Recovering bioactive compounds from Olive oil filter cake by advanced extraction techniques. *International Journal of Molecular Sciences*, **15**, 16 270–16 283.

Martins, S., Mussatto, S. I., Martínez-Avila, G., Montañez-Saenz, J., Aguilar, C. N. & Teixeira, J. A. 2011. Bioactive phenolic compounds: production and extraction by solid-state fermentation. A review. *Biotechnology Advances*, **29**, 365–373.

Moure, A., Cruz, J. M., Franco, D., Dominguez, J. M., Sineiro, J., Dominguez, H., Nunez, M. J. & Parajo, J. C. 2001. Natural antioxidants from residual sources. *Food Chemistry*, **72**, 145–171.

Mukherjee, P. K., Maity, N., Nema, N. K. & Sarkar, B. K. 2011. Bioactive compounds from natural resources against skin aging. *Phytomedicine*, **19**, 64–73.

Nawirska, A. & Kwaśniewska, M. 2005. Dietary fibre fractions from fruit and vegetable processing waste. *Food Chemistry*, **91**, 221–225.

Nguyen, V. T. 2014. Mass proportion, proximate composition and effects of solvents and extraction parameters on pigment yield from cacao pod shell (*Theobroma cacao* L.). *Journal of Food Processing and Preservation*, **39**, 1414–1420.

Nguyen, V. T. & Scarlett, C. J. 2016. Mass proportion, bioactive compounds and antioxidant capacity of carrot peel as affected by various solvents. *Technologies*, **4**, 36.

Perretti, G., Miniati, E., Montanari, L. & Fantozzi, P. 2003. Improving the value of rice by-products by SFE. *The Journal of Supercritical Fluids*, **26**, 63–71.

Sainvitu, P., Nott, K., Richard, G., Blecker, C., Jérôme, C., Wathelet, J. P., Paquot, M. & Deleu, M. 2012. Structure, properties and obtention routes of flaxseed lignan secoisolariciresinol: a review. *Biotechnologie, Agronomie, Societe et Environnement*, **16**, 115–124.

Sánchez-Vioque, R., Bagger, C. L., Rabiller, C. & Guéguen, J. 2001. Foaming properties of acylated rapeseed (*Brassica napus* L.) hydrolysates. *Journal of Colloid and Interface Science*, **244**, 386–393.

Sánchez-Vioque, R., Bagger, C. L., Larré, C. & Guéguen, J. 2004. Emulsifying properties of acylated rapeseed (*Brassica napus* L.) peptides. *Journal of Colloid and Interface Science*, **271**, 220–226.

Santana-Méridas, O., González-Coloma, A. & Sánchez-Vioque, R. 2012. Agricultural residues as a source of bioactive natural products. *Phytochemistry Reviews*, **11**, 447–466.

Shaheda, S. A., S., D., R., N., Anusha, P. & Qudusiya, S. M. 2014. A review on natural bioactive compounds as potential anti-wrinkle agents. *World Journal of Pharmacy and Pharmaceutical Sciences*, **3**.

Tarazona-Díaz, M. P., Viegas, J., Moldao-Martins, M. & Aguayo, E. 2011. Bioactive compounds from flesh and by-product of fresh-cut watermelon cultivars. *Journal of the Science of Food and Agriculture*, **91**, 805–812.

Zanwar, A., Hegde, M. & Bodhankar, S. 2011. Cardioprotective activity of flax lignan concentrate extracted from seeds of *Linum usitatissimum* in isoprenalin induced myocardial necrosis in rats. *Interdisciplinary Toxicology*, **4**, 90–97.

2

Bioactive Compounds and Extraction Techniques

Md. Ariful Alam[1], *Md. Zaidul Islam Sarker*[1], *Kashif Ghafoor*[2], *Rukshana Akter Happy*[3] *and Sahena Ferdosh*[4]

[1] *Department of Pharmaceutical Technology, Faculty of Pharmacy, International Islamic University Malaysia (IIUM), Pahang, Malaysia*
[2] *Department of Food Science and Nutrition, King Saud University, Riyadh, Saudi Arabia*
[3] *Department of Biochemistry and Biotechnology, Faculty of Basic Medical and Pharmaceutical Sciences, University of Science and Technology Chittagong (USTC), Foy's Lake, Chittagong, Bangladesh*
[4] *Department of Plant Science, Faculty of Science, International Islamic University Malaysia (IIUM), Kuantan Campus, Pahang, Malaysia*

2.1 History and Definition of Bioactive Compounds

The chemotherapeutic uses of medicinal plants are the prehistoric evidence of humankind itself. Traditional herbal medicine has been used since ancient times, interrelated with indigenous people's beliefs, concepts and practices. Most of the foods originate from plant sources, and plant foods have been used by primitive peoples to cure myriads of ailments.

One of the most primitive urban civilisations in the world was the Sumerians in Southern Mesopotamia. At least 5000 years ago, they explained the use of herbal remedies in written form on clay slabs: this is considered the oldest evidence in the field of natural medicine (Pan et al., 2014). They defined 12 herbal formulations based on more than 250 plants, including laurel, poppy, caraway, henbane, thyme and mandrake (Falodun, 2010). Archaeological evidence also showed the application of natural medicines about 60 000 years ago in Iraq and 8000 years ago in China (Gourhan, 1975). The history of traditional Chinese medicine dates back to 3000 years (Xutian et al., 2009). *Pen T'Sao* is a Chinese book on roots and grasses transcribed by Emperor Shen Nung circa 2500 BCE. This book mentions drugs made from different parts of herbal plants, most of them still used in modern natural medicines (Houghton, 2007). Another book, *The Devine Farmer's Classic of Herbalism*, written 2000 years ago, cites detailed information about the use of herbs in the development of herbal pharmacopoeias (Petrovska, 2012). In ancient Egypt, medical papyri known as 'Ebers Papyrus' provide many prescriptions of plant species for therapeutic, cosmetic and preservative uses, including pomegranate, castor oil plant, aloe, garlic, onion, fig, willow, coriander, juniper and common centaury (Glesinger, 1954; Petrovska, 2012; Vinatoru, 2001).

Recovering Bioactive Compounds from Agricultural Wastes, First Edition. Edited by Van Tang Nguyen.

In the *Iliad* and *Odyssey*, Homer mentions the pharmacological uses of plant species from Mycenaean Greece and Egyptian Assyria. For example, Artemisia is a genus derived from the Greek word 'Artemis' (Petrovska, 2012). Numerous therapeutic uses of medicinal herbs were also mentioned by Hippocrates, Herodotus, Orpheus and Pythagoras (Petrovska, 2012). Hippocrates categorised medicinal plants according to their process and functional activity: *Artemisia absinthium* L. and *Centaurium umbellatum* were applied for treating fever; *Allium sativum* was used for inhibition of intestinal parasites; *Allium cepa, Apium graveolenes, Petroselinum crispum* and *Asparagus officinalis* were prescribed for diuretics; and *Mandragora officinarum* was recommended for nerve stimulating activity (Petrovska, 2012). The herbalists Dioscorides (c. 40–c. 90 CE) and Galen (131–200 CE) were surgeons in the Roman army who gathered their knowledge of medicinal herbs in the *Materia Medica*, which is still a unique compilation in the field of herbalism (Thorwald, 1991). This book represented a vast trove of information on morphology, aetiology, habitat, sample collection criteria and drug formulation. Drugs formulated from different inland plants for the relief of inflammation, headaches, gastrointestinal problems, diuretics and gynaecological complications, and for use as antipyretics, expectorants, cardiac stimulants and antihydrotic agents were recorded in this book (Petrovska, 1990). Meanwhile, Pliny the Elder compiled *Historia Naturalis*, and recorded thousands of medicinal plants (Petrovska, 2012).

In the 16th century, Theophrastus Bombastus von Hohenheim, better known as Paracelsus, wrote *De Causis Plantarium* (Plant Aetiology) and *De Historia Plantarium* (Plant History), where he classified several medicinal plants still recognised in the field of modern natural medicine.

The application of medicinal plants in the form of herbal drugs or in other formulations was the basis of maintaining health until the 19th century. This period saw the beginning of extraction and isolation techniques for the discovery of bioactive molecules such as glycosides, alkaloids, quinine, saponosides, vitamins and hormones. In the same era, the first institute of herbal medicine was founded in Cluj, Romania (Vinatoru, 2001). This was the end of blind faith regarding in the application of herbs to cure certain ailments and the launching of scientific pharmacy.

Plant species produce a broad range of organic compounds, categorised as either primary or secondary metabolites. Primary metabolites are the principle chemical components essential for sustaining life through photometric synthesis, metabolic cellular respiration, growth and improvement. It includes phytosterols, carbohydrates, proteins, organic acids and lipids (Crozier et al., 2006). Whereas, secondary metabolites have diverse chemical structures capable of producing biological effects on human health. Thus, bioactive compounds can be treated as additional alimentary components naturally formed of minor quantities in plants (Kris-Etherton et al., 2002). The literature refers to bioactive compounds variously as 'phytochemicals' (derived from Greek '*phyto*', meaning plant), 'phytocompounds', 'phytomedicines', 'nutraceuticals' (coined by Stephan DeFelice in 1979), 'secondary products' and, in a broad sense, 'natural products'. Generally, plant-based bioactive compounds are formed in secondary metabolic pathways, and hence are defined as 'secondary metabolites' capable of exhibiting certain pharmacological effects on human health (Bernhoft, 2010). These natural compounds are produced in plants as a result of their own defence mechanisms against predators, pathogens, microbes and abiotic stress (Mazid et al., 2010). Most plant species are capable of producing metabolites for protection and attraction (in terms of communication with other organisms), but not necessarily for growth and development (Bernhoft, 2010).

2.2 Classification and Synthesis of Bioactive Compounds

The precise classification of bioactive compounds is not well established due to the extreme diversity of classes and variations in their chemical properties. However, they can be distinguished in different ways based on their chemical structure (e.g. functional groups, ring structure, precursor molecules, sugar moiety, etc.) and chemical contents. Bioactive compounds can be categorised according to their pharmacological activity, but the way of classification is complicated because of the dissimilarities between chemically interrelated compounds and the similarities between different molecules. The taxonomic source (such as family and genus) of the plants from which they were isolated and identified might be another way to categorise them, as plant species within the same family or genus have similar metabolic chains and produce the same or similar analogues. Plant-based secondary metabolites can be distributed into three broad categories according to their biosynthetic or biochemical pathway: terpenoids, alkaloids and phenolic compounds (Croteau, 2000; Harborne, 1993).

In every species biosynthetic pathways control the biological process of life through the formation of primary metabolites. Further, secondary metabolites are synthesised for use in defence mechanisms, but both are linked through a common metabolic chain. The simplest routes for biosynthesis include: glycosides and polysaccharides, synthesised from pentose; phenolics, tannins and alkaloids, synthesised from shikimic acid; phenolics and alkaloids, synthesised from acetate-malonate; and terpenes, steroids and alkaloids, synthesised from mevalonic acid pathway (Dewick, 2001). The interrelated linkage of secondary metabolites with primary metabolism is depicted in Figure 2.1, which shows how the biosynthetic intermediates are connected with the metabolites of the main biochemical pathways (photosynthesis, glycolysis and the Krebs cycle) in the channel of energy generation. According to Dewick (2001), a large margin of secondary metabolites are formed biosynthetically after the intermediate acetyl coenzyme A (acetyl-CoA), shikimic acid, mevalonic acid and 1-deoxyxylulose 5-phosphate. These mediators are used in the acetate, shikimate, mevalonate and deoxy-xylulose phosphate pathways.

2.3 Extraction of Bioactive Compounds

Medicinal plants have been used for a long time in traditional and modern medicine as nutraceuticals and food supplements due to their adequate source of bioactive compounds. Traditional medicine provides potentially useful new compounds for the development of chemotherapeutic agents (Schafer, 1998). It indicates the massive importance of traditional medicines in the industrial processing of herbal plants for pharmaceutical formulations and chemical precursors.

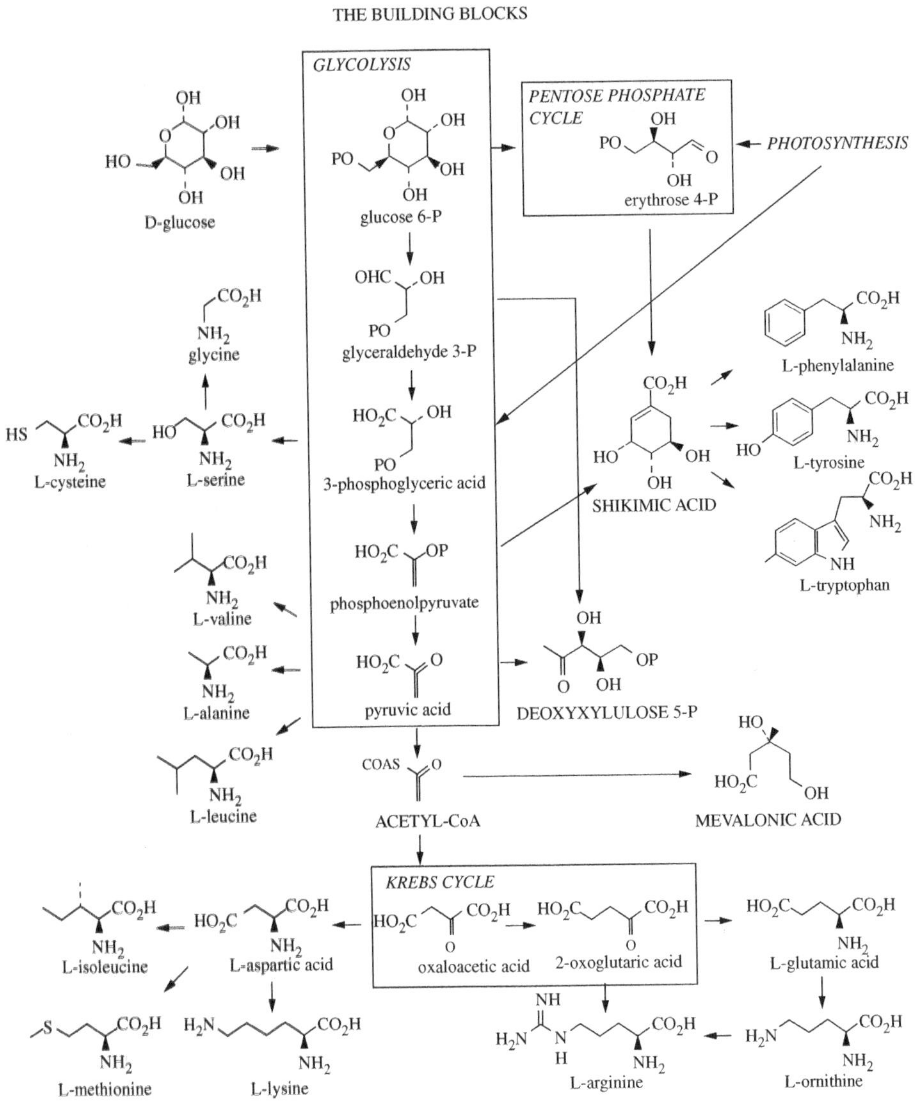

Figure 2.1 Common pathways for secondary metabolite formation from primary metabolism. *Source:* Dewick (2001). Reproduced with permission of John Wiley & Sons.

Thus, the pharmacological effects of precise phytochemicals have been extensively studied in the medical and nutraceutical industries (Ao et al., 2008). However, the industrial processing of such compounds is complicated and involves several crucial steps, such as extraction, pharmacological screening, isolation and characterisation of the bioactive compound, toxicological evaluation and clinical evaluation. Another important thing is to maintain the antioxidant properties of bioactive compounds during processing, because it scavenges free radicals in the human body and minimise oxidative stress (Lobo et al., 2010). Improper physiological function between free radicals and antioxidants is associated with the formations of molecular species, which triggers a number of human diseases (Lobo et al., 2010; McCord, 2000). Therefore, antioxidants from natural sources are used to interrupt the cellular damage, mainly through free radical scavenging activity (Amarowicz et al., 2004; Ghasemzadeh & Jaafar, 2013). However, the bioactive constituents of medicinal plants influenced by some external factors such as environmental conditions, year and season, including the plant's age and leaf maturity. The secondary metabolic processes, as well as the antioxidant activities of bioactive compounds characteristically go hand in hand with age-associated improvements, as reported by Fritz et al. (2001) and Pasko et al. (2009).

Due to the growing interest in natural products, there is a trend towards the development of technologies for the renovation of biomasses into potential natural products in the pharmaceutical, nutraceutical and chemical industries. Hence, technological advances in extraction, separation and identification have been developed in an attempt to produce natural products with potent biological activity (Van Lanen and Shen 2006; Wang and Weller, 2006b). Selection of an appropriate extraction process is a pivotal step because it influences the characterisation of bioactive compounds (Smith, 2003). This process often affects the biological activity, and sometimes the breakdown or damage of synergistic effects of analogue components. On the other hand, the rate of extraction depends upon the distribution constant, temperature, pressure, time and the relative volumes of the phases; therefore, the background knowledge is very important (Smith, 2003). Analysis of thermodynamic process and experimental design can help ensure maximal reproducibility of the extraction process.

Different extraction technologies have been extensively studied as non-conventional methods to enhance the extraction output and the release of bioactive compounds from natural sources, including supercritical fluid extraction (SFE) (Ghafoor et al., 2010; Wang et al., 2008), subcritical water extraction (SWE) (Luque-Rodriguez et al., 2006), microwave-assisted extraction (MAE) (Moreira et al., 2012; Simsek et al., 2012), ultrasound-assisted extraction (UAE) (Chukwumah et al., 2009; Ghafoor et al., 2009), pulsed electric field (PEF) extraction (Luengo et al., 2015; Segovia et al., 2015), enzyme-assisted extraction (EAE) (Wijesinghe & Jeon, 2012), counter-current extraction (Solana et al., 2016), pressurised fluid extraction (PFE) (Herrero et al., 2015), accelerated solvent extraction (ASE) (Zhang et al., 2013), coupled SFE–supercritical fluid chromatography/gas or liquid chromatography (GC/LC) (Patil & Shettigar, 2010) and phytonic extraction (Ghosh et al., 2011).

2.3.1 Supercritical Fluid Extraction

The unique characteristic of supercritical fluids in dissolving low vapour-pressure solid materials was first reported by Hannay & Hogarth (1879). The industrial decaffeination of coffee using supercritical fluid was patented by Zosel (1964).

The foundations of this novel extraction process have been recognised since the early work of Stahl et al. (1980). The application of SFE is broadly studied for its unique physical properties, including gas-like flow, low viscosity and liquid-like densities (Huang et al., 2013; Liu et al., 2009). The extraction of bioactive compounds from various natural sources has been reported in extensive research (del Valle et al., 2012; Wagner et al., 2013).

A substance turns to supercritical state when temperature and pressure increase beyond its critical point, then it manifests some distinctive properties, which enhance the mass transfer from any natural produces (Herrero et al., 2006). In this state, the substance has the properties of both a gas and a liquid (Wang and Weller, 2006a). This fluid has higher diffusivity than liquid and a comparative viscosity to gas (Meisen, 1988). This allows supercritical fluids to diffuse natural solid matrices with less effort than liquids. Thus, SFE is considered a high-value extraction technology for natural products due to the short extraction time, limited solvent use and suitability for thermolabile compounds (Meisen, 1988).

Carbon dioxide is the proper solvent for SFE. It is superior to organic solvents because less expensive, nonhazardous, non-explosive, highly diffuse to targeted compounds and can be easily removed from the finished products (Wang & Weller, 2006a; Wang et al., 2008). In the industrial sectors, SFE is tremendously popular for targeted separation of phytoconstituents from natural sources (Kitzberger et al., 2009). Low and high critical temperatures (T_c) and selectivities are the other key factors that must be considered for an SFE solvent. In case of CO_2, the critical parameters are T_c of 31 °C and P_c of 7.4 MPa; thus, SFE extraction can be favourably carried out without causing thermal damage to the bioactive compounds. In the extraction of highly polar compounds from solid matrices, SFE is less effective (Herrero et al., 2006). The weak solubility of polar molecules into SFE shows their inability to obtain a good separation from natural species (Daintree et al., 2008). In order to overcome this problem, it is crucial to maintain the critical temperature and pressure. The addition of co-solvents or modifiers in small quantities increases the solubility of polar compounds (Sihvonen et al., 1999); this is often called the 'co-solvent effect'. Ethanol is recommended as a co-solvent in SFE, due to its reduced toxicity and miscibility in CO_2 (Liza et al., 2010). Abundant research has been conducted to find out the effect of polar modifiers on SFE from natural sources, along with the solubility of the improvements in SFE (Zuo et al., 2008).

There are two major operations involving in SFE: extraction of bioactive compounds from solid materials by the supercritical fluid solvent and separation of the compounds from the solvent by expansion. Usually, an extractor vessel contains raw material with a temperature and pressure regulator to maintain the optimal conditions. A tank consists of CO_2 as mobile phase, pressurised by a pump to flow into the separator, plus another vessel containing co-solvent, also pressurised to flow and mix with the carrier CO_2 (Figure 2.2). The upcoming fluid with dissolved bioactive compounds from the separator is collected by a tap vessel and the fluid is regenerated and cycled (Zuo et al., 2008).

The industrial application of SFE is quite different from laboratory research, because the extraction is carried out at high pressure. Great effort and expense have been expended in designing productive and economically feasible SFE technology (Herrero et al., 2006).

2.3.2 Subcritical Water Extraction

Subcritical water extraction (SWE) is an environmental friendly process often known as 'hot water extraction under pressure'. It provides rapid, high extraction output from solid substratum, with minimal use of solvent compared to traditional extraction methods (Plaza et al., 2010). Formerly, this method was used for the purification of

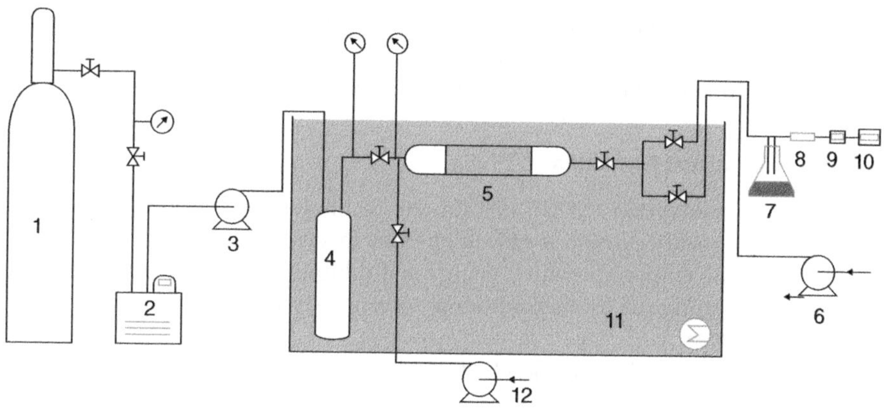

1. CO_2 cylinder
2. Refrigerated bath
3. High-pressure pump of CO_2
4. Supply tank
5. Extractor
6. Peristaltic pump
7. Collection flask
8. Porapak trap Q
9. Gas flow meter
10. Volume totalizer
11. Heater
12. High-pressure pump of ethanol/water

Figure 2.2 Supercritical CO_2 extraction system. *Source:* Garmus et al. (2014). Reproduced with permission of Elsevier.

environmental samples and to extract active compounds from herbal plants (Latawiec and Reid, 2010). Research into SWE has focused mainly on ecological aspects (Okuda et al., 2009) and on the utility of targeted separation of bioactive compounds from natural sources (Zaibunnisa et al., 2009). SWE is suitable for the extraction of antioxidant compounds for nutraceutical applications (Hassas-Roudsari et al., 2009; Ibanez et al., 2003). To date, many natural sources have been used for the extraction of bioactive compounds using SWE, such as antioxidants from rosemary (Ibanez et al., 2003), laurel (Fernandez-Perez et al., 2000), canola meal (Hassas-Roudsari et al., 2009), olive leaves (Ghoreishi & Shahrestani, 2009), bitter melon (*Momordica charantia*) (Budrat & Shotipruk, 2009), grape seeds (García-Marino et al. 2006), onion skin (Ko et al., 2011) and *Citrus unshiu* peel (Cheigh et al., 2012). The phenolic compounds from flax shive (Kim & Mazza, 2006), nutraceuticals such as glycyrrhetic acid, catechins and epicatechin from tea leaves and grape seeds (Pineiro et al., 2004), flavones, anilines and phenols from orange peel (Lamm & Yang, 2003) and caffeine from tea waste (Shalmashi et al., 2010) have been also investigated.

SWE has been used for the separation of essential oils, by changing the influential parameters (Basile et al., 1998). This process is carried out with hot water from 100 to 373 °C, just below the critical temperature, and under high pressure from 10 to 60 bars, in order to retain the water in its liquid form (Herrero et al., 2006). The enhancement of the water temperature slightly decreases the water's dielectric constant, viscosity and surface tension, but increases its molecular diffusion rate (Smith, 2002). The dielectric constant of the water can be decreased to 30 at 250 °C, as this value is similar to that of some other solvents; the same is true in the case of solubility. Hence, this method can be used for the extraction of nonpolar natural bioactive compounds (Adil et al., 2007), without requiring the use of organic solvents. Besides the temperature, other influential factors have a substantial impact on extraction yields, including the extraction time, particle size, solid-to-solvent ratio and pressure (Smith, 2002).

2.3.3 Microwave-Assisted Extraction

Microwave assisted extraction (MAE) is a unique technology which allows the high production of good-quality extract, is easy to operate, involves a short extraction duration, uses minimal solvent, temperature, energy and inhibits degradation of thermolabile compounds, as compared to the traditional solvent-extraction methods (Ma et al., 2012; Wang et al., 2011).

Microwaves are a kind of electromagnetic radiation, with wavelengths between 1 mm and 1 m whereas frequencies ranging from 300 MHz (100 cm) to 300 GHz (0.1 cm).

There are two basic mechanisms of microwave heating: transfer of the energy into the medium by dipole rotation and ionic conduction through reversal of the dipoles and dislocation of the ionised particles present in the solutes and the solvent (Eskilsson & Bjorklund, 2000). These two modes of action take place in various types of application. An electrophoretic migration of ions occurs when an electromagnetic field is applied to a medium, and the resistance of the medium to the flow of ions causes friction and heat. There are two oscillating fields – electric and magnetic both are perpendicular to one another. The transmitted microwaves are absorbed by polar materials and transform this radiation into thermal energy via ionic conduction and dipole rotation mechanisms (Zhang et al., 2011). This evolved energy which accelerates the temperature of moisture within the matrices and vaporises it, generating elevated pressure. As a result, morphological changes in the natural materials soften the tissues by cleaving the cell walls and developing the permeability of the biomaterials. This is the basic principle of the dissolution and extraction of components from biomaterials in MAE.

The extraction conditions are very important to the reproducibility of MAE. The most investigated factors include solvent type, solvent concentration, solvent–sample ratio, irradiation and extraction times, temperature, microwave power and the moisture content of biomaterials. Selection of the solvent system is crucial and should be based on the power of solubilisation of the compounds, the solvent diffusion activity, the dielectric constant (Chen et al., 2008) and the mass transfer kinetics of the process (Spigno & De Faveri, 2009). The solvent is used in the MAE must be capable of absorbing microwave energy; this causes the temperature to increase (Eskilsson & Bjorklund, 2000). For example, hexane is a microwave-transparent solvent, whereas polar solvents like ethanol, methanol and water are heated by absorbing microwave energy (Brachet et al., 2002). MAE is now extensively used in industrial applications for the extraction of bioactive compounds, phytonutrients, functional foods and active pharmaceutical ingredients from biomaterials (Wang and Weller, 2006a). Phenolic compounds have been extracted from a variety of plant sources using MAE (Martino et al., 2006). Sutivisedsak et al. (2010) showed that MAE extracts have high antioxidant activity.

A modified method, open-vessels microwave extraction (OVME), is used for the watery extracts of senna leaves (*Senna folium*), to allow the examination of sennosides A and B. Another fast and proficient MAE method has been used for the targeted extraction of embelin from *Embelia ribes* (Latha, 2007). A continuous microwave system was used to the production level by Terigar et al. (2011) for the investigation of soybean and rice bran oil extraction. Different classes of bioactive compounds have been extracted from various types of medicinal plants (Asghari et al., 2011). Several experiments have reported hypolipidemic, antioxidant and anti-inflammatory activity on the part of guggulsteron. This compound, obtained from *Commiphoria mukul*, has been studied by MAE and compared with traditional extraction methods (Deng, 2007). *Labisia pumila* Benth was investigated for the extraction of flavonoids, isoflavonoids and phenolic compounds by MAE (Karimi & Jaafar, 2011).

2.3.4 Ultrasound-Assisted Extraction

Ultrasound assisted extraction (UAE) technology is an early method, but several advances has been made in recent years (Soria & Villamiel, 2010) and considered as an important process in food engineering operations (Bhaskaracharya et al., 2009). Ultrasound is a sound wave with frequency above 20 kHz and mechanical vibrations in solid, liquid and gas. It has the same properties as regular sound waves and travels through a medium via expansion and contraction cycles. During expansion, bubbles form and create negative pressure, before collapsing (Wang and Weller, 2006b). UAE is comparatively cheap, easy to perform and requires low extraction temperatures, and a competent alternative to conventional extraction methods with high reproducibility and rapid kinetics properties.

In UAE, the surface moisture vaporisation rate upsurges due to the impact of ultrasound waves and produces oscillating velocities and microstreaming at the interfaces, which affect the diffusion boundary layer and cause a sequence of contractions and expansions of the material, affecting mass transfer (Garcia-Perez et al., 2006). It has been used for the extraction of proteins (Qu et al., 2012), sugars (Karki et al., 2010), polysaccharide–protein complexes and oil (Adam et al., 2012). The antioxidant activity of the extracts significantly influenced by the UAE parameters such as time, temperature and frequency (Chukwumah et al., 2009; Ghafoor et al., 2009). In addition, UAE caused less degradation of phenolic compounds during extraction as compared to the other methods (Dobias et al., 2010).

2.3.5 Pulsed Electric Field Extraction

Pulsed electric field (PEF) assisted extraction is a rapid, nonthermal and advanced extraction method for the intracellular components from plant matrices. In PEF, extracting materials are treated for a few microseconds with high-energy electric transmission in an electric field (Angersbach et al., 2000). A polarisation effect on intact membrane induces electropenetrability, altering the cellular membrane, which enhances mass transfer with minimal damage to the cellular tissue (Huang et al., 2012). Electroporation is a phenomenon involving the application of a series of short, high-voltage electric pulses that causes electromechanical compression and hence the formation of pores on the membrane. This electroporation is a dynamic process and can be reversible or irreversible depending on the intensity of the PEF treatment, due to solute diffusion from eukaryotic cellular tissue (Wouters et al., 2001). The irreversible electroporation of the cellular membrane in the plant matrices delivers a broad range of process operations, including expression, extraction or diffusion of cell material (Angersbach et al., 2000). Cell membrane permeability can occur under moderate electric fields (<10 kV/cm) and with low specific energies (<10 kJ/kg) (Puértolas et al., 2013). PEF extraction has been used for selective extraction of polyphenols and anthocyanin from grape peels. The yield of naringin and hesperidin from orange peels was considerably increased at 1–7 kV/cm (Luengo et al., 2013; Medina-Meza & Barbosa-Cánovas, 2015). PEF-assisted extraction was used for bioactive compounds from olive karnel (Roselló-Soto et al., 2015), anthocyanin from red cabbage (Gachovska et al., 2010) and betanin from red beets (Chalermchat et al., 2004).

2.3.6 Enzyme-Assisted Extraction

Enzyme assisted extraction (EAE) of phytoconstituents from plant sources is considered a useful, eco-friendly alternative to conventional solvent-extraction methods. EAE has different advantages, including rapid extraction, maximal yield, minimal solvent usage and low energy consumption as compared to nonenzymatic methods. Intracellular bioactive compounds are maintained intact in the cell cytoplasm and are often found in polysaccharide-lignin chain complexes formed by hydrophobic interactions; these components are not easily extractable through traditional solvent-extraction methods. Enzymatic treatment of plant materials prior to extraction is a suitable way to enhance mass transfer and the discharge of bounded secondary metabolites (Rosenthal et al., 1996). Enzymes are perfect catalysing agents, which ease the breakdown and hydrolysis of complex materials on cell walls and membranes during extraction, causing alteration or synthesis of bioactive constituents of plant origin (Puri et al., 2012). Several enzymes, such as pectinase, cellulase and hemi-cellulases, are extensively used in juice processing and in the separation of suspended solids from beer during brewing (Puri et al., 2012). EAE can be performed in two ways: (i) enzyme-assisted aqueous extraction (EAAE) and (ii) enzyme-assisted cold pressing (EACP). Basically, EAAE methods have been used for the separation of oils from various seeds (Rosenthal et al., 1996), while in EACP processes, enzymes are used for the hydrolytic cleavage of oilseed cell walls (Concha et al., 2004). EAE has been used for the extraction of carotenoids from marigold flower and tomato peel (Barzana et al., 2002; Dehghan-Shoar et al., 2011), glucovanillin from vanilla pods (Ruiz-Terán et al., 2001), oil from grape seeds (Passos et al., 2009) and polyphenolic compounds (Yang et al., 2010), which suggests the prospective application of this method for other plant-originated materials. Enzymatic pretreatment has been used to enhance the separation of flavonoids from citrus peel with minimal use of solvent and heat (Kaur et al., 2010). Application of cellulases and pectinases in EAE significantly increased (206%) the extraction yield of lycopene from tomato peel under optimised parameters (Choudhari & Ananthanarayan, 2007). Pectinase enzyme has been used for the industrial extraction of fruit juices, the removal of suspended solids, the condensation of extract (Nakkeeran et al., 2011) and the separation of pectin (Ptichkina et al., 2008), oil and pigments (Mishra et al., 2005).

2.3.7 Pressurised Liquid Extraction/Pressurised Fluid Extraction/High-Pressure Solvent Extraction/Accelerated Solvent Extraction/Enhanced Solvent Extraction

Pressurised liquid extraction (PLE) is an advanced solvent-extraction technique, which usually operates at high temperature and pressure before reaching the critical point that enhances the efficiency of extraction as compared to normal temperature and pressure

(Camel, 2001). This method was first coined by Richter et al (1996) and also known as other symmetrical names, such as pressurised fluid extraction/high-pressure solvent extraction/accelerated solvent extraction/enhanced solvent extraction (Mustafa & Turner, 2011; Nieto et al., 2010). If water is used as a solvent in PLE, then it is known as 'pressurised hot water extraction' (PHWE), 'subcritical water extraction' or 'superheated water extraction' (Mustafa & Turner, 2011). PLE has several advantages over conventional extraction, including rapid extraction, reduced solvent consumption and the ability to use the filtration process in the extraction cell, whereas automatic instrumentations are the advanced option of PLE-based extraction systems. The PLE technique requires less solvent, because of the operation beyond the atmospheric boiling point promotes solubility and mass transfer properties, as well as decreasing the viscosity and surface tension, therefore accelerated extraction rate (Azmir et al., 2013). Typically, PLE operates with a high temperature (up to 200 °C) and high pressure (up to 200 bar) in order to obtain analytes using a minimal volume of solvent. PLE is composed of two main operating procedures: static and dynamic instruments. The static system comprises a single or a few extraction cycles, without supplying any solvent between cycles. In the dynamic system, solvent is uninterruptedly supplied through the extraction cell using a high-pressure pump. A schematic diagram of a PLE system is presented in Figure 2.3.

PLE requires reduced extraction time and less solvent consumption as compared to Soxhlet extraction (Richter et al., 1996). Selective separation of polar compounds is higher in PLE as compared to SFE (Kaufmann & Christen, 2002). It allows the recovery of a broad range of nonpolar to polar compounds. Certain extraction parameters significantly affect PLE, including type of solvent, temperature, pressure, extraction time, particle size and moisture content (Mustafa & Turner, 2011). PLE is becoming popular for the extraction of nutraceutical components and other phytonutrients. Previously PLE has been used to extract polyphenols from apple and peach pomaces (Adil et al., 2007). The optimisation of factors influencing the yield and bioactive contents in PLE and PHWE has been examined in several studies (Hartonen et al., 2007; Kronholm et al., 2004; Pineiro et al., 2006). The ultimate

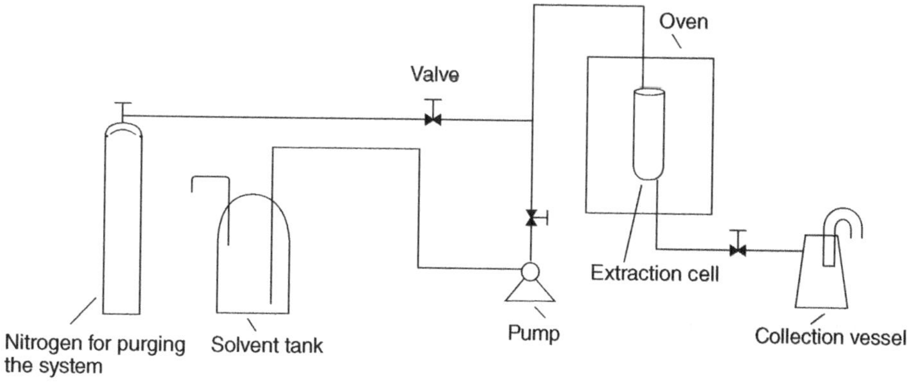

Figure 2.3 Pressurised liquid extraction (PLE) system. *Source:* Richter et al. (1996). Reproduced with permission of American Chemical Society.

findings of this research showed that PLE was the most promising techniques among the convectional methods, with superior recovery rate of catechin and epicatechin from various matrices under optimised conditions in comparison with magnetic stirring and UAE (Pineiro et al., 2004). The systematic approach was applied in the optimisation of PHWE for the recovery of naringenin and other main flavonoids from knotwood of aspen, which obtained higher recoveries than UAE and Soxhlet extraction (Hartonen et al., 2007). Optimal isoflavones were also extracted using PLE from different types of soybeans and soy-based products in comparison with other extraction methods such as vortexing, Soxhlet and sonication (Luthria et al., 2007). PLE has been used to extract lignans from *Schisandra chinensis* (Lee & Kim, 2010) and carotenoids from algae (Denery et al., 2004), herbal teas and vegetables (Loranty et al., 2010).

2.3.8 Phytonic Process

The patented name for phytonic extraction is 'Phytonic Process' (PP). This is one of the newest solvent-based techniques for the extraction of essential oils, fragment components of essential oils and pharmacological extracts using non-chlorofluorocarbons (non-CFCs) (Handa et al., 2008). It is also known as 'florasol extraction' (after Florasol, R134a, one of the refrigerants used to replace Freon). The oils extracted from PP exist in the purest form and include few to no external particles. PP offers significant eco-friendly aspects and has little impact on the environment. In addition, the extraction is carried out below SFE operating temperature, so thermolabile compounds do not degrade. Certain groups of phytoconstituents can be selectively extract in PP by using modified solvent HFC-134a (Handa et al., 2008). The final products of PP are considered safe and healthy due to the low trace of solvent, hence this technique may have application in the industrial processing of herbal formulations, essential oils, fragrances and biologically active natural products.

2.4 Conclusion

The quality and biological activity of the extracts largely depends upon the type of extraction and separation techniques employed in the entire sample preparation process. Certain influential parameters need to be optimised during extraction process in order to get maximum output. Modern extraction technique often linked with advanced chromatograhic and mass spectroscopic method to get a complete metabolite profile of extracted materials. Green approach in extraction technique lead to the development of advanced technology for the reduction of excess solvent evaporation and thus minimisation of huge burden on environment.

Acknowledgement

The study was supported partially by the Research Initiative Grant Scheme (RIGS16-303-0467). The authors extend their appreciation to the International Scientific Partnership Program ISPP at King Saud University, Riyadh, Saudi Arabia, for partially funding this research work through ISPP# 0026.

References

Adam, F., Abert-Vian, M., Peltier, G. & Chemat, F. 2012. 'Solvent-free' ultrasound-assisted extraction of lipids from fresh microalgae cells: a green, clean and scalable process. *Bioresource Technology*, **114**, 457–465.

Adil, İ. H., Çetin, H. İ., Yener, M. E. & Bayindirli, A. 2007. Subcritical (carbon dioxide + ethanol) extraction of polyphenols from apple and peach pomaces, and determination of the antioxidant activities of the extracts. *Journal of Supercritical Fluids*, **43**, 55–63.

Amarowicz, R., Pegg, R. B., Rahimi-Moghaddam, P., Barl, B. & Weil, J. A. 2004. Free-radical scavenging capacity and antioxidant activity of selected plant species from the Canadian prairies. *Food Chemistry*, **84**, 551–562.

Angersbach, A., Heinz, V. & Knorr, D. 2000. Effects of pulsed electric fields on cell membranes in real food systems. *Innovative Food Science & Emerging Technologies*, **1**, 135–149.

Ao, C. W., Li, A. P., Elzaawely, A. A., Xuan, T. D. & Tawata, S. 2008. Evaluation of antioxidant and antibacterial activities of Ficus microcarpa L. fil. extract. *Food Control*, **19**, 940–948.

Asghari, J., Ondruschka, B. & Mazaheritehrani, M. 2011. Extraction of bioactive chemical compounds from the medicinal Asian plants by microwave irradiation. *Journal of Medicinal Plants Research*, **5**, 495–506.

Azmir, J., Zaidul, I. S. M., Rahman, M. M., Sharif, K. M., Mohamed, A., Sahena, F., Jahurul, M. H. A., Ghafoor, K., Norulaini, N. A. N. & Omar, A. K. M. 2013. Techniques for extraction of bioactive compounds from plant materials: a review. *Journal of Food Engineering*, **117**, 426–436.

Barzana, E., Rubio, D., Santamaria, R. I., Garcia-Correa, O., Garcia, F., Ridaura Sanz, V. E. & Lopez-Munguia, A. 2002. Enzyme-mediated solvent extraction of carotenoids from marigold flower (Tagetes erecta). *Journal of Agricultural and Food Chemistry*, **50**, 4491–6

Basile, A., Jimenez-Carmona, M. M. & Clifford, A. A. 1998. Extraction of rosemary by superheated water. *Journal of Agricultural and Food Chemistry*, **46**, 5205–5209.

Bernhoft, A., 2010. A brief review on bioactive compounds in plants. Proceedings from a symposium held at The Norwegian Academy of Science and Letters, Oslo, Norway.

Bhaskaracharya, R. K., Kentish, S. & Ashokkumar, M. 2009. Selected applications of ultrasonics in food processing. *Food Engineering Reviews*, **1**, 31–49.

Brachet, A., Christen, P. & Veuthey, J. L. 2002. Focused microwave-assisted extraction of cocaine and benzoylecgonine from coca leaves. *Phytochemical Analysis*, **13**, 162–169.

Budrat, P. & Shotipruk, A. 2009. Enhanced recovery of phenolic compounds from bitter melon (*Momordica charantia*) by subcritical water extraction. *Sepation and Purification Technology*, **66**: 125–129.

Camel, V. 2001. Recent extraction techniques for solid matrices-supercritical fluid extraction, pressurized fluid extraction and microwave-assisted extraction: their potential and pitfalls. *Analyst*, **126**, 1182–93.

Chalermchat, Y., Fincan, M. & Dejmek, P. 2004. Pulsed electric field treatment for solid-liquid extraction of red beetroot pigment: mathematical modelling of mass transfer. *Journal of Food Engineering*, **64**, 229–236

Cheigh, C. I., Chung, E. Y. & Chung, M. S. 2012. Enhanced extraction of flavanones hesperidin and narirutin from Citrus unshiu peel using subcritical water. *Journal of Food Engineering*, **110**, 472–477.

Chen, L. G., Song, D. Q., Tian, Y., Ding, L., Yu, A. M. & Zhang, H. Q. 2008. Application of on-line microwave sample-preparation techniques. *TrAC – Trends in Analytical Chemistry*, **27**, 151–159.

Choudhari, S. M. & Ananthanarayan, L. 2007. Enzyme aided extraction of lycopene from tomato tissues. *Food Chemistry*, **102**, 77–81.

Chukwumah, Y. C., Walker, L. T., Verghese, M. & Ogutu, S. 2009. Effect of frequency and duration of ultrasonication on the extraction efficiency of selected isoflavones and trans-resveratrol from peanuts (*Arachis hypogaea*). *Ultrasonics Sonochemistry*, **16**, 293–299.

Concha, J., Soto, C., Chamy, R. & Zúñiga, M. E. 2004. Enzymatic pretreatment on rose-hip oil extraction: hydrolysis and pressing conditions. *Journal of the American Oil Chemists' Society*, **81**, 549–552.

Croteau, R., Kutchan, T.M., Lewis, N.G., 2000. Natural products (secondary metabolites). In: Buchanan, B., Gruissem, W. & Jones, R. (eds), *Biochemistry and Molecular Biology of Plants.* American Society of Plant Physiologists, Rockville, MD, pp. 1250–1318.

Crozier, A., Jaganath, I. B. and Clifford, M. N. 2006. Phenols, polyphenols and tannins: an overview. In: Crozier, A., Clifford, M. N. & Ashihara, H. (eds), *Plant Secondary Metabolites: Occurrence, Structure and Role in the Human Diet.* Blackwell, Oxford.

Daintree, L. S., Kordikowski, A. & York, P. 2008. Separation processes for organic molecules using SCF technologies. *Advanced Drug Delivery Reviews*, **60**, 351–372.

Dehghan-Shoar, Z., Hardacre, A. K., Meerdink, G. & Brennan, C. S. 2011. Lycopene extraction from extruded products containing tomato skin. *International Journal of Food Science & Technology*, **46**, 365–371.

Del Valle, J. M., Glatzel, V. & Martinez, J. L. 2012. Supercritical CO2 extraction of allicin from garlic flakes: Screening and kinetic studies. *Food Research International*, **45**, 216–224.

Denery, J. R., Dragull, K., Tang, C. S. & Li, Q. X. 2004. Pressurized fluid extraction of carotenoids from Haematococcus pluvialis and Dunaliella salina and kavalactones from Piper methysticum. *Analytica Chimica Acta*, **501**, 175–181.

Deng, R. 2007. Therapeutic effects of guggul and its constituent guggulsterone: cardiovascular benefits. *Cardiovascular Drug Reviews*, **25**, 375–390.

Dewick, P. M. 2001. Secondary metabolism: the building blocks and construction mechanisms. In: *Medicinal Natural Products: A Biosynthetic Approach* (second edition). John Wiley & Sons, Chichester.

Dobias, P., Pavlikova, P., Adam, M., Eisner, A., Benova, B. & Ventura, K. 2010. Comparison of pressurised fluid and ultrasonic extraction methods for analysis of plant antioxidants and their antioxidant capacity. *Central European Journal of Chemistry*, **8**, 87–95.

Eskilsson, C. S. & Bjorklund, E. 2000. Analytical-scale microwave-assisted extraction. *Journal of Chromatography A*, **902**, 227–250.

Falodun, A. 2010. Herbal medicine in Africa: distribution, standardization and prospects. *Research Journal of Phytochemistry*, **4**, 154–161.

Fernandez-Perez, V., Jimenez-Carmona, M. M. & De Castro, M. D. L. 2000. An approach to the static-dynamic subcritical water extraction of laurel essential oil: comparison with conventional techniques. *Analyst*, **125**, 481–485.

Fritz, R. S. H., Lewkiewicz, D. A., Bothwell, S. & Orians, C. M. 2001. Seedling herbivoryby slugs in a willow hybrid system: Developmental changes in damage, chemical defense, and plant performance. *Oecologia*, **129**, 87–97.

Gachovska, T., Cassada, D., Subbiah, J., Hanna, M., Thippareddi, H. & Snow, D. 2010. Enhanced anthocyanin extraction from red cabbage using pulsed electric field processing. *Journal of Food Science*, **75**, E323–E329.

García-Marino, M., Rivas-Gonzalo, J. C., Ibanez, E. & Garcia-Moreno, C. 2006. Recovery of catechins and proanthocyanidins from winery by-products using subcritical water extraction. *Analytica Chimca Acta*, **563**, 44–50.

Garcia-Perez, J. V., Carcel, J. A., De La Fuente-Blanco, S. & De Sarabia, E. R. F. 2006. Ultrasonic drying of foodstuff in a fluidized bed: Parametric study. *Ultrasonics*, **44**, 539–543.

Garmus, T. T., Paviani, L. C., Queiroga, C. L., Magalhães, P. M. & Cabral, F. A. 2014. Extraction of phenolic compounds from pitanga (*Eugenia uniflora* L.) leaves by sequential extraction in fixed bed extractor using supercritical CO_2, ethanol and water as solvents. *Journal of Supercritical Fluids*, **86**, 4–14.

Ghafoor, K., Choi, Y. H., Jeon, J. Y. & Jo, I. H. 2009. Optimization of ultrasound-assisted extraction of phenolic compounds, antioxidants, and anthocyanins from grape (*Vitis vinifera*) seeds. *Journal of Agricultural and Food Chemistry*, **57**, 4988–4994.

Ghafoor, K., Park, J. & Choi, Y. H. 2010. Optimization of supercritical fluid extraction of bioactive compounds from grape (*Vitis labrusca* B.) peel by using response surface methodology. *Innovative Food Science & Emerging Technologies*, **11**, 485–490.

Ghasemzadeh, A. & Jaafar, H. Z. E. 2013. Profiling of phenolic compounds and their antioxidant and anticancer activities in pandan (*Pandanus amaryllifolius* Roxb.) extracts from different locations of Malaysia. *BMC Complementary and Alternative Medicine*, **13**, 1–9.

Ghoreishi, S. M. & Shahrestani, R. G. 2009. Subcritical water extraction of mannitol from olive leaves. *Journal of Food Engineering*, **93**, 474–481.

Ghosh, U., Haq, M. B. & Chakrabarty, S., 2011. Application of systematic technologies for the extraction of novel phytoconstituents from pharmacologically important plants. *International Journal of Chemical Analytical Science*, **2**, 1153–1158.

Glesinger, L. 1954. Medicine through centuries. *Zagreb: Zora*, 21–38.

Gourhan, A. L. 1975. The flowers found with Shanidar IV, a Neanderthal burial in Iraq. *Science China-Technological Sciences*, **190**, 562–564.

Handa, S., Khanuja, S., Longo, G. & Rakesh, D. 2008. *Extraction Technologies for Medicinal and Aromatic Plants*. Earth, Environmental and Marine Sciences and Technologies, Trieste.

Hannay, J. B. & Hogarth, J. 1879. On the solubility of solid in gases. *Proceeding of the Royal Society of London*, **29**, 324–326.

Harborne, J. 1993. Advances in chemical ecology. *Natural Product Reports*, **10**, 327–348.

Hartonen, K., Parshintsev, J., Sandberg, K., Bergelin, E., Nisula, L. & Riekkola, M. L. 2007. Isolation of flavonoids from aspen knotwood by pressurized hot water extraction and comparison with other extraction techniques. *Talanta*, **74**, 32–8.

Hassas-Roudsari, M., Chang, P. R., Pegg, R. B. & Tyler, R. T. 2009. Antioxidant capacity of bioactives extracted from canola meal by subcritical water, ethanolic and hot water extraction. *Food Chemistry*, **114**, 717–726.

Herrero, M., Cifuentes, A. & Ibañez, E. 2006. Sub- and supercritical fluid extraction of functional ingredients from different natural sources: plants, food by-products, algae and microalgae: A review. *Food Chemistry*, **98**, 136–148.

Herrero, M., Del Pilar Sánchez-Camargo, A., Cifuentes, A. & Ibáñez, E., 2015. Plants, seaweeds, microalgae and food by-products as natural sources of functional ingredients obtained using pressurized liquid extraction and supercritical fluid extraction. *TrAC – Trends in Analytical Chemistry*, **71**, 26–38.

Houghton, P. 2007. Ethnopharmacology of medicinal plants: Asia and the Pacific. *British Journal of Clinical Pharmacology*, **64(2)**, 248.

Huang, K., Tian, H., Gai, L. & Wang, J. 2012. A review of kinetic models for inactivating microorganisms and enzymes by pulsed electric field processing. *Journal of Food Engineering*, **111**, 191–207.

Huang, Z., Li, J. H., Li, H. S., Miao, H., Kawi, S. & Goh, A. H. 2013. Effect of the polar modifiers on supercritical extraction efficiency for template removal from hexagonal mesoporous silica materials: solubility parameter and polarity considerations. *Separation and Purification Technology*, **118**, 120–126.

Ibanez, E., Kubatova, A., Senorans, F. J., Cavero, S., Reglero, G. & Hawthorne, S. B. 2003. Subcritical water extraction of antioxidant compounds from rosemary plants. *Journal of Agricultural and Food Chemistry*, **51**, 375–382.

Karimi, E. & Jaafar, H. Z. 2011. HPLC and GC-MS determination of bioactive compounds in microwave obtained extracts of three varieties of *Labisia pumila* Benth. *Molecules*, **16**, 6791–6805.

Karki, B., Lamsal, B. P., Jung, S., Van Leeuwen, J., Pometto, A. L., Grewell, D. & Khanal, S. K. 2010. Enhancing protein and sugar release from defatted soy flakes using ultrasound technology. *Journal of Food Engineering*, **96**, 270–278.

Kaufmann, B. & Christen, P. 2002. Recent extraction techniques for natural products: microwave-assisted extraction and pressurised solvent extraction. *Phytochemical Analysis*, **13**, 105–113.

Kaur, A., Singh, S., Singh, R. S., Schwarz, W. H. & Puri, M. 2010. Hydrolysis of citrus peel naringin by recombinant α-L-rhamnosidase from Clostridium stercorarium. *Journal of Chemical Technology & Biotechnology*, **85**, 1419–1422.

Kronholm, J., Revilla-Ruiz, P., Porras, S. P., Hartonen, K., Carabias-Martinez, R. & Riekkola, M. L. 2004. Comparison of gas chromatography-mass spectrometry and capillary electrophoresis in analysis of phenolic compounds extracted from solid matrices with pressurized hot water. *Journal of Chromatogr A*, **1022**, 9–16.

Kim, J. W. & Mazza, G. 2006. Optimization of extraction of phenolic compounds from flax shives by pressurized low-polarity water. *Journal of Agricultural and Food Chemistry*, **54**, 7575–7584.

Kitzberger, C. S. G., Lomonaco, R. H., Michielin, E. M. Z., Danielski, L., Correia, J. & Ferreira, S. R. S. 2009. Supercritical fluid extraction of shiitake oil: curve modeling and extract composition. *Journal of Food Engineering*, **90**, 35–43.

Ko, M. J., Cheigh, C. I., Cho, S. W. & Chung, M. S. 2011. Subcritical water extraction of flavonol quercetin from onion skin. *Journal of Food Engineering*, **102**, 327–333.

Kris-Etherton, P. M., Hecker, K. D., Bonanome, A., Coval, S. M., Binkoski, A. E., Hilpert, K. F., Griel, A. E. & Etherton, T. D. 2002. Bioactive compounds in foods: their role in the prevention of cardiovascular disease and cancer. *American Journal of Medicine*, **113**, 71–88.

Lamm, L. J. & Yang, Y. 2003. Off-line coupling of subcritical water extraction with subcritical water chromatography via a sorbent trap and thermal desorption. *Analytical Chemistry*, **75**, 2237–2242.

Latawiec, A. E. & Reid, B. J. 2010. Sequential extraction of polycyclic aromatic hydrocarbons using subcritical water. *Chemosphere*, **78**, 1042–1048.

Latha, C. 2007. Microwave-assisted extraction of embelin from Embelia ribes. *Biotechnology Letters*, **29**, 319–322.

Lee, H. J. & Kim, C. Y. 2010. Simultaneous determination of nine lignans using pressurized liquid extraction and HPLC-DAD in the fruits of Schisandra chinensis. *Food Chemistry*, **120**, 1224–1228.

Liu, G. M., Xu, X., Hao, Q. F. & Gao, Y. X. 2009. Supercritical CO2 extraction optimization of pomegranate (*Punica granatum* L.) seed oil using response surface methodology. *LWT – Food Science and Technology*, **42**, 1491–1495.

Liza, M. S., Rahman, R. A., Mandana, B., Jinap, S., Rahmat, A., Zaidul, I. S. M. & Hamid, A. 2010. Supercritical carbon dioxide extraction of bioactive flavonoid from Strobilanthes crispus (Pecah Kaca). *Food and Bioproducts Processing*, **88**, 319–326.

Lobo, V., Patil, A., Phatak, A. & Chandra, N. 2010. Free radicals, antioxidants and functional foods: impact on human health. *Pharmacognosy Reviews*, **4**, 118–26.

Loranty, A., Rembiałkowska, E., Rosa, E. A. S. & Bennett, R. N. 2010. Identification, quantification and availability of carotenoids and chlorophylls in fruit, herb and medicinal teas. *Journal of Food Composition and Analysis*, **23**, 432–441.

Luengo, E., Álvarez, I. & Raso, J. 2013. Improving the pressing extraction of polyphenols of orange peel by pulsed electric fields. *Innovative Food Science & Emerging Technologies*, **17**, 79–84.

Luengo, E., Martínez, J. M., Coustets, M., Álvarez, I., Teissié, J., Rols, M.-P. & Raso, J. 2015. A comparative study on the effects of millisecond-and microsecond-pulsed electric field treatments on the permeabilization and extraction of pigments from chlorella vulgaris. *Journal of Membrane Biology*, **248**, 883–891.

Luque-Rodriguez, J. M., Perez-Juan, P. & De Castro, M. D. L. 2006. Extraction of polyphenols from vine shoots of Vitis vinifera by superheated ethanol-water mixtures. *Journal of Agricultural and Food Chemistry*, **54**, 8775–8781.

Luthria, D. L., Biswas, R. & Natarajan, S. 2007. Comparison of extraction solvents and techniques used for the assay of isoflavones from soybean. *Food Chemistry*, **105**, 325–333.

Ma, C. H., Yang, L., Zu, Y. G. & Liu, T. T. 2012. Optimization of conditions of solvent-free microwave extraction and study on antioxidant capacity of essential oil from Schisandra chinensis (Turcz.) Baill. *Food Chemistry*, **134**, 2532–2539.

Martino, E., Ramaiola, I., Urbano, M., Bracco, F. & Collina, S. 2006. Microwave-assisted extraction of coumarin and related compounds from *Melilotus officinalis* (L.) Pallas as an alternative to Soxhlet and ultrasound-assisted extraction. *Journal of Chromatography A*, **1125**, 147–151.

Mazid, M., Khan, T. A., Mohammad, F. 2010. Role of secondary metabolites in defense mechanisms of plants. *Biology and Medicine*, **3(2)**.

Mccord, J. M. 2000. The evolution of free radicals and oxidative stress. *American Journal of Medicine*, **108**, 652–659.

Medina-Meza, I. G. & Barbosa-Cánovas, G. V. 2015. Assisted extraction of bioactive compounds from plum and grape peels by ultrasonics and pulsed electric fields. *Journal of Food Engineering*, **166**, 268–275.

Meisen, A. 1988. *Supercritical Fluid Extraction: Principles and Practice.* Butterworth, Stoneham, MA.

Mishra, D., Shukla, A. K., Dixit, A. K. & Singh, K. 2005. Aqueous enzymatic extraction of oil from Mandarin peels. *Journal of Oleo Science*, **54**, 355–359.

Moreira, M. M., Morais, S., Barros, A. A., Delerue-Matos, C. & Guido, L. F. 2012. A novel application of microwave-assisted extraction of polyphenols from brewer's spent grain with HPLC-DAD-MS analysis. *Analytical and Bioanalytical Chemistry*, **403**, 1019–1029.

Mustafa, A. & Turner, C. 2011. Pressurized liquid extraction as a green approach in food and herbal plants extraction: a review. *Analytica Chimica Acta*, **703**, 8–18.

Nakkeeran, E., Umesh-Kumar, S. & Subramanian, R. 2011. *Aspergillus carbonarius* polygalacturonases purified by integrated membrane process and affinity precipitation for apple juice production. *Bioresource Technology*, **102**, 3293–3297.

Nieto, A., Borrull, F., Pocurull, E. & Marcé, R. M. 2010. Pressurized liquid extraction: a useful technique to extract pharmaceuticals and personal-care products from sewage sludge. *TrAC – Trends in Analytical Chemistry*, **29**, 752–764.

Okuda, T., Yamashita, N., Tanaka, H., Matsukawa, H. & Tanabe, K. 2009. Development of extraction method of pharmaceuticals and their occurrences found in Japanese wastewater treatment plants. *Environment International*, **35**, 815–820.

Pan, S. Y., Litscher, G., Gao, S. H., Zhou, S. F., Yu, Z. L., Chen, H. Q., Zhang, S. F., Tang, M. K., Sun, J. N. & Ko, K. M. 2014. Historical perspective of traditional indigenous medical practices: the current renaissance and conservation of herbal resources. *Evidence-Based Complementary and Alternative Medicine*, 525340.

Pasko, P. B., Zagrodzki, H., Gorinstein, P., Fo, S., TA, M., Zachwieja, Z. 2009. Anthocyanins, total polyphenols and antioxidant activity in amaranth and quinoa seeds and sprouts during their growth. *Food Chemistry*, **115**, 994–998.

Passos, C. P., Yilmaz, S., Silva, C. M. & Coimbra, M. A. 2009. Enhancement of grape seed oil extraction using a cell wall degrading enzyme cocktail. *Food Chemistry*, **115**, 48–53.

Patil, P.S. & Shettigar, R., 2010. An advancement of analytical techniques in herbal research. *Journal of Advanced Science Research*, **1**(**1**), 8–14.

Petrovska, B. B. 2012. Historical review of medicinal plants' usage. *Pharmacognosy Review*, **6**, 1–5.

Pineiro, Z., Palma, M. & Barroso, C. G. 2004. Determination of catechins by means of extraction with pressurized liquids. *Journal of Chromatography A*, **1026**, 19–23.

Pineiro, Z., Palma, M. & Barroso, C. G. 2006. Determination of trans-resveratrol in grapes by pressurised liquid extraction and fast high-performance liquid chromatography. *Journal of Chromatography A*, **1110**, 61–5.

Plaza, M., Amigo-Benavent, M., Del Castillo, M. D., Ibanez, E. & Herrero, M. 2010. Facts about the formation of new antioxidants in natural samples after subcritical water extraction. *Food Research International*, **43**, 2341–2348.

Ptichkina, N. M., Markina, O. A. & Rumyantseva, G. N. 2008. Pectin extraction from pumpkin with the aid of microbial enzymes. *Food Hydrocolloids*, **22**, 192–195.

Puértolas, E., Cregenzán, O., Luengo, E., Álvarez, I. & RASO, J. 2013. Pulsed electric field-assisted extraction of anthocyanins from purple-fleshed potato. *Food Chemistry*, **136**(**3**), 1330–1336.

Puri, M., Sharma, D. & Barrow, C. J. 2012. Enzyme-assisted extraction of bioactives from plants. *Trends Biotechnology*, **30**, 37–44.

Qu, W. J., Ma, H. L., Jia, J. Q., He, R. H., Luo, L. & Pan, Z. L. 2012. Enzymolysis kinetics and activities of ACE inhibitory peptides from wheat germ protein prepared with SFP ultrasound-assisted processing. *Ultrasonics Sonochemistry*, **19**, 1021–1026.

Richter, B. E., Jones, B. A., Ezzell, J. L., Porter, N. L., Avdalovic, N. & Pohl, C. 1996. Accelerated solvent extraction: a technique for sample preparation. *Analytical Chemistry*, **68**, 1033–1039

Roselló-Soto, E., Barba, F.J., Parniakov, O., Galanakis, C.M., Lebovka, N., Grimi, N. & Vorobiev, E. 2015. High voltage electrical discharges, pulsed electric field, and ultrasound assisted extraction of protein and phenolic compounds from olive kernel. *Food and Bioprocess Technology*, **8(4)**, 885–894.

Rosenthal, A., Pyle, D. L. & Niranjan, K. 1996. Aqueous and enzymatic processes for edible oil extraction. *Enzyme and Microbial Technology*, **19**, 402–420.

Ruiz-Terán, F., Perez-Amador, I. & López-Munguia, A. 2001. Enzymatic extraction and transformation of glucovanillin to vanillin from vanilla green pods. *Journal of Agricultural and Food Chemistry*, **49**, 5207–5209.

Schafer, K. 1998. Accelerated solvent extraction of lipids for determining the fatty acid composition of biological material. *Analytica Chimica Acta*, **358**, 69–77.

Segovia, F. J., Luengo, E., Corral-Pérez, J. J., Raso, J. & Almajano, M. P. 2015. Improvements in the aqueous extraction of polyphenols from borage (*Borago officinalis* L.) leaves by pulsed electric fields: pulsed electric fields (PEF) applications. *Industrial Crops and Products*, **65**, 390–396.

Shalmashi, A., Abedi, M., Golmohammad, F. & Eikani, M. H. 2010. Isolation of caffeine from tea waste using subcritical water extraction. *Journal of Food Process Engineering*, **33**, 701–711.

Sihvonen, M., Jarvenpaa, E., Hietaniemi, V. & Huopalahti, R. 1999. Advances in supercritical carbon dioxide technologies. *Trends in Food Science & Technology*, **10**, 217–222.

Simsek, M., Sumnu, G. & Sahin, S. 2012. Microwave-assisted extraction of phenolic compounds from sour cherry pomace. *Separation Science and Technology*, **47**, 1248–1254.

Smith, R. M. 2002. Extractions with superheated water. *Journal of Chromatography A*, **975**, 31–46.

Smith, R. M. 2003. Before the injection – modern methods of sample preparation for separation techniques. *Journal of Chromatography A*, **1000**, 3–27.

Solana, M., Teel, J., Hojilla-Evangelista, M., Bertucco, A. & Eller, F., 2016. Counter-current carbon dioxide extraction of fat from soy skim. *Journal of Supercritical Fluids*, **113**, 106–111.

Soria, A. C. & Villamiel, M. 2010. Effect of ultrasound on the technological properties and bioactivity of food: a review. *Trends in Food Science & Technology*, **21**, 323–331.

Spigno, G. & De Faveri, D. M. 2009. Microwave-assisted extraction of tea phenols: a phenomenological study. *Journal of Food Engineering*, **93**, 210–217.

Stahl, E., Schuetz, E. & Mangold, H. K. 1980. Extraction of seed oils with liquid and supercritical carbon dioxide. *Journal of Agricultural and Food Chemistry*, **28**, 1153–1157.

Sutivisedsak, N., Cheng, H., Willett, J., Lesch, W., Tangsrud, R. & Biswas, A. 2010. Microwave-assisted extraction of phenolics from bean (*Phaseolus vulgaris* L.). *Food Research International*, **43**, 516–519.

Terigar, B. G., Balasubramanian, S., Sabliov, C. M., Lima, M. & Boldor, D. 2011. Soybean and rice bran oil extraction in a continuous microwave system: From laboratory to pilot scale. *Journal of Food Engineering*, **104**, 208–217.

Thorwald, J. 1991. *Power and Knowledge of Ancient Physicians.* August Cesarec, Zagreb.

Van Lanen, S. G. & Shen, B. 2006. Microbial genomics for the improvement of natural product discovery. *Current Opinion in Microbiology*, **9**, 252–260.

Vinatoru, M. 2001. An overview of the ultrasonically assisted extraction of bioactive principles from herbs. *Ultrasonics Sonochemistry*, **8**, 303–13.

Wagner, M. E., French, J. & Rizvi, S. S. H. 2013. Supercritical fluid extraction of oil from potato chips: Two scale comparison and mathematical modeling. *Journal of Food Engineering*, **118**, 100–107.

Wang, L. & Weller, C. L. 2006a. Recent advances in extraction of nutraceuticals from plants. *Trends in Food Science & Technology*, **17**, 300–312.

Wang, L. J. & Weller, C. L. 2006b. Recent advances in extraction of nutraceuticals from plants. *Trends in Food Science & Technology*, **17**, 300–312.

Wang, L., Weller, C. L., Schlegel, V. L., Carr, T. P. & Cuppett, S. L. 2008. Supercritical CO2 extraction of lipids from grain sorghum dried distillers grains with solubles. *Bioresource Technology*, **99**, 1373–1382.

Wang, S. Y., Yang, L., Zu, Y. G., Zhao, C. J., Sun, X. W., Zhang, L. & Zhang, Z. H. 2011. Design and performance evaluation of ionic-liquids-based microwave-assisted environmentally friendly extraction technique for camptothecin and 10-hydroxycamptothecin from samara of camptotheca acuminata. *Industrial & Engineering Chemistry Research*, **50**, 13620–13627.

Wijesinghe, W.A.J.P. & Jeon, Y.J., 2012. Enzyme-assistant extraction (EAE) of bioactive components: a useful approach for recovery of industrially important metabolites from seaweeds: a review. *Fitoterapia*, **83(1)**, 6–12.

Wouters, P. C., Alvarez, I. & Raso, J. 2001. Critical factors determining inactivation kinetics by pulsed electric field food processing. *Trends in Food Science & Technology*, **12**, 112–121.

Xutian, S., Zhang, J. & Louise, W. 2009. New exploration and understanding of traditional Chinese medicine. *American Journal of Chineese Medicine*, **37**, 411–26.

Yang, Y. C., Li, J., Zu, Y. G., Fu, Y. J., Luo, M., Wu, N. & Liu, X. L. 2010. Optimisation of microwave-assisted enzymatic extraction of corilagin and geraniin from Geranium sibiricum Linne and evaluation of antioxidant activity. *Food Chemistry*, **122**, 373–380.

Zaibunnisa, A. H., Norashikin, S., Mamot, S. & Osman, H. 2009. An experimental design approach for the extraction of volatile compounds from turmeric leaves (*Curcuma domestica*) using pressurised liquid extraction (PLE). *LWT – Food Science and Technology*, **42**, 233–238.

Zhang, H. F., Yang, X. H. & Wang, Y. 2011. Microwave assisted extraction of secondary metabolites from plants: Current status and future directions. *Trends in Food Science & Technology*, **22**, 672–688.

Zhang, Y., Liu, C., Qi, Y., Li, S. & Wang, J., 2013. Application of accelerated solvent extraction coupled with counter-current chromatography to extraction and online isolation of saponins with a broad range of polarity from Panax notoginseng. *Separation and Purification Technology*, **106**, 82–89.

ZOSEL, K. 1964. German Patent No. **1493**, 190.

Zuo, Y. B., Zeng, A. W., Yuan, X. G. & Yu, K. T. 2008. Extraction of soybean isoflavones from soybean meal with aqueous methanol modified supercritical carbon dioxide. *Journal of Food Engineering*, **89**, 384–389.

3

Recovering Bioactive Compounds from Tea, Coffee, Cacao and Cashew Wastes

Van Tang Nguyen

School of Environmental and Life Sciences, Faculty of Science and Information Technology, University of Newcastle, Ourimbah, NSW, Australia
Department of Food Technology, Faculty of Food Technology, Nha Trang University, Nha Trang, Khanh Hoa, Vietnam

3.1 Introduction

Bioactive compounds, or phytochemical constituents, are of great interest due to their potential application in the nutraceutical, medical, pharmaceutical and cosmetic industries. Therefore, a number of valuable bioactive compounds have been exploited from different materials, particularly inexpensive materials from agriculture, forestry and aquaculture. Tea, coffee and cacao are three of the most popular drinks in the world, with annual worldwide consumption approximately 4.0, 7.0 and 3.5 million tonnes, respectively (Goksu & Poyrazoglu, 2013; Hii et al., 2009; Jalil & Ismail, 2008; Nuhu, 2014; Rodrigues & Bragagnolo, 2013). They are rich in phenolics and alkaloids. Cashew nut, meanwhile, is rich in lutein, β-carotene and unsaturated fatty acids, such as oleic acid and linoleic acid, and has an annual worldwide consumption of about 3.5 million tonnes (Trox et al., 2010). The wastes and byproducts of tea, coffee, cacao and cashew processing and production also have potential in creating higher value-added products, particularly bioactive compounds with potential application in the nutraceutical, medical, pharmaceutical and cosmetic industries. Therefore, a wide range of tea, coffee, cacao and cashew wastes and their major bioactive compounds are presented in this chapter, in order to examine their potential in different industrial fields. In addition, a number of traditional and advanced techniques for the extraction, isolation and purification of valuable bioactive compounds from tea, coffee, cacao and cashew wastes are also discussed.

3.2 Recovering Bioactive Compounds from Tea Wastes

Tea is a valuable industrial tree with wide distribution across the world, including in China, India, Kenya, Sri Lanka, Turkey, Vietnam, Iran, Indonesia, Argentina and Japan. Total harvested tea area and tea production worldwide in 2013 were estimated at about 3.52 million ha and 5.35 million tonnes, respectively, with a gross production value approximately US$22.22 billion (FAOSTAT, 2015). Tea and its products are rich

Recovering Bioactive Compounds from Agricultural Wastes, First Edition. Edited by Van Tang Nguyen.

sources of phenolic compounds (Goksu & Poyrazoglu, 2013), particularly catechins, which possess a wide range of potent biological properties; for example, the antioxidant activity of Oolong tea is comparable to that of vitamin C (Nguyen et al., 2011), and it also has antihypertensive, anti-aging and anticancer activities (Cabrera et al., 2006). Tea wastes include tea fruit, flower, old tea leaves (OTL) and tea stalk during harvesting of leaves, as well as tea dust, stalk and fibre during tea production (Farhoosh et al., 2007).

Figure 3.1 indicates the main steps in green and black tea production and the wastes discarded from these processes, which include OTL and tea stalk during grading of fresh tea leaves; green tea stalk, fibre and dust during sorting of green tea; and black tea stalk, fibre and dust during sorting of black tea.

Yang et al. (2007) determined that tea flower was a rich source of flavones, polyphenols, catechins and caffeine (36.1, 298.0, 134.0 and 34.4 mg/g dried weight), with the ethyl acetate fraction having the highest levels of flavones, polyphenols and catechins with the greatest antioxidant capacity against free radicals. Xu et al. (2012) extracted phenolic antioxidants from tea fruit peel and found that the highest total phenolic content (47.5 mg gallic acid equivalents (GAE)/g dry peel) obtained under optimised parameters of 43% ethanol, 60 °C and 33 minutes ranged from 2.71 to 38.22 mg/g dry peel, with gallocatechin and epigallocatechin the major catechins in tea fruit peel. These compounds displayed potent ferric-reducing antioxidant power. Lee & Yen (2006) isolated and identified two novel compounds in methanol extract from tea seed oil – sesamin and 2,5-bis-benzo[1,3]dioxol-5-yl-tetrahydro-furo [3,4-d][1,3]dioxine (compound B) – which exhibited remarkable antioxidant activity (Figure 3.2).

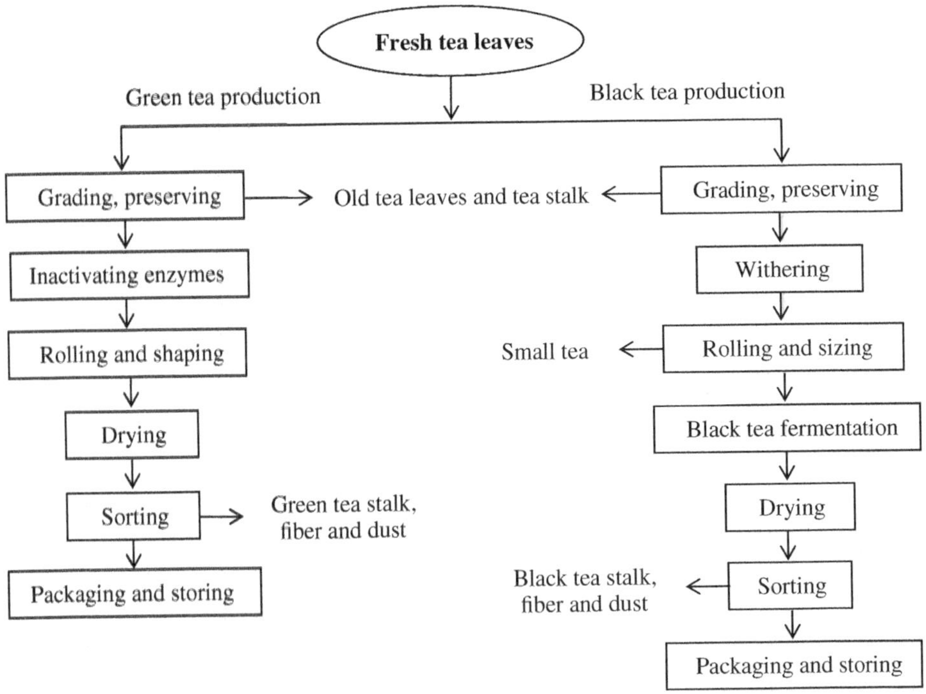

Figure 3.1 Main steps in green and black tea production.

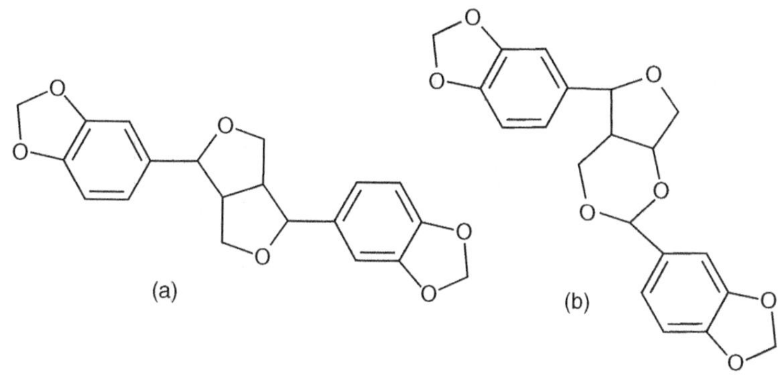

Figure 3.2 Chemical structure of (a) sesamin and (b) compound B isolated from methanol extract of tea seed oil. *Source:* Lee & Yen (2006). Reproduced with permission of American Chemical Society.

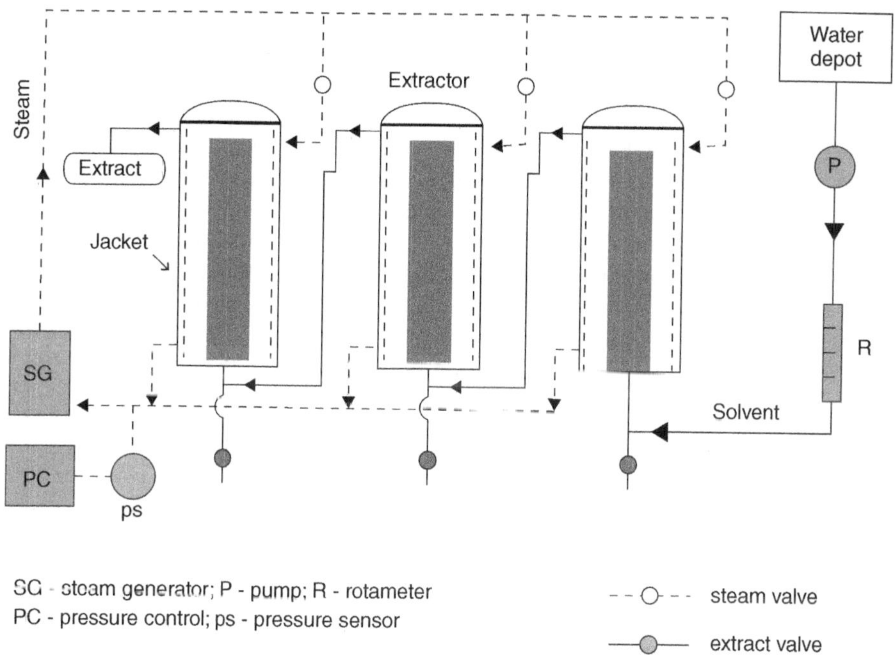

Figure 3.3 Experimental battery plant for solid–liquid extraction (SLE) of caffeine from tea waste with three extractors. *Source:* Senol & Aydin (2006). Reproduced with permission of Elsevier.

Senol & Aydin (2006) optimised the extraction process for caffeine from tea waste obtained during black tea production using a battery-type extractor. They found that water and chloroform were effective solvents for solid–liquid extraction (SLE) of caffeine from tea waste (2% by dry weight) at isothermal conditions of 370 K for water and 293 K for chloroform in the extraction plant, with either three or five extractors (Figure 3.3). Kondo et al. (2007) determined the total extractable phenolics, total extractable tannins and condensed tannins in green tea waste (GTW) and black tea waste (BTW) and found they were 11.3, 9.57 and 2.42 and 7.84, 6.08 and 0.75% by dry weight, respectively.

Farhoosh et al. (2007) studied the extraction of antioxidants from OTL and BTW using various solvents (hot water, ethyl acetate and methanol) and indicated that hot water achieved the highest extraction yields (27 and 24% by dry weight for BTW and OTL, respectively) as compared to methanol and ethyl acetate (12 and 7% and 3.5 and 2% by dry weight for OTL and BTW, respectively). On the other hand, antioxidant capacity based on the reducing power of hot water extracts from both OTL and BTW was lower than that from methanolic extracts and much lower than that from ethyl acetate extracts. Ye et al. (2009) reported that tea stalk was an inexpensive adsorbent for separating catechins from decaffeinated tea (146.7 mg total catechins/g tea) for further application.

The maximum yield of caffeine from tea stalk and fibre wastes was extracted by supercritical carbon dioxide (SC-CO_2) without consolvent (extraction time 7 hours, temperature 60 °C, CO_2 flow rate 11 g/min, pressure 250 bar, particle size 0.202 mm) and with ethanol as co-solvent (flow ratio 5.23 g ethanol/100 g CO_2 (10 g CO_2/min), extraction time 3 hours, temperature 65 °C, pressure 250 bar, particle size 0.202 mm). Yields were 14.9, 19.2 and 14.95, 18.92 mg/g dry weight and 1.49, 1.92 and 1.49, 1.89% by dry weight, respectively; this was 61.9, 65.5 and 62.5, 63.1% greater than with chloroform extraction (0.92 and 1.16% by dry weight) (Gürü & İçen, 2004; İçen & Gürü, 2009, 2010), with a decrease of operational time from 7 to 3 hours when using ethanol as co-solvent in the supercritical extraction (Figure 3.4). Martono (2010) extracted phenolic compounds from industrial GTW and BTW in order to apply an ultraviolet (UV) filter and indicated that the total phenolic content in BTW (35.77 mg GAE/g dry weight) was much higher than that in GTW (13.18 mg GAE/g dry weight), leading to a much stronger UV filter effectiveness but reduced antioxidant activity. These data reveal that the UV filter effectiveness and antioxidant capacity of tea extracts are greatly influenced by the phenolic constituents in tea wastes, where GTW are rich in catechins (non-oxidised process, with high antioxidant capacity), while BTW is rich in theaflavin and thearubigin (oxidised process, with high effectiveness against UV).

Shalmashi et al. (2010) isolated caffeine from tea waste using a subcritical water extraction system and indicated that a water temperature of 175 °C, pressure of 20 bar,

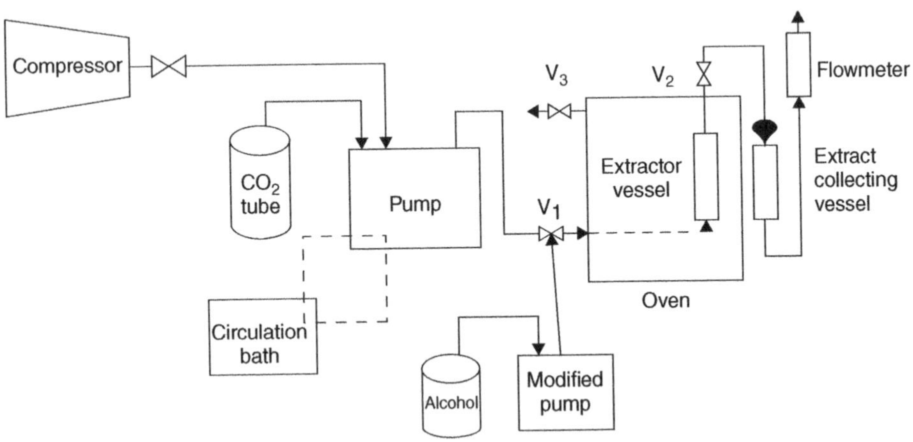

Figure 3.4 Flow diagram of the supercritical extraction set-up. *Source:* İçen & Gürü (2010). Reproduced with permission of Elsevier.

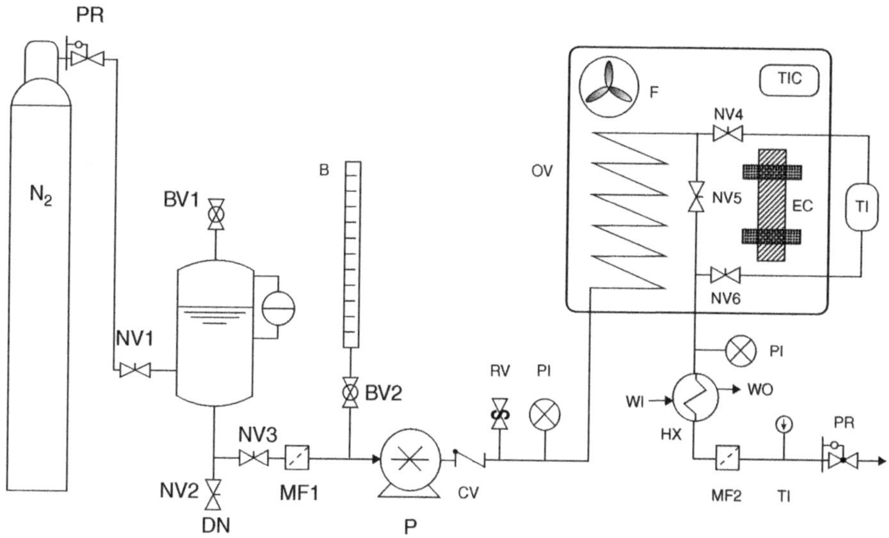

Figure 3.5 Schematic diagram of subcritical water extraction system. B, burette; CV, check valve; DN, drain; EC, extraction cell; F, fan; HX, heat exchanger; MF, microfilter; NV, needle valve; OV, oven; P, pump; PI, pressure indicator; PR, pressure regulator; RV, relief valve; TI, temperature indicator; TIC, temperature indicator controller; WI, water inlet; WO, water outlet. *Source:* Shalmashi et al. (2010). Reproduced with permission of John Wiley & Sons.

water flow rate of 4 g/min, extraction time of 1.5 hours and particle size of 0.5 mm were optimal for recovery of 0.77% caffeine from tea waste (Figure 3.5). Taheri et al. (2011) determined the phenolic compounds and antioxidant capacities of OTL and tea plant barks and found that total phenolic content in a mixture of tea wastes was 17.5 mg GAE/g dry weight and that antioxidant capacity was comparable with that of black tea (0.40 and 0.42 mmol Fe/ml of ferric-reducing antioxidant power, and 40.4 and 62.6% DPPH radical scavenging capacity for tea wastes and black tea, respectively).

Figure 3.6 shows the chemical structures of some major constituents of tea products and wastes (Cabrera et al., 2006).

Heo et al. (2014) applied accelerated solvent extraction under optimal conditions (80% ethanol, 160 °C extraction temperature, 5-minute extraction time, two extractions, 1500 psi) for the extraction of lutein from green tea byproducts following the production of green tea beverages, and found that the lutein content of these byproducts was 18.18 mg/100 g dry weight, which was equivalent to 57.57% of that in green tea (32.67 mg/100 g dry weight). Recently, Nguyen (2017b) studied the extraction of phenolic compounds from black tea stalk and fibre (Figure 3.7) and indicated that ethanol concentration, extraction temperature, extraction time, ratio of solvent to sample and number of extractions had great effects on the extraction efficiency of phenolic compounds. Total phenolic content in black tea stalk and fibre ranged from 208 to 844 and from 227 to 1137 mg GAE/g dry weight, respectively. The optimal parameters for the effective extraction of phenolic compounds from black tea stalk and fibre were found to be 80% ethanol, 50 °C extraction temperature, 3-hour extraction time, 15 ml/g solvent to sample ratio and two extractions.

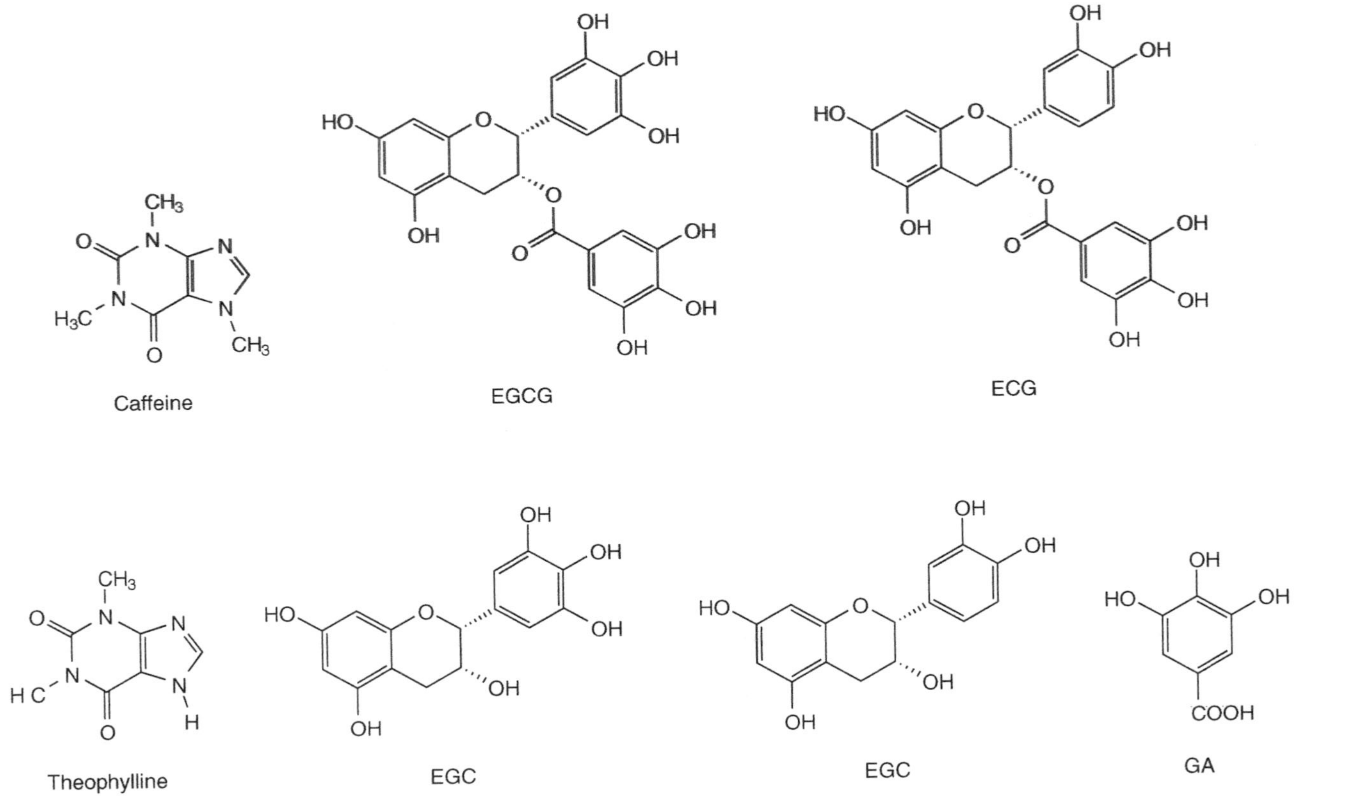

Figure 3.6 Chemical structures of caffeine, theophylline, gallic acid and four major catechins in tea products and wastes. GA, gallic acid; EGCG, (−)-epigallocatechin-3-gallate; EGC, (−)-epigallocatechin; ECG, (−)-epicatechin-3-gallate; EC, (−)-epicatechin. *Source:* Cabrera et al. (2006). Reproduced with permission of Taylor & Francis.

(a)

(b)

Figure 3.7 (a) Black tea stalk and (b) fibre obtained from black tea production.

3.3 Recovering Bioactive Compounds from Coffee Wastes

Coffee is an important industrial tree in many countries, including Brazil, Vietnam, Indonesia, Colombia, Ethiopia, India, Peru, Honduras, Mexico and Guatemala. Total harvested coffee area and green coffee production worldwide in 2013 were estimated at about 10.14 million ha and 8.92 million tonnes, respectively, with a gross production value of approximately US$16.37 billion (FAOSTAT, 2015). Coffee contains a significant amount of bioactive compounds (Nuhu, 2014; Rodrigues & Bragagnolo, 2013), particularly caffeine, which is demonstrated to have stimulatory effects on the central nervous, gastrointestinal, cardiovascular, renal and respiratory systems, resulting in energy arousal and increased motivation to work, alertness and cognitive function (de Sena et al., 2011). Coffee production results in a large amount of wastes, including outer skin/pericarp, pulp/mucilage, parchment/hull/husk and silver skin from the green coffee industry (Esquivel & Jimenez, 2012; Murthy & Naidu, 2010) and spent coffee grounds from the soluble coffee industry (Acevedo et al., 2013; Bravo et al., 2012; Ramalakshmi et al., 2009; Zuorro & Lavecchia, 2012). The ratios of outer skin to pulp, mucilage and parchment per whole coffee fruit are 43.2, 11.8 and 6.1% w/w, respectively (Esquivel & Jimenez, 2012). Oliveira et al. (2006) also consider defective coffee beans, such as black, brown (sour), immature and broken beans, coffee byproducts; these form up to 20% of total coffee production.

3.3.1 Bioactive Compounds from Green Coffee Production

As already mentioned, approximately 60% of coffee fruit is not used for the production of commercial green coffee, and is therefore discarded during processing. There can be harmful effects on the environment if this waste is not treated properly. Studies indicate that coffee pulp is a rich source of bioactive compounds. Rathinavelu & Graziosi (2005) reported that coffee mucilage, particularly the pulp, contains a number of polyphenols, anthocyanins, proanthocyanins, cyanidins, bioflavonoids, tannins, caffeine and chlorogenic acids, while Esquivel & Jimenez (2012) found catechin, epicatechin, chlorogenic acid (5-caffeoylquinic acid), 3,4-, 3,5-, 4,5-dicaffeoylquinic acids, rutin, protocatechuic acid, ferulic acid, 5-feruloylquinic acid, cyanidin-3-rutinoside, cyanidin-3-glucoside, aglycone, proanthocyanidins (condensed tannins) and caffeine in coffee pulp.

Figure 3.8 shows the main steps in green coffee production and the wastes discarded from this process, including coffee husk, hull and silver skin from dry processing and outer skin, pulp/mucilage, parchment/hull and silver skin from wet processing.

Murthy & Naidu (2010) reported that the levels of chlorogenic acid and total phenolics in coffee pulp, husk and silver skin were 2.4, 2.5, 3.0% and 1.48, 1.22, 1.32% GAE, respectively, while Bouafou et al. (2011) reported that the caffeine contents in coffee hull and pulp were 0.9 and 0.8% by dry weight, respectively, and that the content of tannins in coffee pulp was 0.92% by dry weight. Esquivel & Jimenez (2012) indicated that the caffeine content in the outer skin of coffee fruit was much lower than that in coffee bean. The chemical structures of some major bioactive compounds found in coffee products and wastes are shown in Figure 3.9 (Nuhu, 2014).

3.3.2 Bioactive Compounds from Soluble Coffee Production

Ground or insoluble coffee is produced from green coffee beans through two major steps: roasting and grinding. This is also a main step in the production of soluble or instant coffee. The residues left after extraction of soluble compounds from ground coffee are called 'spent coffee grounds' (Figure 3.10). These grounds possess a relevant amount of bioactive compounds, including phenolics, hydrophilic antioxidants (e.g. 3-, 4-, 5-monocaffeoylquinic acids and 3,4-, 3,5-, 4,5-dicaffeoylquinic acids), caffeine and browned compounds (melanoidins). The caffeine (3.59–8.09 mg/g dry matter) and total

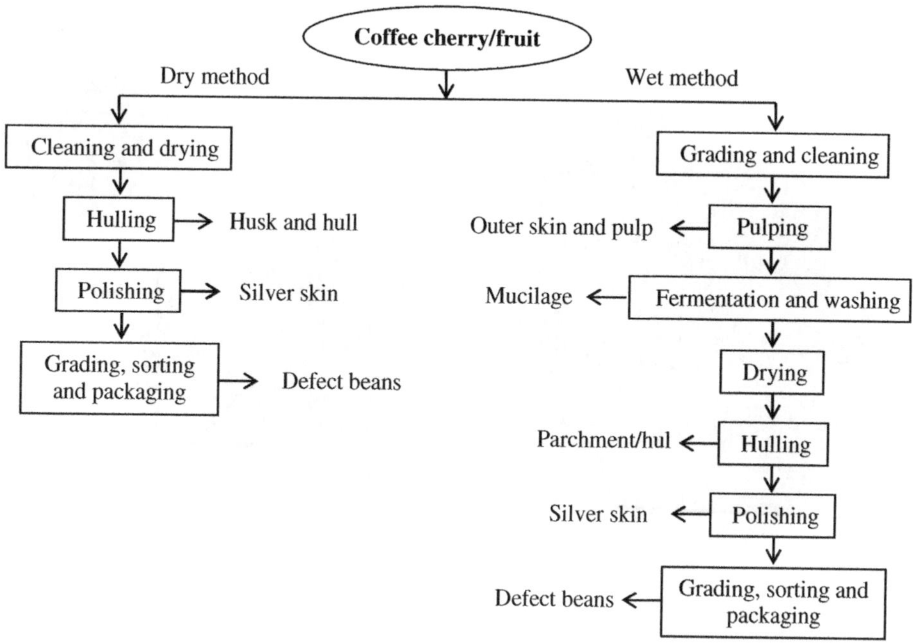

Figure 3.8 Main steps in green coffee production.

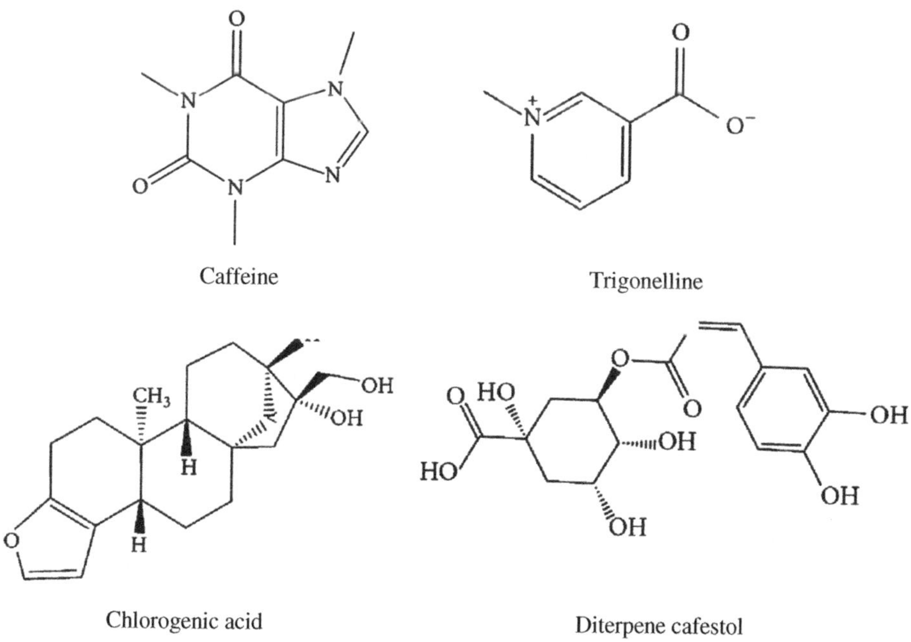

Figure 3.9 Chemical structures of caffeine, trigonelline, chlorogenic acid and diterpene cafestol in coffee products and wastes. *Source:* Adapted from Nuhu (2014).

Figure 3.10 Spent coffee grounds discarded from soluble coffee production.

caffeoylquinic acid (6.22–13.24 mg/g dry matter; particularly dicaffeoylquinic acid, 3.31–5.79 mg/g dry matter) content in spent coffee grounds is much higher than that in coffee brews (Bravo et al., 2012). Total phenolic content ranges from 17.75 to 21.56 mg GAE/g dry matter (Zuorro & Lavecchia, 2012), while polyunsaturated fatty acids (50%; mainly linoleic and palmitic acids) and diterpenes (2.14 mg/g for kahweol and 4.66 mg/g for cafestol) range from 255 to 273 mg GAE/g dry matter, as reported by Acevedo et al. (2013). Levels of chlorogenic acid and total phenolics are 2.3 and 1.02% GAE, respectively (Murthy & Naidu, 2010). Importantly, these compounds have direct links to potential biological activity, such as antioxidant, antidiabetec, anticardiovascular, antitumour, anticancer and anti-Parkinson's disease activities (Ludwig et al., 2014; Ramalakshmi et al., 2009).

Figure 3.11 illustrates the main steps in soluble coffee production.

A number of solvents and extraction techniques have been applied to extract the bioactive compounds from coffee wastes. For instance, Ramalakshmi et al. (2009) used methanol and the Soxhlet system for the extraction of bioactive compounds from spent coffee grounds and found that the contents of caffeine, chlorogenic acids and total phenolics were 1.0–2.5 and 4.87–5.62% and 4.81–6.32 mg GAE/g of the extracts, respectively. Zuorro & Lavecchia (2012) used aqueous ethanol and the conventional method (water bath) to extract phenolic compounds, and found that the highest total phenolic content (17.75–21.56 mg GAE/g dry matter) was achieved under the optimal conditions of 70% ethanol, 50 °C, 2-hour extraction time and 40 ml/g solvent–sample ratio.

Bravo et al. (2012) applied petroleum ether and the Soxhlet system to defat, then used water and a filter coffeemaker at 90 °C for 6 minutes to extract the hydrophilic bioactive compounds. The phenolic compounds (255–273 mg GAE/g dry matter) and

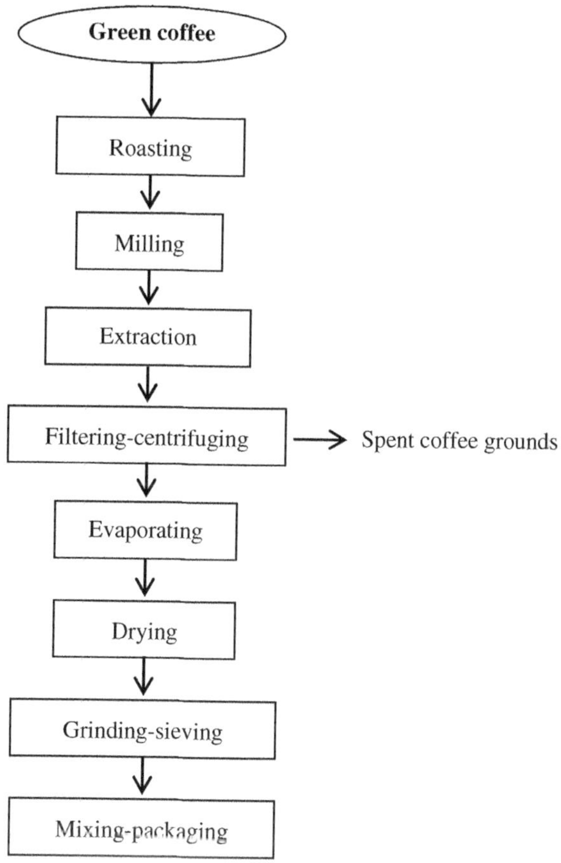

Figure 3.11 Main steps in soluble coffee production.

diterpene compounds (2.14 mg/g for kahweol and 4.66 mg/g for cafestol) were isolated by direct saponification, saponification in oil using the Soxhlet system and supercritical fluid extraction (SFE) (Acevedo et al., 2013). Murthy & Naidu (2010) used a mixture of isopropanol and water (60 : 40) and the Soxhlet apparatus to isolate chlorogenic acid and phenolic compounds from coffee pulp, husk, silver skin and spent coffee grounds, and indicated that the bioactive yield of enzyme-treated samples (1% viscozyme for 1 hour) was higher than that of untreated and steam-treated samples (121 °C for 20 minutes).

Pereira & Meireles (2010) described the application of SFE for the extraction of valuable bioactive compounds and the removal of undesirable compounds (Figure 3.12). They found that SFE is an advanced technique: it is a fast, efficient and clean method, and is able to produce end-products without toxic residues with no degradation of active constituents and high purity. In addition, SFE has the important advantage that the extraction conditions can easily be changed in terms of temperature, pressure and co-solvents, in order to obtain specific compounds.

Basically, a simple SFE consists of two major steps: (i) extraction of soluble substances from solid material using supercritical fluid (solvent) and (ii) separation of these compounds from solvent after extraction. Solvent is first fed and uniformly distributed into the extractors (extractor 1 or 2) by solvent pump and heat exchange. During the

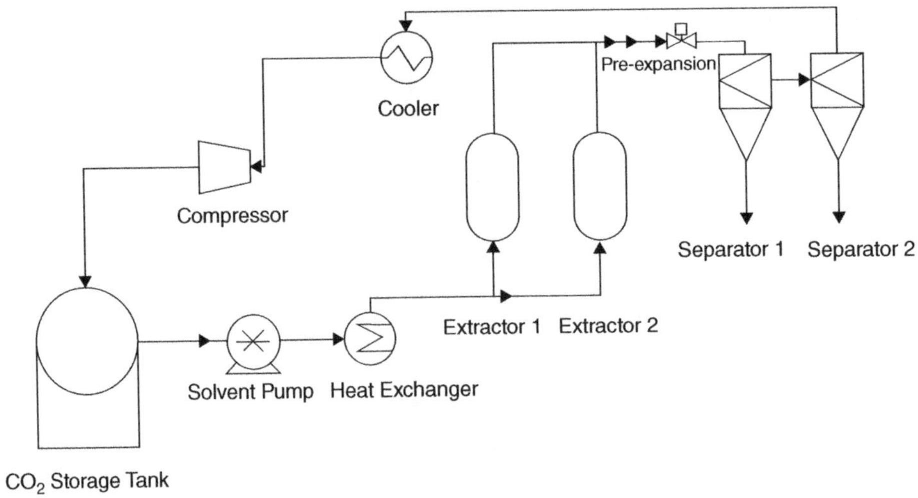

Figure 3.12 Flow diagram of a simplified supercritical fluid extraction (SFE) process. *Source:* Pereira & Meireles (2010). Reproduced with permission of Springer.

extraction, the solvent flows through the fixed bed and dissolves the soluble compounds. The solute–solvent mixture is then separated into the separators (separators 1 and 2) by rapid pressure reduction, temperature increase or both. Finally, the solvent is cooled by the cooler and recompressed by the compressor, and returned to the extractors.

3.4 Recovering Bioactive Compounds from Cacao Wastes

Cacao has been considered 'the food of the gods'. It is mainly grown in Africa (Côte d'Ivoire, Ghana, Nigeria and Cameroon), South and Central America (Brazil, Ecuador, Colombia and Mexico), the Caribbean and South-Western Pacific (the Dominican Republic and Papua New Guinea) and South East Asia (Indonesia, Malaysia and Vietnam). Total harvested cacao area and cacao bean production worldwide in 2013 were estimated at about 10.01 million ha and 4.59 million tonnes, respectively, with a gross production value approximately US$6.95 billion (FAOSTAT, 2015). The consumption of cacao and its products have increased continuously in recent decades. Cacao contains a significant amount of bioactive compounds (Jalil & Ismail, 2008), particularly theobromine, an alkaloid possessing stimulatory effects on the central nervous, gastrointestinal, cardiovascular, renal and respiratory systems (de Sena et al., 2011), and polyphenols (12–18% or 40.0–84.2 mg GAE/g dry weight of whole bean), which can be divided into three major groups: catechins (37%), anthocyanins (4%) and proanthocyanidins (58%). The main catechin is (−)-epicatechin, which makes up 35% of total polyphenols (Hii et al., 2009). The wastes from cacao harvesting and processing include cacao pod shell, cacao pod husk, cacao bean pulp/mucilage and cacao bean hull (Awarikabey et al., 2014; Bruna et al., 2009; Chung & Iiyama, 2003; Chung et al., 2003; Hartati, 2010; Nguyen, 2014; Santos et al., 2014; Yapo et al., 2013), as well as cacao leaves (Osman et al., 2004).

3.4.1 Bioactive Compounds from Cacao Bean Production

Cacao bean production creates a large amount of cacao pod shell, cacao pod husk and cacao bean pulp/mucilage (Figure 3.13 and Table 3.1). These wastes are estimated to make up approximately 85% of total cacao pod mass. Annual worldwide production of cacao pod shell and cacao pod husk is around 6.4 and 55.0 million tonnes, respectively (equal to 1.5 and 13.0 times total cacao bean production, respectively) (Nguyen, 2014; Nguyen & Nguyen, 2017a). Yapo et al. (2013) report that cacao pod husk and cacao bean make up 70.6 and 23.6% of fresh pods, respectively.

Kopp & Bradbury (2006) extracted polyphenols, theobromine and caffeine from cacao pod shell using 50% acetone, water eluate and methanol eluate, and found that cacao pod shell contained 1.09, 0.75 and 0.77 g polyphenols, 680, 665 and 2 mg theobromine, and 65, 60 and 2 mg caffeine/100 g dry weight, respectively. Vriesmann et al. (2011) extracted pectin from cacao pod husk (approximately 74% by fresh weight of cacao pods) using hot water and obtained a yield of 7.5–12.6% by dry weight containing 8.3–9.8 g GAE of phenolics/100 g dry weight and a number of monosaccharides, including glucose, uronic acid, xylose, arabinose, galactose, rhammose, mannose and fucose. Santos et al. (2014) found that fractions from cacao pod husk contained phenolics, flavonoids, steroids, terpenes, amino acids and alkaloids, which were effective against the yeast *Saccharomyces cerevisiae*, the basidiomycete *Moniliophthora perniciosa* and bacteria such as *Pseudomonas aeruginosa* and *Salmonella choleraesuis* (Gram-negative) with a

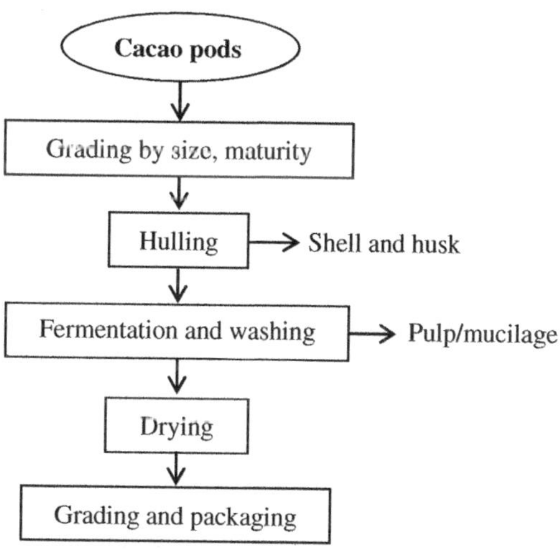

Figure 3.13 Main steps in cacao bean production.

Table 3.1 Mass proportion of cacao pod. *Source:* Nguyen (2014). Reproduced with permission of John Wiley & Sons.

Part of cacao pod (% by fresh weight)		
Shell	**Husk**	**Bean with pulp**
8.15 ± 0.78	70.15 ± 0.67	21.70 ± 0.20

minimum inhibitory concentration (MIC) of 5.0 mg/ml and against *Staphylococcus epidermidis* (Gram-positive) with an MIC of 2.5 mg/ml.

The soluble phenolics and condensed tannins in cacao pod husk were extracted by four different solvents in the order hot ethanol–water, acetone–water, distilled water and hot acidified butanol at 68.93 mg GAE/g dry weight and 52.16 mg/g dry weight, respectively (Yapo et al., 2013). Karim et al. (2014) detected the presence of carboxylic acid, phenolic acid, fatty acid, flavonoids (flavonol and flavones), stilbenoids and terpenoids in cacao pod husk, which possessed good antioxidant activity as compared to ascorbic acid and butylated hydroxytoluene (BHT). Cacao pod husk extract had better action against elastase and collagenase enzymes and higher tyrosinase enzyme inhibition in comparison with ascorbic acid. It also induced proliferation on human fibroblast cells and showed activity as a potential UVB sunscreen agent. Therefore, cacao pod husk extract may be of use as a functional ingredient for anti-wrinkles, skin whitening or sunscreen cosmetic products.

Nguyen (2017c) determined total phenolic content and saponin content from cacao pod husk using various solvents, and indicated that the highest total phenolic content from cacao pod husk was extracted by water, followed by ethanol, methanol, ethyl acetate, acetonitrile and hexane (8.48, 7.32, 6.37, 1.41, 1.07 and 0.61 mg GAE/g dry weight, respectively), while methanol produced the greatest level of saponins (63.30 mg escin equivalents/g dry weight) – significantly higher than that obtained by ethanol, water, ethyl acetate, acetonitrile or hexane (38.04, 25.95, 22.36, 10.39 and 0.66 mg escin equivalents/g dry weight). The extract from cacao pod husk contained (−)-epigalocatechingallate, (−)-epicatechin, caffeic acid and rutin, implying that cacao pod husk is a relevant source of potential antioxidant compounds (Nguyen, 2017c). Ribeiro Dias et al. (2007) indicated the ratio of bean, pulp and husk of cacao pod was 23.1, 26.4 and 50.5% by fresh weight, while the content of soluble phenolics in cacao bean pulp extracted by 50% methanol, methanol and water was 0.13, 0.15 and 0.17% by fresh weight, respectively.

Theobromine is a colourless and odourless substance with a melting point at 357 °C, a slightly bitter taste and slight solubility in water. It is known as the most valuable compound in cacao and cacao products (Hartati, 2010). Nguyen & Nguyen (2017a) determined the theobromine content in cacao pod husk to be 6.35 mg/100 g dried husk. Among various solvents, extraction times and numbers of extractions tested, 70% ethanol achieved the greatest levels of theobromine and the greatest theobromine extraction efficiency (2.23 mg/100 g dry weight and 35.03%) compared to water and chloroform (0.09 and 0.35 mg/100 g dry weight, 1.45 and 5.43%, respectively). Theobromine content and theobromine extraction efficiency were significantly higher with longer extraction times, but were not significantly different with an increase in the number of extractions from once to twice (5.98 and 6.06 mg/100 g dry weight, 92.10 and 95.40%, respectively).

3.4.2 Bioactive Compounds from Chocolate Production

Cacao bean hull is the major waste separated from cacao bean in chocolate production (Figure 3.14), accounting for about 10–12% by dry weight of cacao bean. It contains tannins, steroids, terpenoids, glycosides, flavonoids and purine alkaloids, but no saponins (Awarikabey et al., 2014). Bruna et al. (2009) determined the total phenolic content and antioxidant activity of cacao bean hull using 0.1% HCl and methanol

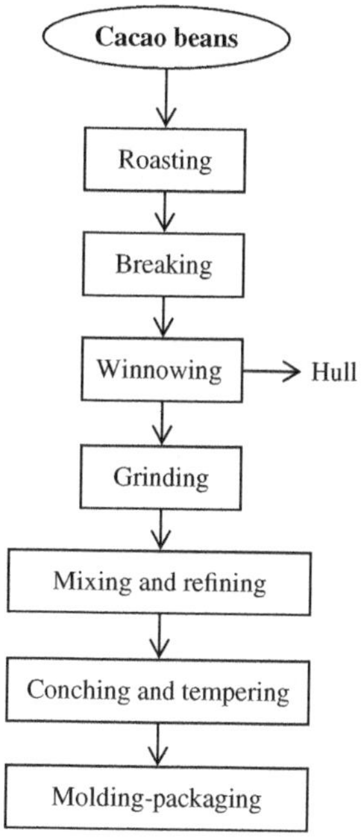

Figure 3.14 Main steps in chocolate production

(15:85 v/v). They found total phenolic content ranged from 2.56 to 4.06 mg GAE/g dry weight, depending on geographical origin and antioxidant activity, as measured by electron spin resonance spectroscopy. Phenolic compounds (lignin) in cacao bean hull were found in the range 35.9–39.1% by dry weight (Chung & Iiyama, 2003). Total soluble phenolics and condensed tannins in the fermented–roasted cacao bean hull extracted by four different solvents (hot ethanol–water, acetone–water, distilled water and hot acidified butanol) were found to be 26.71 mg GAE/g dry weight and 92.43 mg/g dry weight, respectively (Yapo et al., 2013) – much higher than was found with acetone, ethanol, 80% methanol and water using a microwave-assisted extraction (MAE) apparatus (4.85 mg GAE/g dry weight) (Atindana et al., 2012). Chung et al. (2003) indicated that the extract from cacao bean hull was defatted by benzene and extracted twice by 70% acetone, containing polyphenols and alkaloids such as (+)-catechins and caffeine.

Arlorio et al. (2005) extracted cacao bean hull using SC-CO_2 (150/200 bar, 50 °C) and found the phenolic and theobromine contents were 1.82 and 1.29 g/100 g dry weight, respectively – much higher than those obtained from cacao pod husk (Nguyen & Nguyen, 2017a).

Hartati (2010) recommended using hydrotrophy for the extraction of theobromine from cacao bean hull, using ionic organic salts such as sodium benzoate, sodium acetate, sodium alginate and urea as hydrotropic agents – such salts can increase or decrease the solubility in a given solvent, in a process known as 'salting in' or 'salting out' the solute. This technique was successfully applied for the extraction of piperine from black pepper, curcuminoids from *Curcuma longa* and limonin from *Citrus aurantium* seed.

Osman et al. (2004) extracted polyphenols from cocoa shoot and young leaves using methanol, water, hexane and ethyl acetate, and indicated that the total polyphenols in cocoa shoot and young leaves were 19.0 and 28.4 mg catechol/100 mg fresh weight, respectively. The main polyphenols in the extracts were epicatechin (EC), epigallocatechin gallate (EGCG), epigallocatechin (EGC), gallic acid (GA) and epicatechin gallate (ECG). The antioxidant capacity of these polyphenol-enriched extracts was comparable with butylated hydroxyanisole (BHA) in a concentration range of 50–200 ppm. The caffeine contents of cocoa shoot and young leaves were found to be 2.24 and 1.33 mg/100 mg, respectively.

3.5 Recovering Bioactive Compounds from Cashew Wastes

Cashew is as an industrial tree with high value in countries such as Vietnam, Nigeria, India, Côte d'Ivoire and Brazil. Total harvested cashew area and production of cashew nut with shell worldwide in 2013 were estimated at about 5.46 million ha and 4.44 million tonnes, respectively, with a gross production value approximately US$2.82 billion (FAOSTAT, 2015). Cashew nut possesses high amounts of lutein, β-carotene and unsaturated fatty acids (Trox et al., 2010). The wastes from cashew nut production include cashew apple, outer shell and inner skin (da Silva et al., 2014; de Almeida Lopes et al., 2012; de Brito et al., 2007; Gordon et al., 2012; Honorato & Rodrigues, 2010; Marc et al., 2012; Mestres et al., 2009; Queiroz et al., 2011a, 2011b; Rufino et al., 2010) and cashew nut shell (Garkal & Bhande, 2014; Kumar et al., 2009; Patel et al., 2006b; Smith et al., 2003).

Figure 3.15 shows the main steps in cashew nut production and the wastes discarded from this process, including cashew apple, outer shell and inner skin. Cashew apples

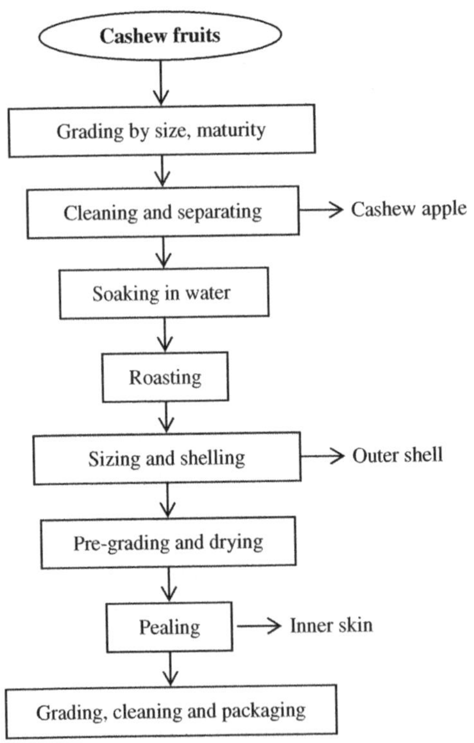

Figure 3.15 Main steps in cashew nut production.

consist of a high ratio of whole cashew fruit – over 60% by fresh weight. Total harvested cashew apple area and cashew apple production worldwide in 2013 were estimated at about 0.69 million ha and 2.00 million tonnes, respectively, with a gross production value approximately US$0.90 billion (FAOSTAT, 2015). Recently, the bioactive compounds in cashew wastes have been identified and quantified. For example, de Brito et al. (2007) determined the flavonoid components of cashew apple by liquid chromatography with diode array detection and electrospray ionisation mass spectrometry (LC-DAD-ESI/MS) and found that methanol–water extract of cashew apple contained one anthocyanin 3-*O*-hexoside of methyl-cyanidin and thirteen glycosylated flavonols, including 3-*O*-galactoside, 3-*O*-glucoside, 3-*O*-rhamnoside, 3-*O*-xylopyranoside, 3-*O*-arabinopyranoside and 3-*O*-arabinofuranoside of quercetin and myricetin, as well as kaempferol 3-*O*-glucoside (Figure 3.16), with total myricetin glycosides, total quercetin glycosides and total glycosylated flavonoids of 0.1511, 0.1139 and 0.2847 mg/g dry weight, respectively. The contents of ascorbic acid, soluble and hydrolysable polyphenols and proanthocyanidins in cashew apple were found to be 163 mg/100 g fresh weight, 12.79 and 18.53 mg GAE/100 g fresh weight and 9.27 mg/100 g fresh weight, respectively (Queiroz et al., 2011a); carotenoids, anthocyanins, flavonoids, polyphenols, β-carotene, 9,12-octadecadienoic acid, 9,12-octadecadienoyl chloride, gallic acid, protocatechuic acid, cinnamic acid (free and conjugate) and citric acid were also identified in cashew apple (da Silva et al., 2014; de Almeida Lopes et al., 2012; Bhakyaraj & Singaravadive, 2012; Gordon et al., 2012; Honorato & Rodrigues, 2010; Mestres et al., 2009; Rufino et al., 2010). The content of cinnamic acid was reduced when cashew apple was cut and stored at 40 °C, whereas the level of 5-hydroxymethylfurfural in cashew apple juice rose with storage at higher temperatures and following injury, revealing non-enzymatic browning (Queiroz et al., 2011b).

Michodjehoun-Mestres et al. (2009) extracted monomeric phenols from cashew apple skins and flesh using acetone/water (60 : 40) and found that the simple phenolics in the skins were much higher than those in flesh, in which flavonol glycosides such as myricetin, quercetin hexosides, pentosides and rhamnosides were major compounds, while anthocyanidin glycosides like peonidin, petunidin and cyanidin 3-*O*-hexosides were detected in skins but absent in flesh. Mestres et al. (2009) also

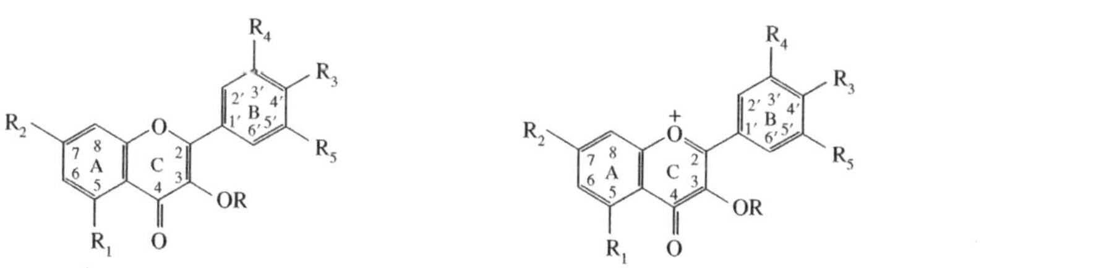

Flavonols (I)

Myricetin (**I-1**): R1=R2=R3=R4=R5=OH, R=H, MW=318

Quercetin (**I-2**): R1=R2=R3=R4=OH, R=H, MW=302

Kaempferol (**I-3**): R1=R2=R3=OH, R=H, MW=286

Rhamnetin (**I-4**): R1=R3=R4=OH, R=H, R2=OMe, MW=316

Isorhamnetin (**I-5**): R1=R2=R3=OH, R=H, R4=OMe, MW=316

Anthocyanidins (II)

Cyanidin (**II-1**): R1=R2=R3=R4=OH, R=R5=H, $[M+H]^+$=287

5-Methylcyanidin (**II-2**): R1=OMe, R2=R3=R4=OH, R=R5=H, $[M+H]^+$=301

Peonidin (**II-3**): R1=R2=R3=OH, R=R5=H, R4=OMe, $[M+H]^+$=301

Delphinidin (**II-4**); R1=R2=R3=R4=R5=OH, R=H, $[M+H]^+$=303

Figure 3.16 Structures of the flavonoids in cashew apple. *Source:* de Brito et al. (2007). Reproduced with permission of Elsevier.

extracted tannins from the skin and flesh of cashew apple using acetone/water (60 : 40) and found that tannins from both skin and flesh had high contents of (−)-epigallocatechin and (−)-epigallocatechin-*O*-gallate, followed by (−)-epicatechin and (−)-epicatechin-3-*O*-gallate; all these compounds were of 2,3–cis configuration, and there were half as many galloylated tannins in skin as in flesh.

Rufino et al. (2010) reported that cashew apples contained 28, 13 and 52 g/kg dry weight extractable polyphenols, ascorbic acid and non-extractable condensed tannins, respectively, when using acidic methanol/water (50 : 50; pH 2), acetone/water (70 : 30) and butanol/HCl/$FeCl_3$ as the solvent system. The anthocyanin, yellow flavonoid, β-carotene and total phenolic contents in cashew apple were found to be 14.74 and 44.91 mg/100 g dry weight, 179.14 μg/100 g dry weight and 6588.41 mg GAE/100 g dry weight, respectively (da Silva et al., 2014). Anacardic acid (30%), cardol (10%), cardanol (50%) and 2-methyl cardol (Figure 3.17) were found in cashew nut shell liquid (CNSL) extracted from cashew nut shell using a Soxhlet extractor and acetone as solvent (Kumar et al., 2009), and displayed potent antioxidant, antibacterial, antifungal and antitumour activities (Kubo et al., 2006; Stasiuk, 2005). Kumar et al. (2009) found the salicylic acid, coumaric acid, flavonoid, flavanols and quercetin contents of CNSL were 1–10, 100–450, 100–600, 30–200 and 50–250 mg/100 g, respectively.

Paramashivappa et al. (2001) reported the procedure for the separation of anacardic acid, cardol and cardanol from CNSL, including two steps where the acid-free CNSL is first treated with liquor ammonia and extracted with hexane/ethyl acetate (98 : 2) to separate cardanol, and the ammonia solution is then extracted with ethyl acetate/hexane (80 : 20) to obtain cardol (Figure 3.18). Smith et al. (2003) separated the CNSL from pericarp of cashew nut using SC-CO_2 and indicated that the yield of CNSL increased to 94% when extracting with one or more intermediate depressurisation steps; notably, most of the oil did not separate from the shell during the depressurisation steps, but was obtained during subsequent repressurisation (Figure 3.19). In addition, Patel et al. (2006b) studied the extraction of CNSL using SC-CO_2 and indicated that the yield of CNSL rose with the increase of pressure, temperature and flow rate of SC-CO_2, but that its composition varied under different operating conditions; in particular, the quality of oil obtained using SC-CO_2 extraction was of better quality than that obtained under the thermal method. Furthermore, Patel et al. (2006a) optimised the process parameters and extraction time for maximum profit and purity of refined CNSL using SFE and found the optimal extraction conditions to be 0.9 hours at 300 bar and 323 K (Figure 3.20). Garkal & Bhande (2014) described the different methods

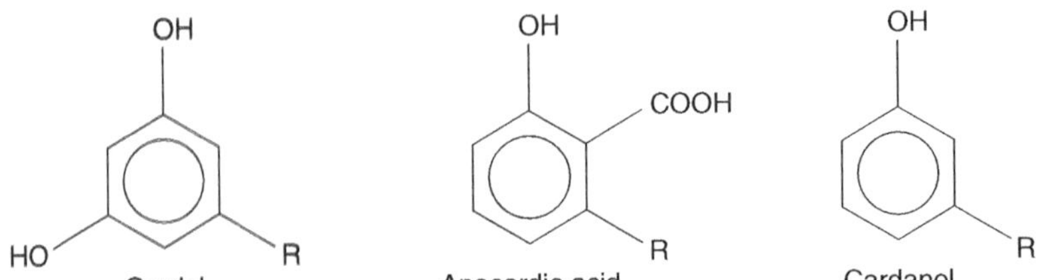

Figure 3.17 Chemical structures of the major phenolic constituents of cashew nut shell liquid (CNSL). *Source:* Paramashivappa et al. (2001). Reproduced with permission of American Chemical Society.

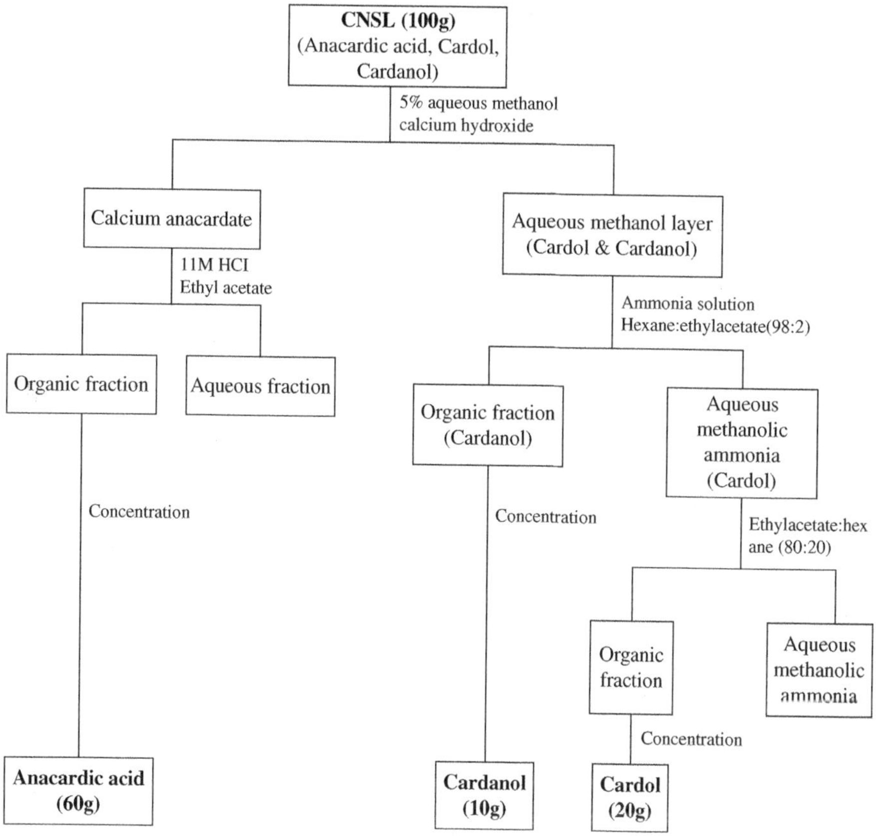

Figure 3.18 Flow chart for the separation of anacardic acid, cardol and cardanol from CNSL. *Source:* Paramashivappa et al. (2001). Reproduced with permission of American Chemical Society.

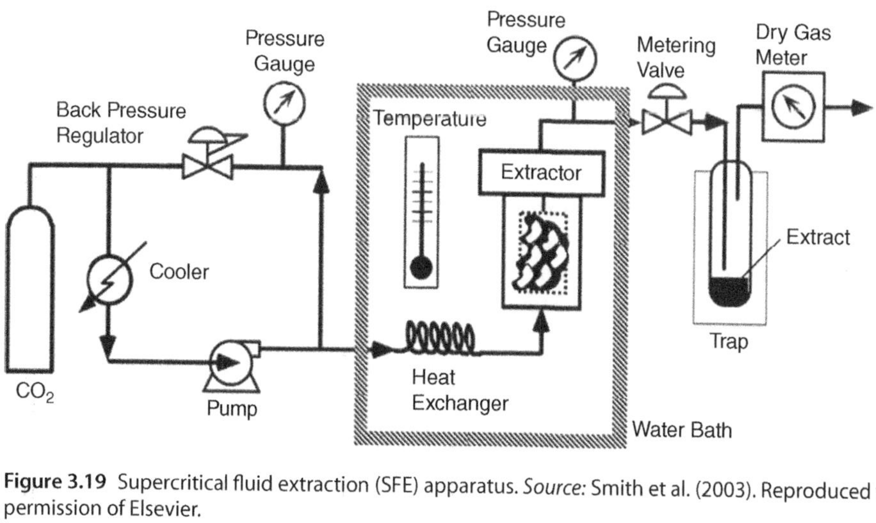

Figure 3.19 Supercritical fluid extraction (SFE) apparatus. *Source:* Smith et al. (2003). Reproduced with permission of Elsevier.

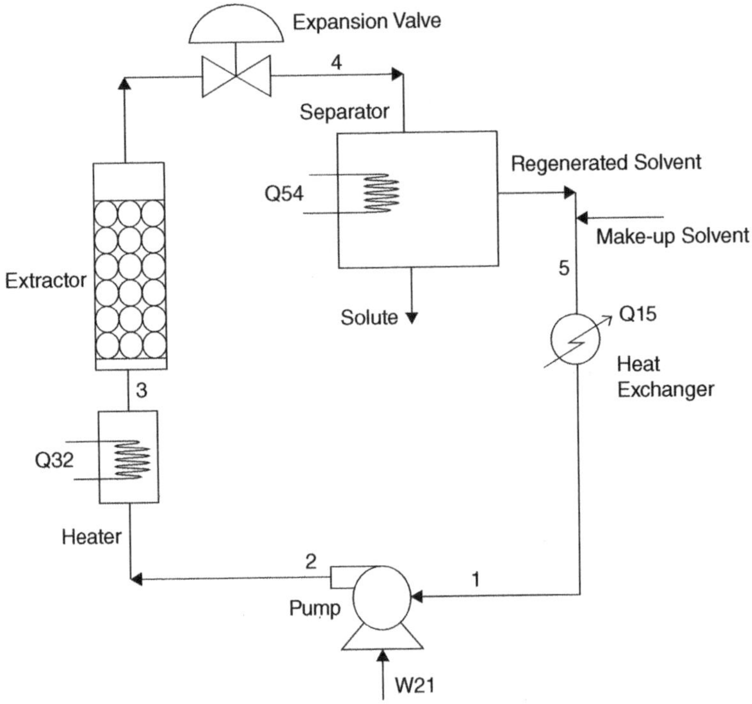

Figure 3.20 Flowchart of the SFE process. *Source:* Patel et al. (2006a). Reproduced with permission of Elsevier.

of extraction of CNSL and the isolation of major constituents of CNSL: thermal, mechanical and solvent extraction. Thermal extraction methods include roasting the nuts at 180–185 °C, recovering 85–90% CNSL; a hot oil bath using steam at 200–250 °C for 2–3 minutes, obtaining 7–12% CNSL; and solar cooking at 225–300 °C, obtaining 50–85% CNSL. Mechanical extraction methods include using screw pressing at high pressure, obtaining 20.7–21.0% CNSL containing 42% cardol, 47% anacardic acid and 3% cardanol, with 85.5–87.8% purity; however, this method has high levels of impurity and viscosity and low levels of thermo-oxidative stabilities and ebullition temperature. Solvent extraction methods include cold and hot solvent extraction using pentane, diethyl ether, carbon tetrachloride, light petroleum, acetone, hexane, ethanol, methanol and toluene as solvents; SC-CO_2 extraction at 250 bar, 400 °C and 16 hours, obtaining 60% CNSL; and vacuum pyrolysis extraction at 5 kPa and 400–6000 °C, achieving 63% oil yield with 86% cardanol containing 5% phenols.

Marc et al. (2012) determined the total phenolics, total flavonoids and anthocyanins in the juices from yellow and red cashew apples, which ranged from 1653.8 to 2374.2 mg/l, 298.6 to 479.8 mg/l and 5.9 to 7.4 mg/l, respectively. The contents of individual phenolic compounds were 1.2–7.9 mg/l quercetin, 4.3–10.3 mg/l naringenin, 20.2–25.9 mg/l caffeic acid, 2.5–7.9 mg/l coumaric acid, 4.6–13.4 mg/l ferulic acid and 1.9–15.4 mg/l gallic acid, indicating that cashew apples are a good source of antioxidant compounds. de Almeida Lopes et al. (2012) investigated the bioactive compounds of cashew apples during ripening and found that the highest levels of ascorbic acid, total anthocyanins and total yellow flavonoids were 279.37, 21.16 and 50.37–56.32 mg/100 g fresh weight, respectively.

Gordon et al. (2012) determined the changes of phenolic composition, ascorbic acid and antioxidant capacity in cashew apple during ripening using a modified pressurised liquid extraction (PLE) method with acetone/water/formic acid (70 : 29 : 1) as solvent and indicated that the amounts of phenolic compounds were highest in unripe cashew apple and decreased during the process of ripening, with myricetin 3-*O*-rhamnoside, quercetin 3-*O*-galactoside and quercetin 3-*O*-rhamnoside the major flavonoids, whereas antioxidant capacity and ascorbic acid content increased during ripening due to the influence of ascorbic acid on antioxidant activity. This showed that the use of polyphenols derived from cashew apple for food purposes should be considered only in unripe fruits.

3.6 Conclusion

The comprehensive information is presented in this chapter indicating high potential of tea, coffee, cacao and cashew wastes from the agricultural production for obtaining the bioactive compounds by using various solvents and different extraction techniques, particularly in consideration, selection and application of cheap, non-toxic and environmentally friendly solvents, such as supercritical fluids and 'green' techniques for extraction, isolation and purification of valuable bioactive compounds. The bioactive compounds have been isolated from these sources to be shown to possess a wide range of relevant biological properties they have therefore high potential for further application in the nutraceutical, medical, pharmaceutical and cosmetic industries for the production of functional products to enhance human health and prevent chronic diseases.

References

Acevedo, F., Rubilar, M., Scheuermann, E., Cancino, B., Uquiche, E., Garcés, M., Inostroza, K. & Shene, C. 2013. Bioactive compounds of spent coffee grounds, a coffee industrial residue. *III Siger*, 1–4.

Arlorio, M., Coïsson, J. D., Travaglia, F., Varsaldi, F., Miglio, G., Lombardi, G. & Martelli, A. 2005. Antioxidant and biological activity of phenolic pigments from *Theobroma cacao* hulls extracted with supercritical CO_2. *Food Research International*, **38**, 1009–1014.

Atindana, J. N., Zhong, F., Mothibe, K. J., Bangoura, M. L. & Lagnika, C. 2012. Quantification of total polyphenolic content and antimicrobial activity of cocoa (*Theobroma cacao* L.) bean shells. *Pakistan Journal of Nutrition*, **11**, 574–579.

Awarikabey, E., Amponsah, I. K. & Woode, M. Y. 2014. The value of the cocoa bean shell (hull) and the effect of various processing methods on the phyto-constituents and antioxidant activity of the nib and shell. *Journal of Natural Products and Plant Resources*, **4**.

Bhakyaraj, R. & Singaravadive, K. 2012. Minerals and bioactive compounds in cashew apple (*Anacardium occidentale* L.). *Journal of Food Resource Science*, **1**, 32–36.

Bouafou, K. G. M., Konan, B. A., Zannou-Tchoko, V. & Kati-Coulibally, S. 2011. Potential food waste and by-products of coffee in animal feed. *Electronic Journal of Biology*, **7**, 74–80.

Bravo, J., Juaniz, I., Monente, C., Caemmerer, B., Kroh, L. W., De Pena, M. P. & CID, C. 2012. Evaluation of Spent Coffee Obtained from the Most Common Coffeemakers as a Source of Hydrophilic Bioactive Compounds. *Journal of Agricultural and Food Chemistry*, **60**, 12 565–12 573.

Bruna, C., Eichholz, I., Rohn, S., Kroh, L. W. & Huyskens-Keil, S. 2009. Bioactive compounds and antioxidant activity of cacao hulls (*Theobroma cacao* L.) from different origins. *Journal of Applied Botany and Food Quality*, **83**, 9–13.

Cabrera, C., Artacho, R. & Giménez, R. 2006. Beneficial effects of green tea – a review. *Journal of the American College of Nutrition*, **25**, 79–99.

Chung, B. Y. & Iiyama, K. 2003. Differences in chemical constituents between vascular bundles and nonvascular bundles of cacao (*Theobroma cacao* L.) hull. *Journal of Wood Science*, **49**, 176–180.

Chung, B. Y., Iiyama, K. & Han, K. W. 2003. Compositional characterization of cacao (*Theobroma cacao* L.) hull. *Agricultural Chemistry and Biotechnology*, **46**, 12–16.

Da Silva, L. M. R., De Figueiredo, E. A. T., Ricardo, N. M. P. S., Vieira, I. G. P., De Figueiredo, R. W., Brasil, I. M. & Gomes, C. L. 2014. Quantification of bioactive compounds in pulps and by-products of tropical fruits from Brazil. *Food Chemistry*, **143**, 398–404.

De Almeida Lopes, M. M., De Miranda, M. R. A., Herbster Moura, C. F. & Enéas Filho, J. 2012. Bioactive compounds and total antioxidant capacity of cashew apples (*Anacardium occidentale* L.) during the ripening of early dwarf cashew clones. *Ciência e Agrotecnologia*, **36**, 325–332.

De Brito, E. S., De Araújo, M. C. P., Lin, L. Z. & Harnly, J. 2007. Determination of the flavonoid components of cashew apple (*Anacardium occidentale*) by LC-DAD-ESI/MS. *Food Chemistry*, **105**, 1112–1118.

De Sena, A. R., De Assis, S. A. & Branco, A. 2011. Analysis of theobromine and related compounds by reversed phase high-performance liquid chromatography with ultraviolet detection: an update (1992–2011). *Food Technology and Biotechnology*, **49**, 413–423.

Esquivel, P. & Jimenez, V. M. 2012. Functional properties of coffee and coffee by-products. *Food Research International*, **46**, 488–495.

FAOSTAT 2015. FAO Statistic Division. http://faostat3.fao.org/home/E (last accessed March 20, 2017).

Farhoosh, R., Golmovahhed, G. A. & Khodaparast, M. H. H. 2007. Antioxidant activity of various extracts of old tea leaves and black tea wastes (*Camellia sinensis* L.). *Food Chemistry*, **100**, 231–236.

Garkal, D. J. & Bhande, R. S. 2014. Review on extraction and isolation of cashew nut shell liquid. *International Journal of Innovations in Engineering Research and Technology*, **1**, 1–8.

Goksu, C. & Poyrazoglu, E. S. 2013. The bioactive compounds of tea and decaffeinated tea (*camellia sinensis*). *International Journal of Chemical, Environmental & Biological Sciences*, **1**, 43–47.

Gordon, A., Friedrich, M., Da Matta, V. M., Herbster Moura, C. F. & Marx, F. 2012. Changes in phenolic composition, ascorbic acid and antioxidant capacity in cashew apple (*Anacardium occidentale* L.) during ripening. *Fruits*, **67**, 267–276.

Gürü, M. & Içen, H. 2004. Obtaining of caffeine from Turkish tea fiber and stalk wastes. *Bioresource Technology*, **94**, 17–19.

Hartati, I. 2010. Hydrotropic extraction of theobromine from cocoa bean shell. *Momentum*, **6**.

Heo, J. Y., Kim, S., Kang, J. H. & Moon, B. K. 2014. Determination of lutein from green tea and green tea by-products using accelerated solvent extraction and UPLC. *Journal of Food Science*, **79**, C816–C821.

Hii, C. L., Law, C. L., Suzannah, S., Misnawi & Cloke, M. 2009. Polyphenols in cocoa (*Theobroma cacao* L.). *Asian Journal of Food and Agro-Industry*, **2**, 702–722.

Honorato, T. L. & Rodrigues, S. 2010. Dextransucrase stability in cashew apple juice. *Food and Bioprocess Technology*, **3**, 105–110.

Içen, H. & Gürü, M. 2009. Extraction of caffeine from tea stalk and fiber wastes using supercritical carbon dioxide. *Journal of Supercritical Fluids*, **50**, 225–228.

Içen, H. & Gürü, M. 2010. Effect of ethanol content on supercritical carbon dioxide extraction of caffeine from tea stalk and fiber wastes. *Journal of Supercritical Fluids*, **55**, 156–160.

Jalil, A. M. M. & Ismail, A. 2008. Polyphenols in cocoa and cocoa products: Is there a link between antioxidant properties and health? *Molecules*, **13**, 2190–2219.

Karim, A. A., Azlan, A., Ismail, A., Hashim, P., Abd Gani, S. S., Zainudin, B. H. & Abdullah, N. A. 2014. Phenolic composition, antioxidant, anti-wrinkles and tyrosinase inhibitory activities of cocoa pod extract. *BMC Complement Altern Med*, **14**, 381.

Kondo, M., Hidaka, M., Kita, K. & Yokota, H. 2007. Ensiled green tea and black tea waste as protein supplement for goats. In: Priolo, A., Bion di, L., Ben Salem, H. & Morand-Fehr, P. (eds). Advanced nutrition and feeding strategies to improve sheep and goat. *Zaragoza:CIHEAM (Option s Méditerran éen n es: Série A. Sémin aires Méditerran éen s; n . 7 4)*, 165–169.

Kopp, G. M. & Bradbury, A. G. 2006. Polyphenol-enriched composition from cocoa shell extraction. United State Patent Application Publication EP1728434 A1.

Kubo, I., Masuoka, N., HA, T. J. & Tsujimoto, K. 2006. Antioxidant activity of anacardic acids. *Food Chemistry*, **99**, 555–562.

Kumar, P. S., Kumar, N. A., Sivakumar, R. & Kaushik, C. 2009. Experimentation on solvent extraction of polyphenols from natural waste. *Journal of Materials Science*, **44**, 5894–5899.

Lee, C.-P. & Yen, G.-C. 2006. Antioxidant activity and bioactive compounds of tea seed (*Camellia oleifera* Abel.) oil. *Journal of Agricultural and Food Chemistry*, **54**, 779–84.

Ludwig, I. A., Clifford, M. N., Lean, M. E. J., Ashihara, H. & Crozier, A. 2014. Coffee: biochemistry and potential impact on health. *Food & Function*, **5**, 1695–1717.

Marc, A., Didier Ange, K., Fabrice Achille, T. & N'Guessan Georges, A. 2012. Phenolic profile of cashew apple juice (*Anacardium occidentale* L.) from Yamoussoukro and Korhogo (Côte d'Ivoire). *Journal of Applied Biosciences*, **49**, 3331–3338.

Martono, Y. 2010. Potency of industrial tea waste: comparison between green and black tea industrial wastes as UV filter for sunscreen. *Indonesian Journal of Cancer Chemoprevention*, **I**, 54–59.

Mestres, L. M., Souquet, J. M., Fulcrand, H., Meudec, E., Reynes, M. & Brillouet, J. M. 2009. Characterisation of highly polymerised prodelphinidins from skin and flesh of four cashew apple (*Anacardium occidentale* L.) genotypes. *Food Chemistry*, **114**, 989–995.

Michodjehoun-Mestres, L., Souquet, J. M., Fulcrand, H., Bouchut, C., Reynes, M. & Brillouet, J. M. 2009. Monomeric phenols of cashew apple (*Anacardium occidentale* L.). *Food Chemistry*, **112**, 851–857.

Murthy, P. S. & Naidu, M. M. 2010. Recovery of phenolic antioxidants and functional compounds from coffee industry by-products. *Food Bioprocess Technology*, **5**, 897–903.

Nguyen, V. T. 2014. Mass proportion, proximate composition and effects of solvents and extraction parameters on pigment yield from cacao pod shell (*Theobroma cacao* L.). *Journal of Food Processing and Preservation*, **39**, 1414–1420.

Nguyen, V. T & Nguyen, N. H. 2017a. Proximate composition, extraction and purification of theobromine from cacao pod husk (*Theobroma cacao* L.). *Technologies*, **5**, 14.

Nguyen, V. T. 2017b. Proximate composition and the effect of extraction parameters on phenolic compounds from black tea by-products (*Camellia sinensis* L.). Unpublished.

Nguyen, V. T. 2017c. Bioactive compounds and antioxidant capacity of cacao pod husk (*Theobroma cacao* L.) as affected by various solvents. Unpublished.

Nguyen, V. T., UENG, J. P. & TSAI, G. J. 2011. Proximate composition, total phenolic content, and antioxidant activity of Seagrape (*Caulerpa lentillifera*). *Journal of Food Science*, **76**, C950–C958.

Nuhu, A. A. 2014. Bioactive micronutrients in coffee: recent analytical approaches for characterization and quantification. *ISRN Nutrition*, Article ID 384230.

Oliveira, L. S., Franca, A. S., Mendonc, J. C. F. & Barros-Júnior, M. C. 2006. Proximate composition and fatty acids profile of green and roasted defective coffee beans. *LWT*, **39**, 235–239.

Osman, H., Nasarudin, R. & Lee, S. L. 2004. Extracts of cocoa (*Theobroma cacao* L.) leaves and their antioxidation potential. *Food Chemistry*, **86**, 41–46.

Paramashivappa, R., Phani Kumar, P., Vithayathil, P. J. & Srinivasa Rao, A. 2001. Novel method for isolation of major phenolic constituents from cashew (*Anacardium occidentale* L.) nut shell liquid. *Journal of Agriculture and Food Chemistry*, **49**, 2548–2551.

Patel, R. N., Bandyopadhyay, S. & Ganesh, A. 2006a. Economic appraisal of supercritical fluid extraction of refined cashew nut shell liquid. *Journal of Chromatography A*, **1124**, 130–138.

Patel, R. N., Bandyopadhyay, S. & Ganesh, A. 2006b. Extraction of cashew (*Anacardium occidentale*) nut shell liquid using supercritical carbon dioxide. *Bioresource Technology*, **97**, 847–853.

Pereira, C. G. & Meireles, M. A. A. 2010. Supercritical fluid extraction of bioactive compounds: Fundamentals, applications and economic perspectives. *Food and Bioprocess Technology*, **3**, 340–372.

Queiroz, C., Lopes, M. L. M., Fialho, E. & Valente-Mesquita, V. L. 2011a. Changes in bioactive compounds and antioxidant capacity of fresh-cut cashew apple. *Food Research International*, **44**, 1459–1462.

Queiroz, C., Da Silva, A. J. R., Lopes, M. L. M., Fialho, E. & Valente-Mesquita, V. L. 2011b. Polyphenol oxidase activity, phenolic acid composition and browning in cashew apple (*Anacardium occidentale* L.) after processing. *Food Chemistry*, **125**, 128–132.

Ramalakshmi, K., Rao, L. J. M., Ishikawa, Y. T. & Goto, M. 2009. Bioactivities of low-grade green coffee and spent coffee in different *in vitro* model systems. *Food Chemistry*, **115**, 79–85.

Rathinavelu, R. & Graziosi, G. 2005. Potential alternative uses of coffee wastes and by-products. International Coffee Organization. Available from: http://www.ico.org/documents/ed1967e.pdf (last accessed 20 March 2017).

Ribeiro Dias, D., Freitas Schwan, R., Sena Freire, E. & Dos Santos Serôdio, R. 2007. Elaboration of a fruit wine from cocoa (*Theobroma cacao* L.) pulp. *International Journal of Food Science and Technology*, **42**, 319–329.

Rodrigues, N. P. & Bragagnolo, N. 2013. Identification and quantification of bioactive compounds in coffee brews by HPLC–DAD–MS^n. *Journal of Food Composition and Analysis*, **32**, 105–115.

Rufino, M. D. S. M., Pérez-Jiménez, J., Tabernero, M., Alves, R. E., De Brito, E. S. & Saura-Calixto, F. 2010. Acerola and cashew apple as sources of antioxidants and dietary fibre. *International Journal of Food Science and Technology*, **45**, 2227–2233.

Santos, R. X., Oliveira, D. A., Sodre, G. A., Gosmann, G., Brendel, M. & Pungartnik, C. 2014. Antimicrobial activity of fermented *Theobroma cacao* pod husk extract. *Genetics and Molecular Research*, **13**, 7725–35.

Senol, A. & Aydin, A. 2006. Solid–liquid extraction of caffeine from tea waste using battery type extractor_Process optimization. *Journal of Food Engineering*, **75**, 565–573.

Shalmashi, A., Abedi, M., Golmohammad, F. & Eikani, M. H. 2010. Isolation of caffeine from teawaste using subcriticalwater extraction. *Journal of Food Process Engineering*, **33**, 701–711.

Smith, R. L. Jr, Malaluan, R. M., Setianto, W. B., Inomata, H. & Arai, K. 2003. Separation of cashew (*Anacardium occidentale* L.) nut shell liquid with supercritical carbon dioxide. *Bioresource Technology*, **88**, 1–7.

Stasiuk, M. 2005. Bioactive compounds from *Anacardium occidentale* cashew nut shell liquid. *Nova Science Publishers*, 87–106.

Taheri, M., Sariri, R., Giahi, M. & Ghafoori, H. 2011. Screening antioxidant activity of extracts from different tea samples. *Pharmacologyonline*, **2**, 674–683.

Trox, J., Vadivel, V., Vetter, W., Stuetz, W., Scherbaum, V., Gola, U., Nohr, D. & Biesalski, H. K. 2010. Bioactive compounds in cashew nut (*Anacardium occidentale* L.) kernels: Effect of different shelling methods. *Journal of Agricultural and Food Chemistry*, **58**, 5341–5346.

Vriesmann, L. C., De Mello Castanho Amboni, R. D. & De Oliveira Petkowicz, C. L. 2011. Cacao pod husks (*Theobroma cacao* L.): Composition and hot-water-soluble pectins. *Industrial Crops and Products*, **34**, 1173– 1181.

Xu, P., Bao, J. S., Gao, J. J., Zhou, T. & Wang, Y. F. 2012. Optimization of extraction of phenolic antioxidants from tea (*Camellia Sinensis* L.) fruit peel biomass using response surface methodology. *Bioresources*, **7**, 2431–2443.

Yang, Z., Xu, Y., Jie, G., He, P. & Tu, Y. 2007. Study on the antioxidant activity of tea flowers (*Camellia sinensis*). *Asia Pacific Journal of Clinical Nutrition*, **16** Suppl. 1, 148–52.

Yapo, B. M., Besson, V., Koubala, B. B. & Koffi, K. L. 2013. Adding value to cacao pod husks as a potential antioxidant-dietary fiber source. *American Journal of Food and Nutrition*, **1**, 38–46.

Ye, J. H., Jin, J., Liang, H. L., Lu, J. L., Du, Y. Y., Zheng, X. Q. & Liang, Y. R. 2009. Using tea stalk lignocellulose as an adsorbent for separating decaffeinated tea catechins. *Bioresource Technology*, **100**, 622–628.

Zuorro, A. & Lavecchia, R. 2012. Spent coffee grounds as a valuable source of phenolic compounds and bioenergy. *Journal of Cleaner Production*, **34**, 49–56.

4

Recovering Bioactive Compounds from Fruit and Vegetable Wastes

Hong Ngoc Thuy Pham

School of Environmental and Life Sciences, Faculty of Science and Information Technology, University of Newcastle, Ourimbah, NSW, Australia
Department of Postharvest Technology, Faculty of Food Technology, Nha Trang University, Nha Trang, Khanh Hoa, Vietnam

4.1 Introduction

Fruit and vegetable wastes and residues account for a high proportion of the total mass, in comparison with their edible parts. These wastes are a rich source of bioactive compounds like antioxidants. For instance, the skin of macadamia, an inedible part, accounts for approximately 40% of the nut's weight; peels and stones are major wastes of mango processing, amounting to 35–60% of the total fruit weight; and banana peels account for 30% of the ripe fruit and possess high levels of phenolics, flavonoids, proanthocyanidins and other antioxidants (Dailey & Vuong, 2015a; Schieber et al., 2001). Guo et al. (2013) indicated that peel and seed fractions of some fruits, such as pomegranate peel, grape seed, hawthorn peel, longan and lychee seeds, exhibit relatively high antioxidant activity and are a rich source of natural antioxidants. In addition, in some materials, the bioactive compound content in byproducts is higher than in their major parts. For example, greater levels of anthocyanins and total phenolic compounds were found in the byproducts of cashew apple (peel and pulp's leftovers) and Surinam cherry (pulp's leftovers) as compared to their pulps (da Silva et al., 2014). Therefore, the wastes of fruit and vegetable processing are a promising source for the recovery of bioactive compounds such as natural antioxidants, which can be used to replace synthetic antioxidants such as butylated hydroxytoluene (BHT) and butylated hydroxyanisole (BHA) as sources of health benefits and functional properties (Llorach et al., 2002; Moure et al., 2001). Consequently, recovery of bioactive compounds from fruit and vegetable byproducts for application in the health food, medical, cosmetic and pharmaceutical industries is essential to enhancing their added value and to reducing the pollution risk for the environment.

4.2 Bioactive Compound Resources in Fruit and Vegetable Wastes

Table 4.1 indicates the diverse bioactive compound resources in the residual parts of the fruit and vegetable processing industries, including leaves, stalks, peels or skins,

Recovering Bioactive Compounds from Agricultural Wastes, First Edition. Edited by Van Tang Nguyen.

Table 4.1 Bioactive compound resources from fruit and vegetable wastes.

Residue	Fruit/vegetable sources	Bioactive compounds	References
Leaves	Papaya, olive	Phenolic compounds, flavonoids, proanthocyanidins, saponins, gallic acid, tyrosol, oleuropein	Goldsmith et al. (2015); Vuong et al. (2013)
Stalks	Grape	Anthocyanins, catechins, flavonol glycosides, phenolic acids, procyanidins	Garcia-Perez et al. (2009)
Peels or skins	Mango, cashew apple, guava, papaya, sapodilla, potato, macadamia	Gallic acid, syringic acid, gentisyl-protocatechuic, mangiferin, ellagic acid, quercetin, anthocyanins, carotenoids, chlorogenic, proto-catechuic, caffeic acids, phenolics, flavonoids, proanthocyanidins	Dailey & Vuong (2015a, 2015b); da Silva et al. (2014); Garcia-Mendoza et al. (2015); Schieber et al. (2001)
Seeds	Olive, grape, papaya, acerola, guava, soursop, passion fruit, apricot	Protocatechuic acid, hydroxybenzoic acid, vanillic acid, caffeic acid, syringic acid, sinapic acid, ferulic acid, *p*-coumaric acid, rutin, hesperidin, quercetin, cinnamic acid, persipan	Alu'datt et al. (2011); da Silva et al. (2014); Schieber et al. (2001)
Hulls	Pistachio	Phenolic compounds	Goli et al. (2005)
Pomace	Olive, grape, apple, tomato	Bioactive polar lipids, anthocyanins, catechins, flavonol glycosides, phenolic acids, alcohols, stilbenes and lycopene, syrigin, procyanidin B2, (−)-epicatechin, phlorizin and quercetin	Bai et al. (2010); Karantonis et al. (2008); Kumcuoglu et al. (2014);Schieber et al. (2001)
Residues	Asparagus	Ferulic acid, kaempferol, quercetin, rutin and isorhamnetin	Fan et al. (2015)

seeds, hulls and pomace. These compounds can be divided into phenolic acids, catechins, flavonoids, proanthocyanidins, saponins, anthocyanins, procyanidins, alcohols and stilbenes.

4.3 Recovering Bioactive Compounds from Fruit and Vegetable Wastes

Figure 4.1 presents various techniques for the recovery of bioactive compounds from fruit and vegetable wastes, including some principal steps as follows: preparation of dried samples, extraction, powdered extract production, isolation and purification. In order to remove undesirable components, the bioactive compound-enriched powder

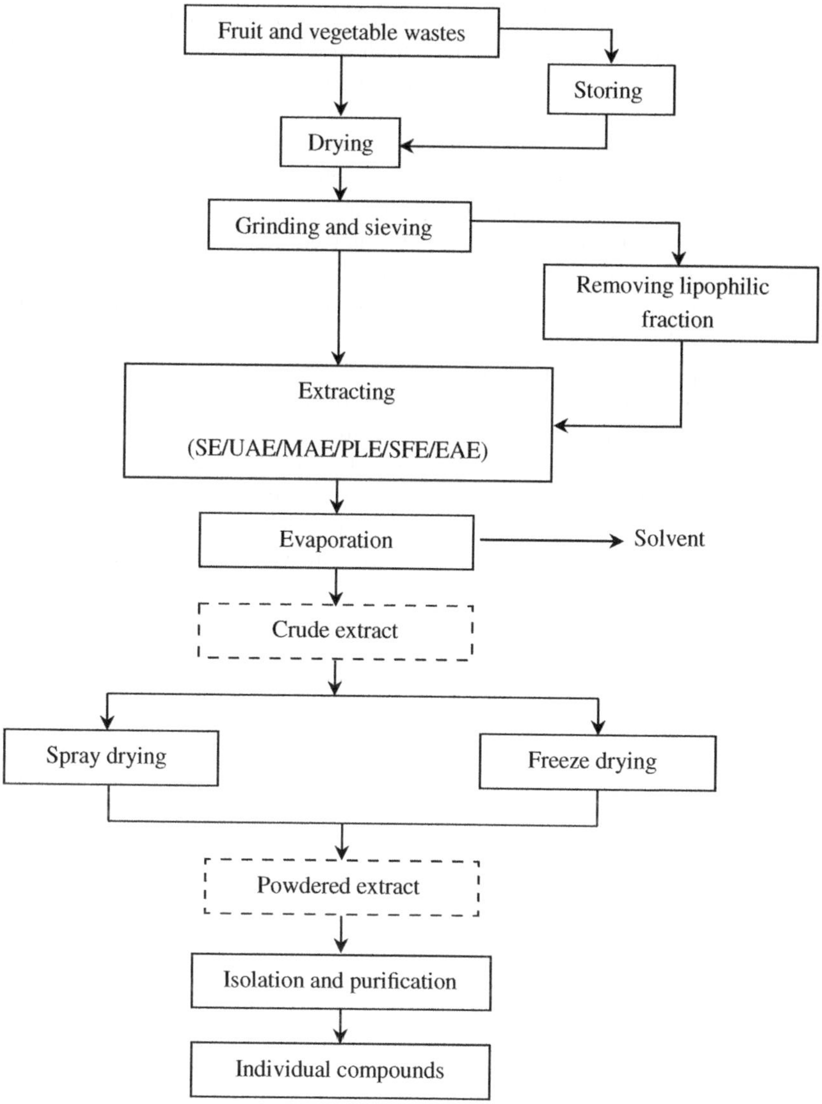

Figure 4.1 General process for the recovery of bioactive compounds from fruit and vegetable wastes. SE, solvent extraction; UAE, ultrasound-assisted extraction; MAE, microwave-assisted extraction; PLE, pressurised liquid extraction; SFE, supercritical fluid extraction; EAE, enzyme-assisted extraction.

is isolated and purified using high-performance liquid chromatography (HPLC) or ion-exchange chromatography (IEC) in order to obtain the target compounds.

4.3.1 Preparation of Dried Materials

Preparation of dried materials has three main steps: drying, grinding and sieving (Kha & Nguyen, 2014) (Figure 4.1). If raw materials are not processed immediately after collecting, they should be stored at below $-18 \pm 2\,^{\circ}C$ in order to reduce oxidation and degradation of bioactive compounds under undesirable conditions such as high temperature, ultraviolet (UV) radiation from sunlight or enzymes. The drying process is an essential step in dried sample preparation, as it inactivates the enzymes responsible for degrading valuable bioactive compounds and decreases the rate of microbial growth (Dorta et al., 2012). Moreover, drying also helps to reduce transportation and preservation costs as it decreases moisture content and so reduces the weight and volume of the raw materials (Pham et al., 2015). Drying treatment has also been found to have a significant effect on bioactive compounds and the antioxidant activity of plant materials and fruit and vegetable wastes (Dorta et al., 2012; Nguyen et al., 2015a; Orphanides et al., 2013; Rabeta & Vithyia, 2013; Sui et al., 2014). Recently, many drying methods have been applied to the preparation of dried materials, such as sun drying, hot air drying, low-temperature air drying, infrared drying, vacuum drying, freeze drying, microwave drying and microwave vacuum drying.

Depending on the raw materials and the characteristics of the bioactive compounds within them, suitable drying conditions may vary. Although sun drying and hot air drying are simple and inexpensive techniques, they can cause a degradation of bioactive compounds such as potential antioxidants in certain materials (Nguyen et al., 2015a; Vega-Gálvez et al., 2009). Freeze drying has been found to be optimal for retaining bioactive compounds that are very sensitive to heat and oxygen, but it is an expensive and time-consuming method (Kha & Nguyen, 2014; Vuong et al., 2015). Therefore, other drying methods can be applied. Microwave drying is an advanced technique in which quick absorption of energy by water molecules leads to a very quick evaporation. Vacuum drying, which allows water to evaporate at a lower temperature without air, can limit the negative impact of high temperature and oxygen (Zhang et al., 2014). A combined microwave and vacuum drying method known as microwave vacuum drying has been shown to be competitive with freeze drying (Wojdyło et al., 2009, 2014; Zhang et al., 2014). An infrared drying method, in which infrared energy is transferred from the heating element to the material surface without heating the surrounding air and is directly absorbed by the material instead of receiving the heat from the air (Sui et al., 2014), has been revealed as a suitable drying method for the retention of polyphenols, proanthocyanidins and antioxidants of wine grape pomace and *Phyllanthus amarus* (Nguyen et al., 2015a; Sui et al., 2014). In conclusion, selecting an optimal drying method is very important in maintaining the high yield of bioactive compounds such as promising antioxidants within fruit and vegetable wastes. However, the economic aspect and the accessibility of drying equipment should be considered at pilot and industrial scales (Kha & Nguyen, 2014).

The next steps are grinding and sieving to give the dried material a suitable particle size for further extraction. Larger particles can cause a longer extraction process, whereas smaller particles result in a larger surface area that increases the contact between the plant matrix and the solvent and thus enhances the extraction yield. However, fine particles may also cause difficulty in maintaining a proper flow rate for supercritical fluid extraction (SFE) (Wang & Weller, 2006). Some materials with a high lipophilic fraction, such as olive cake, should be removed in a clean step, using *n*-hexane as solvent (Lozano-Sánchez et al., 2014).

4.3.2 Extraction Process

Extraction is an essential step in liberating phytochemicals from raw materials. Therefore, it is important to choose a suitable extraction method for a given material from among the applicable techniques, such as solvent extraction, ultrasound-assisted extraction (UAE), microwave-assisted extraction (MAE), pressurised liquid extraction (PLE), supercritical fluid extraction (SFE) and enzyme-assisted extraction (EAE). Different materials may have different optimum extraction methods due to the characteristics of bioactive compounds and the diversity of tissue plant structures. For each extraction method, there are factors which affect the extraction yield of bioactive compounds, and the optimisation of extraction conditions is an important step in obtaining the highest levels of target components.

Table 4.2 summarises the bioactive compounds obtained from fruit and vegetable wastes using different extraction methods in recent years.

4.3.2.1 Solvent Extraction

Solvent extraction, which is a simple and feasible extraction method, can be implemented in a water bath or using a magnetic stirrer. Solvent extraction is usually used for the isolation of antioxidants yielded by extraction, depending on the solvent polarity and extraction method used (Goli et al., 2005). The mechanism of solvent extraction is the liquid phase (the solvent), diffusion inside the material (mass transfer mechanism), then dissolution of solids from the material to the solvent (due to the concentration gradient at the solid–liquid interface) until a constant concentration is reached (Kha & Nguyen, 2014).

The solvent type has been found to have a significant effect on the recovery of bioactive compounds from fruit and vegetable wastes (Dailey & Vuong, 2015a; Pinelo et al., 2005; Rezaie et al., 2015). Some solvents, such as water, ethanol, methanol, ethyl acetate, acetone and acetone nitrile, are commonly used for the extraction of bioactive compounds from fruit and vegetable wastes. Different solvents may be suitable for different raw materials, due to the characteristics of the bioactive compounds and the different dielectric constants and polarities of the solvents.

Goli et al. (2005) revealed that the level of phenolic compounds obtained from pistachio (*Pistachia vera*) hulls by solvent extraction (either in water or in methanol) was as high as UAE, and approximately five times higher than SFE. Solvent extraction, however, usually requires a larger amount of solvents and a longer extraction time than these other methods, and a further filtration step must also be applied.

4.3.2.2 Ultrasound-Assisted Extraction

UAE is a simple, cheap and effective method in comparison with conventional extraction techniques. It utilises the effects of acoustic cavitations generated in solvent under the passage of ultrasonic waves. Gas bubbles are produced and compressed during the application of ultrasound. The compression causes an increase in the pressure and temperature, which leads to the collapse of the bubbles once they reach a critical size (Figure 4.2). Smaller bubbles are created, which either act as new cavitation nuclei or are simply dissolved. Consequently, a resultant 'shockwave' passes through the solvent, enhancing the mixing and resulting in shear disruption of the cell membrane of the material. Ultrasound also accelerates penetration of the solvent into the sample matrix,

Table 4.2 Bioactive compounds recovered from fruit and vegetable wastes using different extraction methods.

Extraction methods and conditions	Fruit/vegetable wastes	Bioactive compounds	References
SE, ethanol as extraction solvent, ratio of solvent to material 10 : 1 ml/g, extraction time 5 hours, ambient temperature	Seed coat of sweet Thai tamarind (*Tamarindus indica* L.).	(−)-epicatechin	Luengthanaphol et al. (2004)
SE, water as extraction solvent, ratio of solvent to dried sample 4 : 1 ml/g, extraction time 6 hours, room temperature	Pistachio (*Pistachia vera*) hull	Phenolic compounds	Goli et al. (2005)
SE, 50% ethanol as extraction solvent, ratio of solvent to sample 30 : 1, extraction temperature 80 °C, extraction time 2 hours	Residues of *Asparagus officinalis* L.	Ferulic acid, kaempferol, quercetin, rutin and isorhamnetin	Fan et al. (2015)
UAE, extraction temperature 40 °C, sonication power 150 W, 80% ethanol, ratio of sample to solvent 0.25 g/ml, extraction time 30 minutes	Orange peel (*Citrus sinensis* L.)	Naringin, hesperidin	Khan et al. (2010)
UAE, extraction temperature 55 °C, extraction time 40 minutes, amplitude 75%, ethanol (99.5%) as extraction solvent	Winter melon seed (*Benincasa hispida*)	Antioxidants	Bimakr et al. (2013)
UAE, extraction temperature 61.03 °C, liquid–solid ratio 21.70 ml/g, sonication time 23.67 minutes, power 400 W, mixture of ethanol (70%, v/v) and hydrochloric acid (0.01%, v/v) as extraction solvent	Blueberry (*Vaccinium ashei*) wine pomace	Phenolic compounds and anthocyanins	He et al. (2016)
MAE, microwave power 650.4 W, extraction time 53.7 seconds, 62.1% ethanol as extraction solvent, ratio of solvent to dried material 22.9:1 ml/g	Apple pomace	Polyphenols (chlorogenic acid, caffeic acid, syrigin, procyanidin B2, (−)-epicatechin, cinnamic acid, coumaric acid, phlorizin and quercetin)	Bai et al. (2010)
MAE, microwave power 121 W, extraction time 6 minutes, irradiation time 30 s/min, 32.6% ethanol as extraction solvent, ratio of solvent to dried material 100 : 1 ml/g	Grape seeds (*Vitis vinifera*)	Phenolic antioxidants	Krishnaswamy et al. (2013)

Table 4.2 (Continued)

Extraction methods and conditions	Fruit/vegetable wastes	Bioactive compounds	References
MAE, microwave power 152 W, extraction time 49 seconds, 66% methanol as extraction solvent, ratio of solvent to dried material 16 ml/g.	Citrus mandarin peels (Kinnow fruit)	Phenolic acids (including gallic acid, ρ-hydroxybenzoic acid, vanillic acid, ρ-coumaric acid and ferulic acid)	Hayat et al. (2009)
PLE, ethanol as extraction solvent, temperature 35 °C, pressure 4 MPa, static extraction time 5 minutes, dynamic extraction time 17 minutes, flow rate 2.4 g/min	Pressed palm fibre	Carotenoids	Cardenas-Toro et al. (2015)
PLE, ethanol as extraction solvent, pressure 100 bar, extraction temperature 60 °C, five static cycles for 5 minutes each, flush volume of 70%, purge with nitrogen gas of 100 seconds at end of each extraction	Sea buckthorn (*Hippophaë rhamnoides* L.) seed	Antimicrobial and antioxidant components	Michel et al. (2012)
SFE, temperature 313 K, pressure 35 MPa, extraction time 240 minutes, ratio of solvent to sample 6.6	Grape seed (*Vitis vinifera* L.)	Linoleic, palmitic, stearic and oleic acids	Prado et al. (2012)
SFE, extraction temperature 50 °C, pressure 27.6 MPa, 0.5 hours in static mode, following by 6 hours in dynamic mode, CO_2 flow rate 1 l/min, restrictor heated at 65 °C, ethanol as modifier	Fruit peel of *Citrus maxima*	Coumarins (imperatorin, meranzin, meranzin hydrate)	Teng et al. (2005)
SFE, supercritical CO_2 containing 10% ethanol (w/w), temperature 313 K, pressure 20 of MPa, extraction time 180 minutes, flow rate 5.88 of g/min, ratio of solvent (CO_2) to sample 47	Grape bagasse (*Vitis vinifera*)	Phenolic compounds (syringic, vanillic, gallic, ρ-hydroxybenzoic, protocatechuic and ρ-coumaric acids and quercetin)	Farias Campomanes et al. (2013)
EAE, Celluzyme MX concentration 1.5% (w/w), ratio of fresh material to aqueous enzyme solution 1 : 8 g/ml, extraction temperature 50 °C, extraction time 3 hours	Grapefruit peel (*Citrus × paradisi*)	Phenolics	Li et al. (2006)

(Continued)

Table 4.2 (Continued)

Extraction methods and conditions	Fruit/vegetable wastes	Bioactive compounds	References
EAE, mixture of Peclyve PR and Cellulyve 50LC (ratio 1 : 1), extraction temperature 30 °C, extraction time 3.18 hours, enzyme load 0.16 kg/kg	Tomato processing waste	Licopene	Zuorro et al. (2011)
EAE, Grindamyl CA 150 or Maxoliva (pectinase) concentration 10%, hydro-ethanolic mixture (75 : 25, v/v) as extraction solvent, extraction temperature 50 °C, extraction time 18 hours	Raspberry waste	Phenolic antioxidants	Laroze et al. (2010)

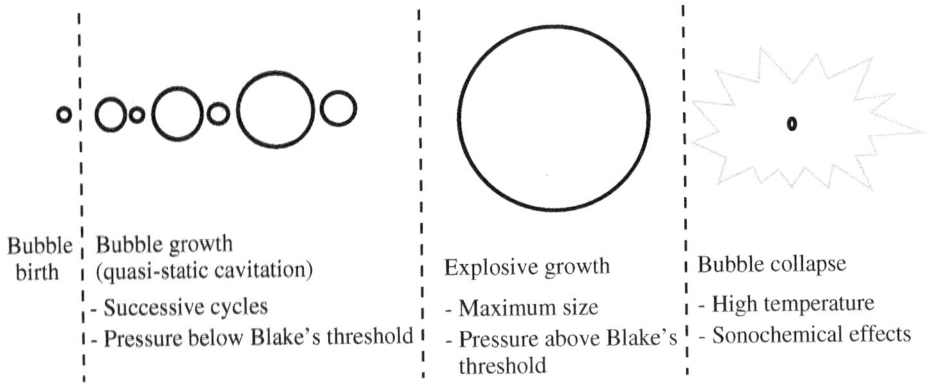

Figure 4.2 Formation, growth and collapse of a cavitation bubble. *Source:* González-García et al. (2010), http://www.mdpi.com/2073-4441/2/1/28/htm. Used under CC BY 3.0 https://creativecommons.org/licenses/by/3.0/.

resulting in an increase in the contact surface area between the solid and liquid phase, and leading to a quick extraction of dissolvable solids from the material into the solvent (Esclapez et al., 2011; Rostagno et al., 2003; Wang et al., 2008).

Ultrasonic baths or probes working in batch or flow mode are the most widely used UAE system at laboratory scale (Figure 4.3). UAE parameters, including ultrasound temperature, power and time, solvent type and ratio of solvent to sample, directly affect the efficiency of extraction of bioactive compounds from the raw materials. Khan et al. (2010) indicated that phenolic compound extraction of orange peel was three times as fast under UAE as when using the conventional method. In addition, Pan et al. (2012) found that flavonoids were extracted from hawthorn seed 91.7% faster and with a 1.32-fold greater yield using UAE compared to conventional flux extraction. Similarly, compared with standard Soxhlet extraction, UAE extracts from winter melon (*Benincasa hispida*) seeds had higher antioxidant activity, as the lower

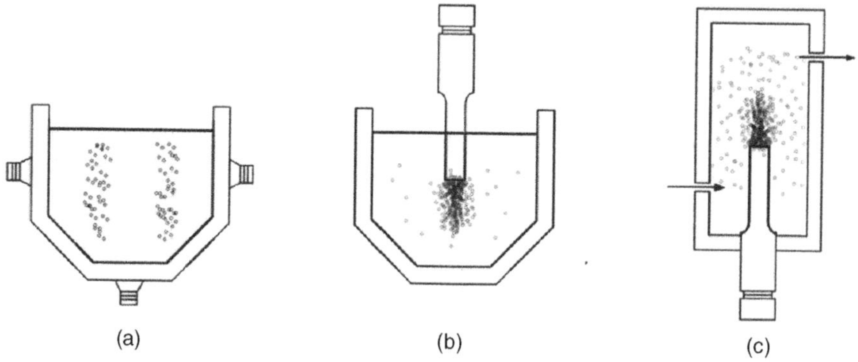

Figure 4.3 Typical reactor configurations: (a) ultrasonic bath, (b) transducer dipped in batch configuration, (c) transducer immersed in a flow reactor. *Source:* González-García et al. (2010), http://www.mdpi.com/2073-4441/2/1/28/htm. Used under CC BY 3.0 https://creativecommons.org/licenses/by/3.0/.

extraction temperature meant a limited thermal degradation of most of the thermolabile compounds and the shorter extraction time meant lower energy consumption (optimal UAE conditions were as follows: amplitude 75%, extraction time 40 minutes, extraction temperature 55 °C) (Bimakr et al., 2013). Thus, UAE is a feasible, effective, energy-saving and upscalable technique for the extraction of bioactive compounds from fruit and vegetable wastes.

4.3.2.3 Microwave-Assisted Extraction

MAE has received great attention recently because of its more rapid rate and similar or better recoveries of bioactive compounds, with either less extraction time or less solvent and energy consumption, compared to conventional methods (Bai et al., 2010). Microwaves are non-ionising electromagnetic waves within the frequency band of 300 MHz to 300 GHz in the electromagnetic spectrum, corresponding to wavelengths between 1 mm and 1 m. Generally, microwave applications are performed at 915 MHz (United States), 896 MHz (United Kingdom) or 2450 MHz (worldwide) for all practical purposes (Kaufmann & Christe, 2002; Kha & Nguyen, 2014). Microwaves possess both electric and magnetic fields, which are perpendicular to each other, due to their electromagnetic nature. The electric field causes heating via two coincident mechanisms in the forms of dipolar rotation and ionic induction. The alignment on the electric field of molecules possessing a dipole moment in either the solvent or the sample causes dipolar rotation. This vibration produces collisions with surrounding molecules, and thus thermal energy is liberated into the medium, which heats the whole sample simultaneously (Kaufmann & Christe, 2002).

Bourasa et al. (2015) applied the MAE technique for the extraction of polyphenols and other antioxidants from *Quercus* bark and found that extraction efficiencies were increased three and two times, respectively, as compared to conventional methods. In addition, Bai et al. (2010) found that the yield of polyphenols using MAE was higher than with reflux extraction or UAE (by 26 and 13%, respectively), and that MAE had an extraction time of only 53.7 seconds, compared to 6 hours and 1 hour, respectively. Therefore, MAE is a rapid and efficient extraction method.

MAE parameters such as microwave power, irradiation time, extraction time, ratio of solvent to sample, solvent concentration (for organic solvents) and particle size have been found to affect directly the extraction yield of bioactive compounds. Therefore, recent studies have attempted to optimise MAE parameters in order to obtain the highest possible levels of bioactive compounds from natural materials (Bai et al., 2010; Bhuyan et al., 2015; Nguyen et al., 2015b, 2016a, 2016b, 2016c; Rezaei et al., 2013; Wen et al., 2015).

4.3.2.4 Pressurised Liquid Extraction

PLE is based on accelerated solvent extraction under high pressure. The temperature of the extraction process can be above or below the boiling point of the solvents, with a pressure of 3.5–20.0 MPa. Organic solvents are used at elevated pressures and temperatures to enhance the extraction process – the high pressure lets the solvents reach temperatures above their atmospheric boiling point while remaining in the liquid state, and the increased temperature speeds the extraction kinetics by increasing the solubility and mass transfer properties. Increasing the temperature results in a decrease in the viscosity and surface tension of the solvents, which leads to greater penetration of the material matrix and weaker interactions between bioactive compounds and the matrix, raising the solubility of bioactive compounds into the solvents; however, this extraction method may cause the degradation of thermolabile compounds (Kaufmann & Christe, 2002; Kha & Nguyen, 2014; Lozano-Sánchez et al., 2014).

Water can be used to replace organic solvents, and this extraction technique is thus also called subcritical water extraction or pressurised hot-water extraction (Tunchaiyaphum et al., 2013). PLE parameters should be tested in order to optimise the extraction process, including solvent type, extraction temperature, pressure, extraction time, flow rate, material particle size and ratio of solid to solvent. Cardenas-Toro et al. (2015) found that with optimal conditions of temperature 55 °C, pressure 4 MPa and flow rate 2.4 g/min, the carotenoid yield from pressed palm fibre with ethanol as solvent was twofold higher with PLE than with Soxhlet extraction (Figure 4.4). Lozano-Sánchez et al. (2014) found suitable operating conditions for the recovery of phenolic and polar compounds from olive oil filter cake were temperature 120 °C, ratio of ethanol to water 50 : 50, extraction time 20 minutes and pressure 1500 psi. However, a preliminary extraction step was carried out to remove the lipophilic fraction from the filter cake using *n*-hexane as solvent at room temperature under a pressure of 1500 psi.

4.3.2.5 Supercritical Fluid Extraction

Every substance has three basic states (solid, liquid and gas). Supercritical state is a distinctive state that can only be attained if a substance is subjected to temperature and pressure beyond its critical point. The critical point is defined as the characteristic temperature and pressure above which distinctive gas and liquid phases do not exist. In supercritical state, the specific properties of a gas or a liquid disappear, which means that supercritical fluid cannot be liquefied by modifying temperature and pressure. Supercritical fluid possesses the gas-like properties of diffusion viscosity and surface tension, as well as liquid-like density and solvation power. These properties make it suitable for extracting bioactive compounds rapidly and with high yield (Azmir et al., 2013).

Carbon dioxide is inert and allows extraction at near room temperature (31 °C), which is good for thermolabile compound extraction and low critical pressures (74 bar).

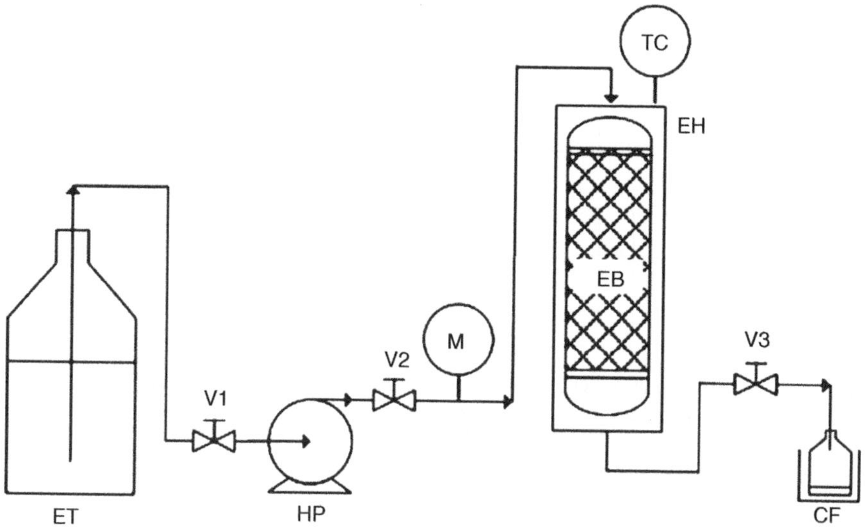

Figure 4.4 Pressurised liquid extraction (PLE) unit. ET, ethanol tank; HP, high-pressure pump; V1, V2, V3, valves; M, manometer; TC, temperature controller; EH, electric heater; EB, extraction bed; CF, collection flask. *Source:* Cardenas-Toro et al. (2015). Reproduced with permission of Elsevier.

It is necessary to use a co-solvent such as ethanol, methanol, *n*-hexane, isopropanol, acetonitrile or dichloromethane in order to increase the polarity of the solvent mixture (CO_2 and co-solvent) when extracting polar compounds, because CO_2 is nonpolar and can only extract lipid, fat and nonpolar compounds. There are many variables which affect the extraction efficiency, such as pressure, extraction time, extraction temperature, solvent flow rate, co-solvent, co-solvent flow rate, particle size, porosity, density, bed diameter and height (Azmir et al., 2013; Bimakr et al., 2012), revealing a limit on the optimisation of extraction conditions.

Preparation of materials is an important step in the SFE process as the high moisture content of the fresh material can cause mechanical difficulties, such as restrictor clogging due to ice formation. Although water is only about 0.3% soluble in supercritical carbon dioxide (SC-CO_2), highly water-soluble solutes would prefer to partition into the aqueous phase, leading to low SFE efficiency. Consequently, it is necessary to add some chemicals in order to retain the moisture in SFE of fresh materials such as Na_2SO_4 and silica gel (Wang & Weller, 2006).

To increase the recovery yield of bioactive compounds, both PLE and SFE can be employed for a sequential two-stage extraction process. For instance, nonpolar flavonoids and carotenoids present in mango peel byproduct were extracted by SFE (pressure 30 MPa, extraction temperature 40 °C, CO_2 flow rate 1.1 l/min, extraction time 7.7 hours) (Figure 4.5). During the SFE process, glass beads were placed inside the extraction cell to maintain proper flow rate of CO_2 in the extraction vessel and to fill the empty spaces of the extractor in order to avoid preferential CO_2 pathways during extraction. Polar polyphenolic compounds were then extracted by PLE under the same operating conditions, using pressurised ethanol for 5.5 hours (Garcia-Mendoza et al., 2015).

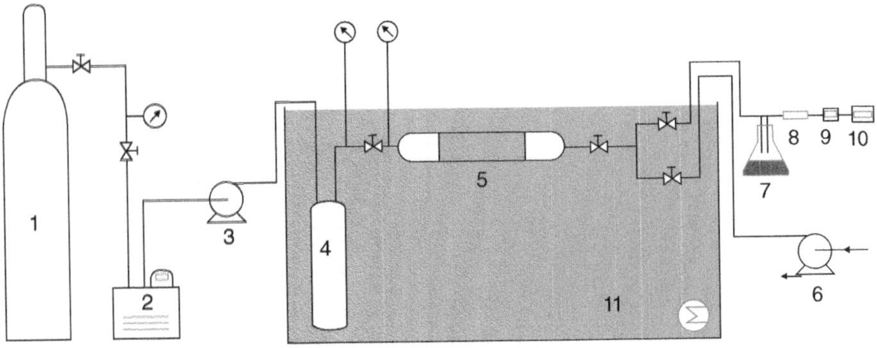

Figure 4.5 SC-CO_2 extraction unit. 1, CO_2 cylinder; 2, refrigeration bath; 3, high-pressure pump; 4, CO_2 supply tank; 5, extractor; 6, peristaltic pump; 7, extract collection flask; 8, Porapak-Q volatiles trap; 9, gas flow metre; 10, volume totaliser; 11, heating bath. *Source:* Garcia-Mendoza et al. (2015). Reproduced with permission of Elsevier.

4.3.2.6 Enzyme-Assisted Extraction

The main mechanism underlying enzyme-assisted extraction (EAE) involves cell wall-degrading enzymes such as glucanases and pectinase weakening or breaking down the cell wall, rendering the intracellular materials more accessible for extraction (Li et al., 2006). Another mechanism may be the direct enzyme-catalysed breakage of the ether and/or ester linkages between bioactive compounds such as phenols and plant cell wall polymers (Pinelo et al., 2008). Different EAE parameters, including extraction temperature, enzyme concentration, types of enzymes and material particle size, directly influence the extraction efficiency of bioactive compounds. An important emerging point is that the effectiveness of the enzyme treatment depends on the material, especially the solute properties and the part used (Zuorro et al., 2011). It has been found that commercial food-grade enzyme preparations with pectinolytic, cellulolytic and hemicellulolytic properties can enhance lycopene recovery from tomato skins (Lavecchia & Zuorro, 2008) and tomato paste (Zuorro & Lavecchia, 2010), even with a short extraction time and around ambient temperature. Moreover, Landbo & Meyer (2001) found that the yield of total phenols from blackcurrant juice press pomace increased significantly after treatment with each of the pectinolytic enzyme preparations Pectinex BE, Macer8 FJ and Macer8 R, as well as with the protease Novozym 89, revealing that each of the pectinases and the acid protease had significantly positive effects on the breakdown of the plant cell wall in blackcurrant pomace. The main limitation on the application of EAE in industrial processes is the high cost of enzymes; however, it is likely we will be able to obtain enzymatic formulations at lower costs and with better quality given current biotechnology advances (Bhat, 2000; Zúñiga et al., 2003).

4.3.3 Production of Powdered Extract

After extraction, the extracts containing bioactive compounds are concentrated using a rotavapor under reduced pressure, then crude powder is obtained by using freeze drying or spray drying to remove the solvent. The important parameters of spray drying are considered inlet temperature, outlet temperature and aspiration, while freeze drying is carried out at negative temperature (usually at −40 °C) under vacuum conditions.

Although freeze drying can generate high-quality powdered extract, it involves a long drying time and high production costs. Spray drying is faster and cheaper, and can also produce products of high quality. Tan et al. (2014) employed both methods to produce bitter melon powder. The working conditions of spray drying were inlet temperature 150 °C, outlet temperature 90 °C and aspiration 100%, while the working conditions of freeze drying were temperature −40 °C and duration 48 hours. The results of this study indicated that spray drying gave a high-quality powder in terms of saponins and antioxidant capacity.

4.3.4 Isolation and Purification of Bioactive Compounds

Isolation and purification of bioactive compounds from powdered extracts containing undesirable components can be carried out using HPLC or IEC. The principle behind the HPLC technique is the mass transfer of analytes between the stationary (immobile packing within the column) and mobile (organic solvent) phases. The bioactive compounds must be dissolved in the mobile phase and then forced to flow via the stationary column. Bioactive compounds can be isolated because of the different equilibrium distribution coefficients of different materials in the stationary and mobile phases. There are two HPLC systems: analytical HPLC and preparative HPLC. Analytical HPLC quantifies, identifies and focuses on all compounds in the mixture, while preparative HPLC isolates, purifies and focuses on one or a few compounds. It is important to choose a particular stationary (column) phase, such as adsorption, reverse phase, size exclusion or IEC, as well as effective mobile phases for preparative chromatography (Kha & Nguyen, 2014).

With regard to these process, Lozano-Sánchez et al. (2014) characterised the phenolic profiles of extracts from olive filter cake using HPLC coupled to a diode array detector and an electrospray time-of-flight mass spectrometer (HPLC-DAD-ESI-TOF/MS). The analytical column used was a Zorbax Eclipse Plus C18 (150 × 4.6 mm, 1.8 μm), which was set a flow rate of 0.80 ml/min and column temperature of 25 °C. The mobile phase was a mixture of 0.25% water (solvent A) and methanol (solvent B). The compounds were detected first in sequence with DAD at 240 and 280 nm, then by mass spectrometer with the microTOF coupled to the HPLC system. External mass spectrometer calibration was performed with sodium acetate clusters (5 mM sodium hydroxide in water/2-propanol at a ratio of 1 : 1 (v/v), with 0.2% acetic acid). All the main phenolic compounds were identified in the sample, including phenolic alcohols or derivatives (hydroxytyrosol and its oxidation product), secoiridoids (decarboxymethylated and hydroxydated forms of oleuropein and ligstroide aglycones), flavones (luteolin and apigenin) and elenolic acid derivatives.

Use of IEC to isolate and purify target compounds in powdered extracts is considered an inexpensive isolation method. Thus, it is an economic alternative when recovering bioactive compounds from residues of fruits and vegetables. It can also be used in conjunction with other processes, such as membrane filtration, to get a higher purity of bioactive compounds. The principle behind this method is mass transfer, in which the extract containing the solute (liquid phase or the adsorbate) is transferred to the solid phase as a result of retention on the surface of the solid or of reaction with the solid (or the adsorbent). The requirements for a good adsorbent are a large contact surface and high porosity. In this method, resin – a synthetic agent – is widely employed as an ion exchanger, either an anion-exchange resin (strong or weak base) or a cation-exchange resin (strong or weak acid). The choice of a certain type of resin depends on the types of ions in the extract (Kha & Nguyen, 2014). Many adsorbents are used in ion-exchange methods, such as Fe_3O_4 magnetic nanoparticles in the isolation of Ginkgolic acids from ginkgo leaves (Li et al., 2014) and rice hulls in the isolation of β-carotene from soybean oil (Chen et al., 2003). The main limitation of this method is the need for recovery of the adsorbate from the adsorbent. Therefore, selection of the most suitable eluent to wash and collect the adsorbate is necessary.

4.4 Conclusion

This chapter presented the principle steps in obtaining bioactive compounds, including preparation of a dried sample, extraction, powdered extract production, isolation and purification. It discussed both traditional (solvent extraction) and novel (UAE, MAE, PLE and SFE) extraction techniques. Every raw material has a particular set of physical characteristics and a particular chemical composition, so the technique used for extraction, isolation and purification should be selected based on the particular material. Recovery of bioactive compounds from fruit and vegetable wastes is intended to enhance the value of the residual materials and to reduce the pollution risk for the environment. The bioactive compounds derived from these sources can be applied in the health food, medical, cosmetic and pharmaceutical industries, instead of using synthetic compounds.

References

Alu'datt, M. H., Alli, I., Ereifej, K., Alhamad, M. N., Alsaad, A. & Rababeh, T. 2011. Optimisation and characterisation of various extraction conditions of phenolic compounds and antioxidant activity in olive seeds. *Natural Product Research*, **25**, 876–889.

Azmir, J., Zaidul, I. S. M., Rahman, M. M., Sharif, K. M., Mohamed, A., Sahena, F., Jahurul, M. H. A., Ghafoor, K., Norulaini, N. A. N. & Omar, A. K. M. 2013. Techniques for extraction of bioactive compounds from plant materials: a review. *Journal of Food Engineering*, **117**, 426–436.

Bai, X. L., Yue, T. L., Yuan, Y. H. & Zhang, H. W. 2010. Optimization of microwave-assisted extraction of polyphenols from apple pomace using response surface methodology and HPLC analysis. *Journal of Separation Science*, **33**, 3751–3758.

Bhat, M. K. 2000. Cellulases and related enzymes in biotechnology. *Biotechnology Advances*, **18**, 355–383.

Bhuyan, D. J., Vuong, Q. V., Chalmers, A. C., Van Altena, I. A., Bowyer, M. C. & Scarlett, C. J. 2015. Microwave-assisted extraction of *Eucalyptus robusta* leaf for the optimal yield of total phenolic compounds. *Industrial Crops and Products*, **69**, 1–10.

Bimakr, M., Rahman, R. A., Ganjloo, A., Taip, F. S., Salleh, M. L. & Sarker, M. Z. I. 2012. Optimization of supercritical carbon dioxide extraction of bioactive flavonoid compounds from spearmint (*Mentha spicata* L.) leaves by using response surface methodology. *Food Bioprocess Technology*, **5**, 912–920.

Bimakr, M., Rahman, R. A., Saleena Taip, F., Adzahan, N. M., Islam Sarker, Z. & Ganjloo, A. 2013. Ultrasound-assisted extraction of valuable compounds from winter melon (*Benincasa hispida*) seeds. *International Food Research Journal*, **20**, 331–338.

Bourasa, M., Chadnia, M., Barbaa, F. J., Grimia, N., Balsa, O. & Vorobieva, E. 2015. Optimization of microwave-assisted extraction of polyphenols from *Quercus* bark. *Industrial Crops and Products*, **77**, 590–601.

Cardenas-Toro, F. P., Alcázar-Alay, S. C., Coutinho, J. P., Godoy, H. T., Forster-Carneiro, T. & Meireles, M. A. A. 2015. Pressurized liquid extraction and low-pressure solvent extraction of carotenoids from pressed palm fiber: experimental and economical evaluation. *Food and Bioproducts Processing*, **94**, 90–100.

Chen, C. Y., Lin, C. I. & Chen, H. K. 2003. Kinetics of adsorption of β-carotene from soy oil with activated rice hull ash. *Journal of Chemical Engineering of Japan*, **36**, 265–270.

Da Silva, L. M. R., De Figueiredo, E. A. T., Ricardo, N. M. P. S., Vieira, I. G. P., De Figueiredo, R. W., Brasil, I. M. & Gomes, C. L. 2014. Quantification of bioactive compounds in pulps and by-products of tropical fruits from Brazil. *Food Chemistry*, **143**, 398–404.

Dailey, A. & Vuong, Q. V. 2015a. Effect of extraction solvents on recovery of bioactive compounds and antioxidant properties from macadamia (*Macadamia tetraphylla*) skin waste. *Cogent Food & Agriculture*, **1**.

Dailey, A. & Vuong, Q. V. 2015b. Optimisation of ultrasonic conditions as an advanced extraction technique for recovery of phenolic compounds and antioxidant activity from Macadamia (*Macadamia tetraphylla*) skin waste. *Technologies*, **3**, 302–320.

Dorta, E., Lobo, M. G. & González, M. 2012. Using drying treatments to stabilise mango peel and seed: Effect on antioxidant activity. *Food Science and Technology*, **45**, 261–268.

Esclapez, M. D., Garcia-Perez, J. V., Mulet, A. & Carcel, J. A. 2011. Ultrasound-assisted extraction of natural products. *Food Engineering Reviews*, **3**, 108–120.

Fan, R., Yuan, F., Wang, N., Gao, Y. & Huang, Y. 2015. Extraction and analysis of antioxidant compounds from the residues of *Asparagus officinalis* L. *Journal of Food Science and Technology*, **52**, 2690–2700.

Farias-Campomanes, A. M., Rostagno, M. A. & Meireles, M. A. A. 2013. Production of polyphenol extracts from grape bagasse using supercritical fluids: yield, extract composition and economic evaluation. *Journal of Supercritical Fluids*, **77**, 70–78.

Garcia-MendOZA, M. P., Paula, J. T., Paviani, L. C., Cabral, F. A. & Martinez-Correa, H. A. 2015. Extracts from mango peel by-product obtained by supercritical CO2 and pressurized solvent processes. *Lwt-Food Science and Technology*, **62**, 131–137.

Garcia-Perez, J. V., Carcel, J. A., Benedito, J., Blasco, M. & Mulet, A. 2009. Extraction of antioxidant compounds from grape stalk dried at different conditions. *Diffusion in Solids and Liquids IV*, **283–286**, 604–609.

Goldsmith, C. D., Vuong, Q. V., Sadeqzadeh, E., Stathopoulos, C. E., Roach, P. D. & Scarlett, C. J. 2015. Phytochemical properties and anti-proliferative activity of *Olea europaea* L. Leaf extracts against pancreatic cancer cells. *Molecules*, **20**, 12 992–13 004.

Goli, A. H., Barzegar, M. & Sahari, M. A. 2005. Antioxidant activity and total phenolic compounds of pistachio (*Pistachia vera*) hull extracts. *Food Chemistry*, **92**, 521–525.

González-García, J., Sáez, V., Tudela, I., Díez-Garcia, M. I., Deseada Esclapez, M. & Louisnard, O. 2010. Sonochemical treatment of water polluted by chlorinated organocompounds. A review. *Water*, **2**, 28–74.

Guo, C., Yang, J., Wei, J., Li, Y., Xu, J. & Jiang, Y. 2013. Antioxidant activities of peel, pulp and seed fractions of common fruits as determined by FRAP assay. *Nutrition Research*, **23**, 1719–1726.

Hayat, K., Hussain, S., Abbas, S., Farooq, U., Ding, B., Xia, S., Jia, C., Zhang, X. & Xia, W. 2009. Optimized microwave-assisted extraction of phenolic acids from citrus mandarin peels and evaluation of antioxidant activity *in vitro*. *Separation and Purification Technology*, **70**, 63–70.

He, B., Zhang, L.-L., Yue, X.-Y., Liang, J., Jiang, J., Gao, X.-L. & Yue, P.-X. 2016. Optimization of ultrasound-assisted extraction of phenolic compounds and anthocyanins from blueberry (*Vaccinium ashei*) wine pomace. *Food Chemistry*, **204**, 70–76.

Karantonis, H. C., Tsantila, N., Stamatakis, G., Samiotaki, M., Panayotou, G., Antonopoulou, S. & Demopoulos, C. A. 2008. Bioactive polar lipids in olive oil, pomace and waste byproducts. *Journal of Food Biochemistry*, **32**, 443–459.

Kaufmann, B. & Christe, P. 2002. Recent extraction techniques for natural products: microwave-assisted extraction and pressurised solvent extraction. *Phytochemical Analysis*, **13**, 105–113.

Kha, T. C. & Nguyen, M. H. 2014. Extraction and isolation of plant bioactives. In: SCARLETT, C. J. & VUONG, Q. V. (eds) *Plant Bioactive Compounds for Pancreatic Cancer Prevention and Treatment*. New York: Nova Science Publishers.

Khan, M. K., Abert-Vian, M., Fabiano-Tixier, A. S., Dangles, O. & Chemat, F. 2010. Ultrasound-assisted extraction of polyphenols (flavanone glycosides) from orange (*Citrus sinensis* L.) peel. *Food Chemistry*, **119**, 851–858.

Krishnaswamy, K., Orsat, V., Gariépy, Y. & Thangavel, K. 2013. Optimization of microwave-assisted extraction of phenolic antioxidants from grape seeds (*Vitis vinifera*). *Food and Bioprocess Technology*, **6**, 441–455.

Kumcuoglu, S., Yilmaz, T. & Tavman, S. 2014. Ultrasound assisted extraction of lycopene from tomato processing wastes. *Journal of Food Science and Technology*, **51**, 4102–4107.

Landbo, A.-K. & Meyer, A. S. 2001. Enzyme-assisted extraction of antioxidative phenols from black currant juice press residues (*Ribes nigrum*). *Journal of Agricultural and Food Chemistry*, **49**, 3169–3177.

Laroze, L., Soto, C. & Zúñiga, M. E. 2010. Phenolic antioxidants extraction from raspberry wastes assisted by-enzymes. *Electronic Journal of Biotechnology*, **13**, 11–12.

Lavecchia, R. & Zuorro, A. 2008. Improved lycopene extraction from tomato peels using cell-wall degrading enzymes. *European Food Research and Technology*, **228**, 153–158.

Li, B. B., Smith, B. & Hossain, M. M. 2006. Extraction of phenolics from citrus peels: II. Enzyme-assisted extraction method. *Separation and Purification Technology*, **48**, 189–196.

Li, R., Shen, Y., Zhang, X., Ma, M., Chen, B. & Van Beek, T. A. 2014. Efficient purification of ginkgolic acids from ginkgo biloba leaves by selective adsorption on Fe_3O_4 magnetic nanoparticles. *Journal of Natural Products*, **77**, 571–575.

Llorach, R., Espín, J. C., Tomás-Barberán, F. A. & Ferreres, F. 2002. Artichoke (*Cynara scolymus* L.) byproducts as a potential source of health-promoting antioxidant phenolics. *Journal of Agriculture and Food Chemistry*, **50**, 3458–3464.

Lozano-Sánchez, J., Castro-Puyana, M., Mendiola, J. A., Segura-Carretero, A., Cifuentes, A. & Ibáñez, E. 2014. Recovering bioactive compounds from olive oil filter cake by advanced extraction techniques. *Molecular Sciences*, **15**, 16270–16283.

Luengthanaphol, S., Mongkholkhajornsilp, D., Douglas, S., Douglas, P. L., Pengsopa, L.-I. & Pongamphai, S. 2004. Extraction of antioxidants from sweet Thai tamarind seed coat – preliminary experiments. *Journal of Food Engineering*, **63**, 247–252.

Michel, T., Destandau, E., Le Floch, G., Lucchesi, M. E. & Elfakir, C. 2012. Antimicrobial, antioxidant and phytochemical investigations of sea buckthorn (*Hippophaë rhamnoides* L.) leaf, stem, root and seed. *Food Chemistry*, **131**, 754–760.

Moure, A., Jose M. Cruz, J. M., Daniel Franco, D., Domínguez, J. M., Jorge Sineiro, J., Domínguez, H., Núnnez, M. J. & Parajó, J. C. 2001. Natural antioxidants from residual sources. *Food Chemistry*, **72**, 145–171.

Nguyen, V. T., Vuong, Q. V., Bowyer, M. C., Van Altena, I. A. & Scarlett, C. J. 2015a. Effects of different drying methods on bioactive compound yield and antioxidant capacity of *Phyllanthus amarus*. *Drying Technology*, **33**, 1006–1017.

Nguyen, V. T., Bowyer, M. C., Vuong, Q. V., Van Altena, I. A. & Scarlett, C. J. 2015b. Phytochemicals and antioxidant capacity of Xao tam phan (*Paramignya trimera*) root as affected by various solvents and extraction methods. *Industrial Crops and Products*, **67**, 192–200.

Nguyen, V. T., Pham, H. N. T., Bowyer, M. C., Van Altena, I. A. & Scarlett, C. J. 2016a. Influence of solvents and novel extraction methods on bioactive compounds and antioxidant capacity of *Phyllanthus amarus*. *Chemical Papers*, **70**, 556–566.

Nguyen, V. T., Bowyer, M. C., Van Altena, I. A. & Scarlett, C. J. 2016b. Optimisation of microwave-assisted extraction from *Phyllanthus amarus* for phenolic compounds-enriched extracts and antioxidant capacity. *Chemical Papers*, **70**, 713–725.

Nguyen, V. T., Vuong, Q. V., Bowyer, M. C., Van Altena, I. A. & Scarlett, C. J. 2016c. Microwave-assisted extraction for saponins and antioxidant capacity from Xao tam phan (*Paramignya trimera*) root. *Journal of Food Processing and Preservation*, **41**, e12851.

Orphanides, A., Goulas, V. & Gekas, V. 2013. Effect of drying method on the phenolic content and antioxidant capacity of Spearmint. *Czech Journal of Food Science*, **31**, 509–513.

Pan, G., Yu, G., Zhu, C. & Qiao, J. 2012. Optimization of ultrasound-assisted extraction (UAE) of flavonoids compounds (FC) from hawthorn seed (HS). *Ultrasonics Sonochemistry*, **19**, 486–90.

Pham, H. N. T., Nguyen, V. T., Vuong, Q. V., Bowyer, M. C. & Scarlett, C. J. 2015. Effect of extraction solvents and drying methods on the physicochemical and antioxidant properties of *Helicteres hirsuta* Lour. leaves. *Technologies*, **3**, 285–301.

Pinelo, M., Rubilar, M., Jerez, M., Sineiro, J. & Nunez, M. J. 2005. Effect of solvent, temperature, and solvent-to-solid ratio on the total phenolic content and antiradical activity of extracts from different components of grape pomace. *Journal of Agriculture and Food Chemistry*, **53**, 2111–2117.

Pinelo, M., Zornoza, B. & Meyer, A. S. 2008. Selective release of phenols from apple skin: Mass transfer kinetics during solvent and enzyme-assisted extraction. *Separation and Purification Technology*, **63**, 620–627.

Prado, J. M., Dalmolin, I., Carareto, N. D. D., Basso, R. C., Meirelles, A. J. A., Oliveira, J. V., Batista, E. A. C. & Meireles, M. A. A. 2012. Supercritical fluid extraction of grape seed: process scale-up, extract chemical composition and economic evaluation. *Journal of Food Engineering*, **109**, 249–257.

Rabeta, M. S. & Vithyia, M. 2013. Effect of different drying methods on the antioxidant properties of *Vitex negundo* Linn. tea. *International Food Research Journal*, **20**, 3171–3176.

Rezaei, S., Rezaei, K., Haghighi, M. & Labbafi, M. 2013. Solvent and solvent to sample ratio as main parameters in the microwave-assisted extraction of polyphenolic compounds from apple pomace. *Food Science & Biotechnology*, **22**, 1–6.

Rezaie, M., Farhoosh, R., Iranshahi, M., Sharif, A. & Golmohamadzadeh, S. 2015. Ultrasonic-assisted extraction of antioxidative compounds from Bene *(Pistacia atlantica* subsp. *mutica*) hull using various solvents of different physicochemical properties. *Food Chemistry*, **173**, 577–583.

Rostagno, M. A., Palma, M. & Barroso, C. G. 2003. Ultrasound-assisted extraction of soy isoflavones. *Journal of Chromatography A*, **1012**, 119–128.

Schieber, A., Stintzing, F. C. & Carle, R. 2001. By-products of plant food processing as a source of functional compounds – recent developments. *Trends in Food Science & Technology*, **12**, 401–413.

Sui, Y., Yang, J., Ye, Q., Li, H. & Wang, H. 2014. Infrared, convective, and sequential infrared and convective drying of wine grape pomace. *Drying Technology*, **32**, 686–694.

Tan, S. P., Vuong, Q. V., Stathopoulos, C. E., Parks, S. E. & Roach, P. D. 2014. Optimized aqueous extraction of saponins from bitter melon for production of a saponin-enriched bitter melon powder. *Journal of Food Science*, **79**, E1372–E1381.

Teng, W. Y., Chen, C. C. & Chung, R. S. 2005. HPLC comparison of supercritical fluid extraction and solvent extraction of coumarins from the peel of *Citrus maxima* fruit. *Phytochemical Analysis*, **16**, 459–462.

Tunchaiyaphum, S., Eshtiaghi, M. N. & Yoswathana, N. 2013. Extraction of bioactive compounds from mango peels using green technology. *International Journal of Chemical Engineering and Applications*, **4**, 194–198.

Vega-Gálvez, A., Di Scala, K., Rodríguez, K., Lemus-Mondaca, R., Miranda, M., López, J. & PEREZ-WON, M. 2009. Effect of air-drying temperature on physico-chemical properties, antioxidant capacity, colour and total phenolic content of red pepper (*Capsicum annuum*, L. var. Hungarian). *Food Chemistry*, **117**, 647–653.

Vuong, Q. V., Hirun, S., Roach, P. D., Bowyer, M. C., Phillips, P. A. & Scarlett, C. J. 2013. Effect of extraction conditions on total phenolic compounds and antioxidant activities of *Carica* papaya leaf aqueous extracts. *Journal of Herbal Medicine*, **3**, 104–111.

Vuong, Q. V., Zammit, N., Munro, B. R., Murchie, S., Bowyer, M. C. & Scarlett, C. J. 2015. Effect of drying conditions on physicochemical and antioxidant properties of *Vitex agnus-castus* leaves. *Journal of Food Processing and Preservation*, **39**.

Wang, L. & Weller, C. L. 2006. Recent advances in extraction of nutra-ceuticals from plants. *Trends in Food Science & Technology*, **17**, 300–312.

Wang, J., Sun, B. G., Cao, Y. P., Tian, Y. & Li, X. H. 2008. Optimisation of ultrasound-assisted extraction of phenolic compounds from wheat bran. *Food Chemistry*, **106**, 804–810.

Wen, Y., Chen, H., Zhou, X., Deng, Q., Zhao, Y., Zhao, C. & Gong, X. 2015. Optimization of microwave-assisted extraction and antioxidant activities of anthocyanins from blackberry using a response surface methodology. *RSC Advances*, **5**, 19686–19695.

Wojdyło, A., Figiel, A. & Oszmianski, J. 2009. Effect of drying methods with the application of vacuum microwaves on the bioactive compounds, color, and antioxidant activity of strawberry fruits. *Journal of Agriculture and Food Chemistry*, **57**, 1337–1343.

Wojdyło, A., Figiel, A., Lech, K., Nowicka, P. & Oszmiański, J. 2014. Effect of convective and vacuum–microwave drying on the bioactive compounds, color, and antioxidant capacity of sour cherries. *Food and Bioprocess Technology*, 7, 829–841.

Zhang, Y., Zeng, H., Huang, Z., Wang, Y. & Zheng, B. 2014. Drying characteristics and kinetics of *Anoectochilus roxburghii* by microwave vacuum drying. *Journal of Food Processing and Preservation*, **38**, 2223–2231.

Zúñiga, M. E., Soto, C., Mora, A., Chamy, R. & Lema, J. M. 2003. Enzymic pre-treatment of Guevina avellana mol oil extraction by pressing. *Process Biochemistry*, **39**, 51–57.

Zuorro, A. & Lavecchia, R. 2010. Mild enzymatic method for the extraction of lycopene from tomato paste. *Biotechnology & Biotechnological Equipment*, **24**, 1854–1857.

Zuorro, A., Fidaleo, M. & Lavecchia, R. 2011. Enzyme-assisted extraction of lycopene from tomato processing waste. *Enzyme and Microbial Technology*, **49**, 567–573.

5

Recovering Bioactive Compounds from Wine Wastes

Van Tang Nguyen

School of Environmental and Life Sciences, Faculty of Science and Information Technology, University of Newcastle, Ourimbah, NSW, Australia
Department of Food Technology, Faculty of Food Technology, Nha Trang University, Nha Trang, Khanh Hoa, Vietnam

5.1 Introduction

Grapes are one of the main agro-economic crops in the world, with total harvested area, total production and total gross production value of 7.16 million ha, 77.18 million tonnes and US$78.50 billion, respectively, in 2013 (FAOSTAT, 2015). Grapes are used to produce several alcoholic beverages (about 80% of all produced grapes), including wine, cognac and Pisco, with wine becoming one of the most consumed alcoholic beverages globally (Putnik et al., 2016). Italy, France and Spain are the three largest wine-producing countries, generating about 46% of total worldwide wine production (Barba et al., 2016). The winemaking industry produces a large quantity of organic residues/wastes as biodegradable solids, which provide a great environmental risk and necessitate expensive treatment (Bustamante et al., 2008; Teixeira et al., 2014). Kammerer & Carle (2008) reported that due to incomplete extraction during pressing and fermentation, grape byproducts are rich sources of phenolic compounds with potent antioxidant capacity, such as anthocyanins, catechins and glycosides of flavonols. Their compositions depend on a variety of factors, such as environmental and processing conditions and the initial phenolic content of the grapes (Schieber et al., 2001). These byproducts can therefore be used as low-cost alternative materials for the production of functional food additives.

Among the wastes discarded from wine production, grape bagasse is the main residue to be produced after pressing, accounting for 20% of total byproducts from Pisco production. Grape bagasse is produced by distillation and consists mainly of stems, skins and seeds (25, 25 and 50% by fresh weight, respectively) (Jin & Kelly, 2009). Grape seeds are considered a disposable waste material by most wineries and are usually discarded, burned or used as animal feed. However, grape seed oil presents health benefits because of its high content of unsaturated fatty acids and antioxidant compounds, such as monomeric flavan-3-ols, phenolic acids and oligomeric proanthocyanidins. It displays antimicrobial activity and is used in the health food, cosmetic and pharmaceutical industries (Fiori, 2010; Prado et al., 2012; Prasain et al., 2009; Yilmaz & Toledo, 2006). Grape pomace (seeds and skins), stems, rachis and lees are other residues left behind after pressing or pressing + fermentation in white and red wine production. They

Recovering Bioactive Compounds from Agricultural Wastes, First Edition. Edited by Van Tang Nguyen.

are also rich in phenolics (phenolic acids, flavonols, flavanols, stilbenes, procyanidins, anthocyanins), resveratrol, lignin and so on (Bertran et al., 2004; Di Lecce et al., 2014; Fontana et al., 2013; Prozil et al., 2014; Ruberto et al., 2007; Sousa et al., 2014; Xiang et al., 2003). Therefore, the use of wine byproducts/residues/wastes as inexpensive, biorenewable and sustainable sources for the recovery of bioactive compounds is a promising trend in the nutraceutical, medical, pharmaceutical and cosmetic industries, increasing the value of grape production and gesturing towards sustainable development.

5.2 Recovering Bioactive Compounds from Wine Wastes

5.2.1 Polyphenols from Wine Wastes

Phenolic compounds are widely distributed in nature, with plants like tea, coffee, grapes and other fruits being rich sources (Farías-Campomanes et al., 2013; Nguyen et al., 2011; Nuhu, 2014; Vuong et al., 2010). Phenolics include not just an ample variety of molecules with a polyphenol structure, such as several hydroxyl groups on aromatic rings, but also molecules with one phenol ring, such as phenolic acids and phenolic alcohols. In general, phenolics in grapes and wine can be classified into three main groups: (i) phenolic acids (mainly benzoic and hydroxycinnamic acids), (ii) simple flavonoids (catechins, flavonols, and anthocyanins) and (iii) tannins and proanthocyanidins (Fontana et al., 2013). Among the phenolic compounds found in grapes, the condensed tannins (proanthocyanidins) are the most abundant in the skins and seeds (Bosso et al., 2016). Phenolic compounds have direct correlation to the colour, astringency, bitterness and structure of wines, and it has been proved that the sensory perception of coarseness rises with increasing galloylation of proantocyanidins (Bosso et al., 2016; Di Lecce et al., 2014). Recently, phenolic compounds have received great interest among researchers because of their potential activity in preventing chronic diseases; for example, both *in vitro* and *in vivo* studies have indicated that phenolic compounds possess a wide range of pharmacological activities, such as antioxidant, anti-allergic, antimutagenic, anti-aging, anti-inflammatory, antimicrobial, anticarcinogenic and anticancer properties (Adámez et al., 2012; Baydar et al., 2007; Cabrera et al., 2006; Fontana et al., 2013; García-Lafuente et al., 2009). Therefore, many studies have been carried out which have attempted to recover phenolic compounds from wine wastes such as grape bagasse/pomace (seeds and skins), leaves, stems/stalks, rachis and lees.

Figure 5.1 provides a flow chart of white and red wine production processes, including the residues/wastes discarded from these processes. Among these, pomace is an important source of polyphenols, accounting for 20–30% of the initial weight of the grape (Bosso et al., 2016).

According to Teixeira et al. (2014), the natural bioactive compounds derived from wine byproducts/wastes are considered to act as health promoters. Among them, grape pomace originates from the production of grape juice (must) by pressing of whole grapes. About 9 million tonnes of this organic residue are produced annually, accounting for 20% (w/w) of the total grapes used for wine production. Regarding its composition, grape pomace is a suitable material for many different processes, particularly the extraction of grape seed oil and polyphenols (mainly anthocyanins, flavonols, flavanols, phenolic acids and resveratrol). The relative proportion of grape

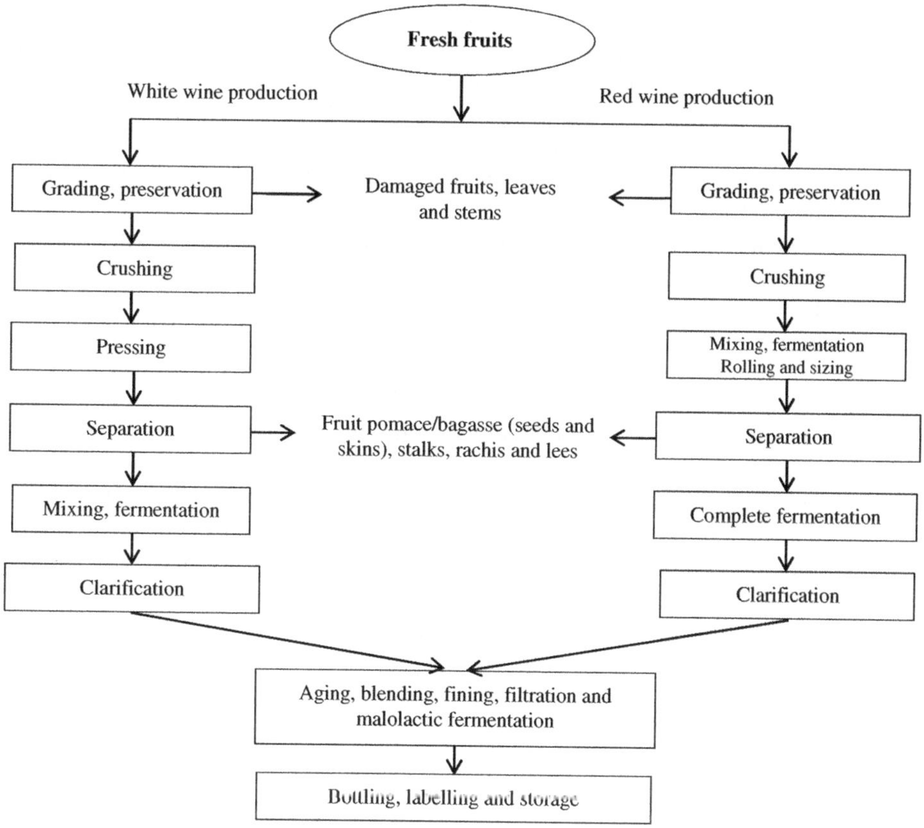

Figure 5.1 Flow chart of while and red wine production.

seeds ranges from 38 to 52% by dry weight, with the phenolic content in seeds varying from 60 to 70% of the total extractable compounds, while grape skin accounts for 65% of the total grape pomace, and is known as a rich source of phenolic compounds. Grape stems (stalks or clusters) are also a residue of the wine making industry and have been partially applied as a source of astringent compounds, mainly proanthocyanidins (condensed tannins). In general, this matter is removed before the vinification steps, to avoid an excessive astringency of the wine or a negative effect on the organoleptic characteristics. The quantity of stems varies between 1.4 and 7.0% of the raw matter processed. Currently, the economic value of grape stems is low, being mainly used as animal feed or soil amendments. Grape stems could therefore be a valuable and interesting source of dietetic fibre and antioxidants. The phenolic composition of grape stems consists of flavan-3-ols, hydroxycinnamic acids, monomeric and oligomeric flavonols and stilbenes, with total phenolics approximately 5.8% by dry weight. Grape leaves are the least studied or valorised residue of grape crops and the wine industry. The bioactive compounds in the leaves consist mainly of phenolic acids, flavonols, tannins, procyanidins, anthocyanins, vitamins, carotenoids and terpenes, which attract considerable interest as a promising source of compounds with nutritional properties and biological potential. Wine lees are the residues formed at the bottom of receptacles containing wine, after fermentation, storage or authorised treatments, as well as the residue obtained following the filtration or centrifugation of wine. Lees are mainly composed of phenolic compounds, such as total phenolic compounds (50 mg gallic acid equivalents (GAE)/g dry weight), anthocyanins (6.0–11.7 mg/g dry weight) and other phenolics (29.8 mg/g dry weight).

Farías-Campomanes et al. (2013) applied supercritical fluid extraction (SFE) to isolate phenolics from grape bagasse (Figure 5.2) and found that it was the most efficient process for recovering phenolic compounds from bagasse in Pisco production (20–23 g/kg of extract at 35 and 20 MPa, respectively), as compared to conventional extraction techniques (1.4 and 1.8 g/kg of extract using agitated bed and Soxhlet extraction, respectively) (Table 5.1). Syringic, vanillic, gallic, *p*-hydroxybenzoic, protocatechuic, *p*-coumaric acids and quercetin were the main compounds identified in the bagasse extract, while rutin and resveratrol were not identified in this extract (Table 5.1 and Figure 5.3). Based on the kinetic parameters and the economic analysis, the best manufacturing cost and extraction yield were achieved under the following conditions: temperature 313 K, pressure 20 MPa, extraction time 180 minutes and solvent/feed ratio 47. The estimated cost of manufacturing and expected phenolic content using a 0.5 m^3 SFE plant were US$133.16/kg of extract and at least 23 g/kg of extract, respectively.

Di Lecce et al. (2014) extracted the phenolic components from the different residue parts of Albariño grapes – skin, pulp and seeds – using a mixture of ethanol and water (80 : 20, v/v) at pH 3.5 (acetic acid), then quantified their content using hybrid

Table 5.1 Operational conditions and phenolic compounds in extracts from grape bagasse. *Source:* Farías-Campomanes et al. (2013). Reproduced with permission of Elsevier.

		Extraction methods			
		Agitated bed	Soxhlet	SFE	
	Pressure (MPa)	0.1	0.1	20	35
	Temperature (K)	333	352	313	313
	Solvent	EtOH	EtOH	CO_2 + EtOH	CO_2 + EtOH
	S/F	10	20	80	115
	Global yield (%)	8.2 ± 0.2	10.4 ± 0.5	5.5 ± 0.1	5.6
		Phenolic concentration of extracts (g/kg of extract)			
1	**Gallic acid**	0.29 ± 0.08	0.4 ± 0.1	4.48 ± 0.01	3.0 ± 0.1
2	**Protocatechuic acid**	0.11 ± 0.04	0.13 ± 0.04	0.91 ± 0.02	1.12 ± 0.03
3	**Vanillic acid**	0.29 ± 0.02	0.4 ± 0.2	4.3 ± 0.1	4.2 ± 0.02
4	***p*-hydroxybenzoic acid**	0.11 ± 0.04	0.12 ± 0.04	1.75 ± 0.01	1.43 ± 0.02
5	**Syringic acid**	0.58 ± 0.04	0.66 ± 0.2	10.78 ± 0.03	9.5 ± 0.1
6	***p*-coumaric acid**	–	0.06 ± 0.01	0.53 ± 0.03	0.7 ± 0.01
7	**Quercetin**	0.04 ± 0.01	0.08 ± 0.02	0.27 ± 0.01	0.15 ± 0.01
	Total	1.4 ± 0.2	1.8 ± 0.6	23.0 ± 0.2	20.0 ± 0.3

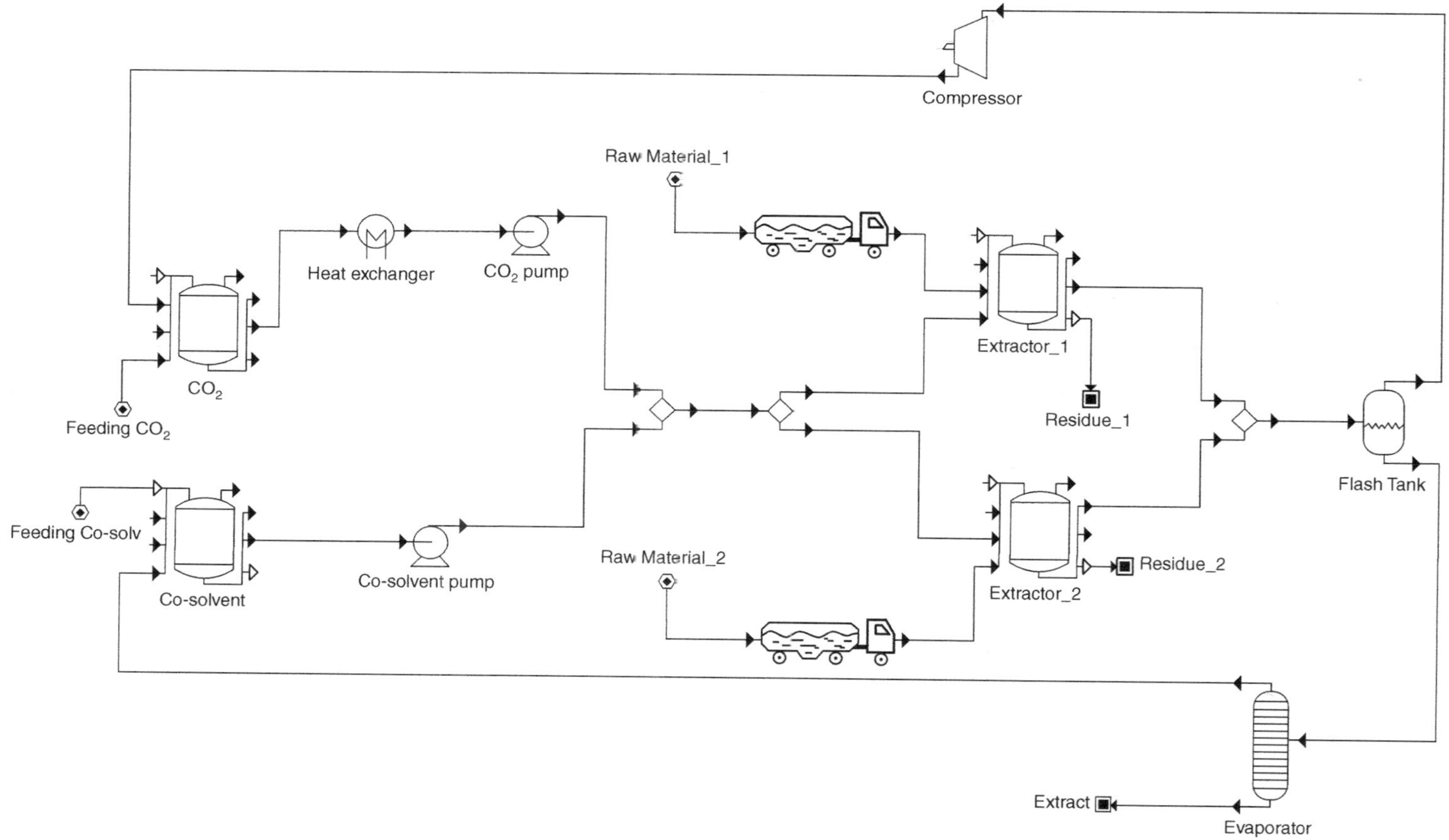

Figure 5.2 Flow chart of supercritical fluid extraction (SFE) of polyphenols from grape bagasse. *Source:* Farías-Campomanes et al. (2013). Reproduced with permission of Elsevier.

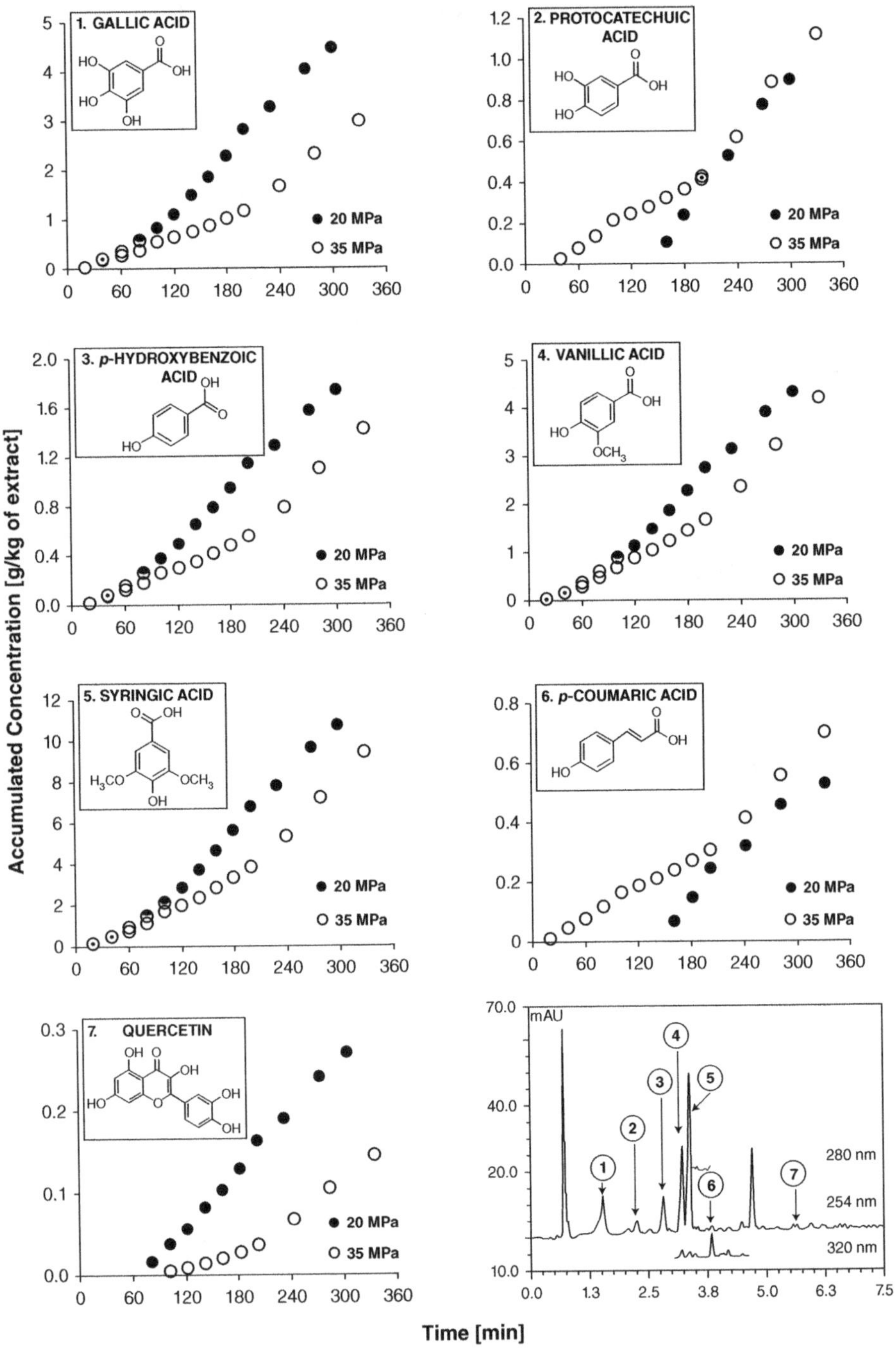

Figure 5.3 Accumulated concentration of phenolic compounds in extracts from grape bagasse. *Source:* Farías-Campomanes et al. (2013). Reproduced with permission of Elsevier.

quadrupole time-of-flight and triple-quadrupole mass spectrometry. The results indicated that the most abundant classes of phenols in Albariño grape berries were the monomeric and oligomeric forms of flavan-3-ols, which appeared in hypodermal layers of the skin and in the soft parenchyma of the seed. The total flavanol content in grape seed was 611 mg/100 g of fresh weight, in that the concentrations of (+)-catechin and (−)-epicatechin were the highest among the phenolic compounds identified (106.50 and 77.51 mg/100 g of fresh weight, respectively) (Table 5.2). Flavanols, flavonols and hydroxycinnamates were also found in Albariño grape skin in considerable amounts (32.90, 31.45 and 18.98 mg/100 g fresh weight, respectively), with the flavanols present in glycoside form, such as 3-glucosides, rutinoside and glucuronide flavonols. The content ranged between 0.39 and 12.4 mg/100 g of fresh weight for quercetin-3-O-rutinoside and quercetin-3-O-glucoside, respectively. Among the hydroxycinnamic acids in grape skin, cis-coutaric acid was the major acid, followed by caftaric isomers. The phenolic content in the pulp was very low, with hydroxycinnamic acids and flavanols the most representative compounds (1.63 and 1.35 mg/100 g of fresh weight, respectively), and catechin, epicatechin and procyanidin B3 also detected at small amounts.

González-Centeno et al. (2014) applied ultrasound-assisted extraction (UAE) using water as solvent for the separation of phenolics and flavonoids from grape pomace (*Vitis vinifera* L.) and found that the optimal UAE conditions for obtaining the highest levels of phenolics, flavonoids and antioxidant capacity were frequency 40 kHz, power density 150 W/l and extraction time 25 minutes. Under these conditions, the total phenolics and total flavonols achieved were 32.31 mg GAE/100 g fresh weight and 2.04 mg quercetin/100 g fresh weight, respectively, and the maximum values of antioxidant capacity were predicted to be 53.47 and 43.66 mg trolox equivalents/100 g fresh weight for cupric ion-reducing antioxidant capacity (CUPRAC) and ferric-reducing antioxidant power (FRAP), respectively. When compared to UAE using organic solvents, this research used only water as solvent for the extraction, and applied it at lower temperature and with a shorter extraction time, decreasing production costs in terms of solvent and energy expenses and reducing the pollution risks for both people and the environment.

Pedroza et al. (2015) applied microwave-assisted extraction (MAE) for phenolic compounds from dried grape skins as a waste and found that the phenolic yield, antioxidant capacity and energy consumption were comparable with those using solid–liquid extraction (SLE) at 60 °C for 2 hours, while the extraction time was shorter. As in SLE, an equivalent yield of total phenolics was achieved with an MAE extraction time of 1033 seconds (equivalent to 83 seconds of irradiation at 900 W, saving 83% extraction time with 70% energetic efficiency as compared to SLE). To enhance the phenolic extraction efficiency, pre-maceration of samples and solvent preheating have been proposed for large-scale industrial processes. This study also indicated the importance of an experimental design based on factors (such as temperature, irradiation time, mixing, vessel geometry and vessel covering) that greatly affect the total phenolic yield, regardless of the extraction method.

Melo et al. (2015) optimised extraction conditions using aqueous ethanol as solvent and combining conventional and ultrasound extraction (for 20 and 15 minutes, respectively) and compared the phenolic compositions and antioxidant capacities of extracts from winery byproducts (approximately 20 million tons were discarded in 2012), and found that the optimal conditions for the extraction of phenolic compounds

Table 5.2 Content of phenolic compounds in the different residual parts of Albariño grape (mg/100 g of fresh weight). *Source:* Di Lecce et al. (2014). Reproduced with permission of Elsevier.

Compounds	Skin	Pulp	Seed
Gallic acid	1.19	nd	1.92
Gallic acid hexose I	nd	nd	0.79
Gallic acid dihexose	1.13	nd	1.25
Protocatechuic acid-*O*-hexoside	nd	nd	0.45
Gallic acid hexose II	nd	nd	1.36
Protocatechuic acid- *O*-hexoside	nd	nd	1.54
Hydroxybenzoic acids	2.32 ± 0.1	nd	7.31 ± 0.3
Cis-caftaric acid	2.53	0.11	nd
Trans-caftaric acid	4.04	0.37	nd
Cis-coutaric acid	6.23	Nd	nd
Trans-coutaric acid	0.27	Nd	nd
Coumaric acid- *O*-hexoside	2.27	1.03	nd
Fertaric acid	1.68	0.12	nd
p-coumaric acid	1.96	Nd	nd
Hydroxycinnamic acids	18.98 ± 0.8	1.63 ± 0.1	nd
(Epi)gallocatechin-(epi)catechin	nd	nd	3.58
Procyanidin trimer C1	nd	nd	12.65
(Epi)gallocatechin-3-gallate	nd	nd	1.54
Epigallocatechin	2.09	nd	nd
(Epi)catechin-3-hexose	nd	nd	1.49
(Epi)catechin-3-hexose	nd	nd	3.52
Procyanidin B3	nd	0.57	44.65
Procyanidin B1	nd	nd	3.09
(+)-catechin	11.45	0.55	106.5
Procyanidin trimer I	nd	nd	31.43
Procyanidin trimer II	nd	nd	18.54
Procyanidin B4	8.04	nd	58.39
Procyanidin B2	8.65	nd	64.53
(−)-epicatechin	2.67	0.23	77.51
(Epi)catechin-(epi)catechingallate I	nd	nd	26.76
Procyanidin trimer III	nd	nd	13.54
(Epi)catechin-(epi)catechingallate II	nd	nd	23.73
(Epi)catechin-(epi)catechingallate III	nd	nd	21.43
(−)-epicatechin-3-*O*-gallate	nd	nd	76.54
Dimer digallater	nd	nd	21.43
Flavanols	32.9 ± 2.7	1.35 ± 0.1	610.8 ± 35.8
Quercetin-3-*O*-rutinoside	0.42	nd	nd
Quercetin-3-*O*-glucuronide	0.98	nd	nd
Quercetin-3-*O*-glucoside	12.43	nd	nd

Table 5.2 (Continued)

Compounds	Skin	Pulp	Seed
Dihydroquercetin-3-*O*-rhamnoside	5.65	nd	nd
Quercetin-3-*O*-pentoside	0.23	nd	nd
Kaempferol-3-*O*-glucuronide	3.21	nd	nd
Kaempferol-3-*O*-glucoside	8.43	nd	nd
Flavonols	31.45 ± 1.6	nd	nd
Trans-piceid	6.93	nd	nd
Trans-resveratrol	1.43	nd	nd
Stilbenes	8.36 ± 0.4	nd	nd
Total	94.21 ± 5.1	2.98 ± 0.2	618.1 ± 36.1

nd, not detected.

were ethanol concentration 43% (petit verdot and syrah rachis) or 57% (chenin blanc rachis and chenin blanc, petit verdot and syrah pomace) and temperature 96 °C. Gallic acid, syringic acid, procyanidins B1 and B2, catechin, epicatechin, epicatechin gallate, quercetin 3-β-d-glucoside, delfinidin 3-glucoside, peonidin 3-*O*-glucoside and malvidin 3-glucoside were found to be the main phenolic compounds in these extracts, depending on the particular pomace and rachis (Table 5.3). The antioxidant capacity of the extracts also depended on the type of pomace and rachis, with rachis extracts showing significantly higher antioxidant capacity than pomace extracts in terms of free radicals (DPPH and ABTS) and reactive oxygen species (ROS) (peroxyl, superoxide and hypochlorous acid), especially for chenin blanc (except superoxide radical and hypochlorous acid) (Table 5.4).

Bosso et al. (2016) used two different ternary solvent mixtures – acetone/ethanol/water and acetone/ethanol/ethyl acetate – in different proportions to extract condensed tannins or proanthocyanidins from grape pomace seeds, focusing on two binary aqueous mixtures (acetone–water and ethanol–water). The outcomes indicated that the ternary solvent mixtures greatly influenced total flavonoids and total polyphenols from grape pomace seeds, in that the highest amounts of total flavonoids (12.60 or 2.85 mg/g dry weight) and total polyphenols (8.78 or 2.86 mg GAE/g dry weight) were achieved using mixtures of acetone/ethanol/water (50 : 0 : 50) and acetone/ethanol/ethyl acetate (50 : 50 : 0), respectively, followed by acetone/ethanol/water (33 : 33 : 33) and acetone/ethanol/ethyl acetate (16.5 : 66.0 : 16.5), while the lowest amounts were obtained by acetone/ethanol/water (100 : 0 : 0) and acetone/ethanol/ethyl acetate (0 : 0 : 100) (Table 5.5). Total flavonoid content from grape pomace seeds was significantly affected by the binary aqueous mixtures in terms of acetone–water and ethanol–water, with the greatest total flavonoid contents (5.38 and 7.98 mg/g dry weight) obtained by mixtures of ethanol–water (70 : 30) and acetone–water (50 : 50), respectively, followed by ethanol–water (50 : 50) and acetone–water (75 : 25).

Xu et al. (2016) determined the phenolic compound content and antioxidant and antibacterial properties of pomace extracts from four Virginia-grown grape varieties using 80% (v/v) aqueous acetone (solvent to material ratio of 4 ml/g) as solvent at room

Table 5.3 Phenolic compounds of extracts from chenin blanc, petit verdot and syrah pomace and rachis. *Source:* Melo et al. (2015). Reproduced with permission of Elsevier.

Phenolic compound (μg/mg)	Winery byproducts						Validation parameters		
	Chenin blanc Rachis	Pomace	Petit verdot Rachis	Pomace	Syrah Rachis	Pomace	LOD (μg)	LOQ (μg)	Linearity (R^2)
Hydroxybenzoic acid									
Gallic acid	0.15 ± 0.02	0.05 ± 0.0	0.14 ± 0.01	0.84 ± 0.04	0.99 ± 0.0	0.43 ± 0.01	0.0052	0.0160	0.997
Syringic acid	nd	nd	nd	0.13 ± 0.01	nd	0.04 ± 0.0	0.0007	0.0022	0.998
Proanthocyanidin									
Procyanidin B1	5.93 ± 0.58	0.21 ± 0.01	3.04 ± 0.15	1.93 ± 0.06	2.09 ± 0.02	0.84 ± 0.06	0.0046	0.0140	1.000
Procyanidin B2	nd	0.15 ± 0.01	nd	1.89 ± 0.15	nd	0.77 ± 0.04	0.0059	0.0180	0.999
Flavan-3-ol									
Catechin	4.11 ± 0.03	0.24 ± 0.10	2.49 ± 0.03	7.43 ± 0.4	2.57 ± 0.07	3.43 ± 0.26	0.0380	0.110	0.983
Epicatechin	0.17 ± 0.01	0.32 ± 0.05	0.04 ± 0.0	5.81 ± 0.22	0.24 ± 0.01	3.11 ± 0.12	0.0087	0.0260	0.999
Epicatechin gallate	nd	0.08 ± 0.01	nd	nd	nd	Nd	0.0032	0.0098	0.999
Flavonol									
Quercetin 3-β-D-glucoside	nd	0.09 ± 0.0	nd	nd		0.44 ± 0.02	0.0068	0.0210	0.999
Anthocyanin									
Delfinidin 3-glucoside	nd	nd	nd	0.07 ± 0.0	nd	0.05 ± 0.0	0.0012	0.0035	0.999
Peonidin 3-*O*-glucoside	nd	nd	nd	0.04 ± 0.0	nd	0.44 ± 0.01	0.0012	0.0035	0.999
Malvidin 3-glucoside	nd	nd	nd	1.66 ± 0.08	nd	1.71 ± 0.07	0.0075	0.023	0.999
TPC	124.96 ± 6.40^{A}	20.17 ± 0.6^{E}	96.77 ± 6.6^{B}	66.69 ± 7.0^{C}	50.63 ± 2.11^{D}	39.30 ± 2.37^{D}	nd	nd	nd
TMA	nd	nd	0.22 ± 0.01^{C}	0.99 ± 0.06^{B}	0.09 ± 0.0^{D}	1.41 ± 0.03^{A}	nd	nd	nd

Each value is expressed as mean (triplicate) ± standard deviation (SD). The same letters in a row are not significantly different at the 0.05 level according to ANOVA.
nd, not detected; LOD, limit of detection; LOQ, limit of quantification; TPC, total phenolic compounds, expressed as μg gallic acid equivalent per mg lyophilised winery byproduct; TMA, total monomeric anthocyanin pigment, expressed as μg malvidin equivalent per mg lyophilised winery byproduct.

Table 5.4 Antioxidant capacities of extracts from chenin blanc, petit verdot and syrah pomace and rachis. *Source:* Melo et al. (2015). Reproduced with permission of Elsevier.

	Synthetic free radical		Reactive oxygen species		
Parameter	DPPH· (μmol/g)[a)]	ABTS·+ (μmol/g)[a)]	Peroxyl radical (ROO·) (μmol/g)[a)]	Superoxide radical ($O_2^{\cdot-}$) (EC_{50}, mg/ml)[b)]	Hypochlorous acid (HOCl) (EC_{50}, mg/ml)[b)]
Chenin blanc					
Rachis	1086 ± 30^A	2419 ± 113^A	3718 ± 97^A	0.17 ± 0.05^C	0.021 ± 0.001^C
Pomace	191 ± 10^E	218 ± 17^C	739 ± 29^D	2.16 ± 0.02^A	0.128 ± 0.012^A
Petit verdot					
Rachis	685 ± 13^B	860 ± 14^B	2939 ± 269^B	0.28 ± 0.01^B	0.035 ± 0.001^{BC}
Pomace	540 ± 9^C	626 ± 9^B	2796 ± 75^{BC}	0.19 ± 0.01^C	0.017 ± 0.001^C
Syrah					
Rachis	473 ± 8^C	805 ± 31^B	2191 ± 140^C	0.26 ± 0.01^B	0.042 ± 0.001^B
Pomace	310 ± 7^D	653 ± 34^B	1363 ± 79^D	0.24 ± 0.01^{BC}	0.031 ± 0.001^{BC}
Gallic acid	–	–	$13\,783 \pm 1332$	0.07 ± 0.0	$8.46 \times 10^{-3} \pm 0.0$
Epicatechin	–	–	$32\,163 \pm 1748$	0.23 ± 0.01	$8.41 \times 10^{-3} \pm 0.0$

Each value is expressed as mean (triplicate) ± standard deviation (SD). The same letters in a column are not significantly different at the 0.05 level according to ANOVA.

a) Expressed as μmol trolox equivalent per gram of lyophilised winery byproduct.

b) EC_{50} is the mean quantity (mg/ml) of winery byproduct extract required to quench 50% of reactive oxygen species (ROS) in the assay.

temperature for 24 hours (Tables 5.6, 5.7 and 5.8) and found that the extraction yields of white grape pomaces (24.9 and 20.6% for viognier and vidal blanc, respectively) were higher than those of red grape pomaces (12.4 and 5.3% for cabernet Franc and chambourcin, respectively). Total phenolic content (TPC), total flavonoid content (TFC), total anthocyanin content (TAC), tannins, condensed tannins and antioxidant capacities in terms of DPPH and ABTS radical scavenging were significantly different among the four pomace extracts. Cabernet Franc had the highest TPC, TFC, tannins and condensed tannins (153.8 mg GAE, 91.7 mg catechin equivalents, 152.2 mg GAE/g dry extract and 50.5% leucocyanidin equivalents, respectively) and the greatest ABTS radical scavenging capacity (1013 μmol trolox equivalents/g dry extract), indicating that ABTS radical scavenging capacity is positively correlated with TPC, TFC, tannins and condensed tannins. DPPH radical scavenging capacity, meanwhile, had a positive correlation with TAC. Nine flavonoid compounds were identified in the grape pomace extracts, of which catechin and epicatechin were the two most abundant (except epicatechin in vidal blanc), with their contents ranged from 214 to 910 and 56.9 to 427 mg/g dry weight, respectively. All grape pomace extracts exhibited antibacterial activity against *L. monocytogenes* ATCC 7644 and *S. aureus* ATCC 29213, but not against *E. coli* O157:H7 ATCC 3510 or *S. typhimurium* ATCC 14028. Of the four, cabernet Franc and viognier pomaces present with the most potential as sources of natural antioxidant and antimicrobial agents.

Table 5.5 Concentrations of total flavonoids (mg/g dried weight) and total polyphenols (mg GAE/g dried weight) extracted by two ternary solvent mixtures in different proportions. *Source:* Bosso et al. (2016). Reproduced with permission of Elsevier.

Run order[a]	Acetone	Ethanol	Water	Total flavonoids	Total polyphenols (GAE)	Acetone	Ethanol	Ethyl acetate	Total flavonoids	Total polyphenols (GAE)
14	100	0	0	0.50	0.92	100	0	0	0.50	0.92
15	0	100	0	1.84	2.14	0	100	0	1.84	2.14
9	0	0	100	0.72	0.91	0	0	100	0.27	0.58
10	50	50	0	2.64	2.94	50	50	0	2.64	2.94
7	50	0	50	12.60	8.78	50	0	50	0.37	0.61
1	0	50	50	4.92	4.18	0	50	50	1.56	1.49
2	0	50	50	4.90	4.35	0	50	50	1.42	1.45
4	50	0	50	12.07	8.32	50	0	50	0.34	0.50
5	50	50	0	2.85	2.86	50	50	0	2.85	2.86
3	66	16.5	16.5	7.67	6.05	66	16.5	16.5	1.31	1.48
6	16.5	66	16.5	6.38	4.68	16.5	66	16.5	2.29	2.48
8	16.5	16.5	66	7.74	5.75	16.5	16.5	66	0.98	1.29
13	33	33	33	9.80	7.64	33	33	33	2.13	2.37
11	33	33	33	9.71	7.61	33	33	33	2.06	2.18
12	33	33	33	10.15	7.83	33	33	33	2.21	2.43

a) The order of execution of the extraction tests.

Table 5.6 Total yield and phenolic, flavonoid and anthocyanin contents in extracts from four grape pomaces. *Source:* Xu et al. (2016). Reproduced with permission of John Wiley & Sons.

Variety	Extraction yield (%)	TPC (mg GAE/g extract)	TFC (mg CE/g extract)	TAC (mg Cyd-3-glu equivalent/g extract)
Viognier	24.9 ± 1.00^{A}	99.1 ± 0.29^{B}	75.0 ± 0.42^{B}	0.02 ± 0.01^{C}
Vidal blanc	20.6 ± 2.57^{B}	55.5 ± 0.87^{B}	32.8 ± 0.41^{D}	0.06 ± 0.01^{C}
Cabernet Franc	12.4 ± 0.55^{C}	153.8 ± 1.83^{A}	91.7 ± 1.00^{A}	1.38 ± 0.03^{B}
Chambourcin	5.3 ± 0.61^{D}	92.0 ± 2.16^{C}	38.9 ± 0.74^{C}	10.7 ± 0.05^{A}

TPC, total phenolic content; TFC, total flavonoid content; TAC, total anthocyanin content; GAE, gallic acid equivalent; CE, catechin equivalent.
Data are expressed as mean ± standard deviation ($n = 3$).
Means followed by the same letter within a column indicate no significant ($P > 0.05$) difference among samples.

Table 5.7 Tannins, condensed tannins and antioxidant activities in extracts from four grape pomaces. *Source:* Xu et al. (2016). Reproduced with permission of John Wiley & Sons.

Varieties	Tannins (mg GAE/g extract)	Condensed tannins (% leucocyanidin equivalent)	DPPH (μmol TE/g extract)	ABTS (μmol TE/g extract)
Viognier	98.6 ± 0.17^{B}	25.7 ± 2.56^{C}	3.54 ± 0.06^{A}	951.0 ± 44.4^{B}
Vidal blanc	54.5 ± 0.85^{D}	8.61 ± 0.95^{D}	7.71 ± 0.02^{B}	334.0 ± 7.39^{D}
Cabernet Franc	152.2 ± 2.56^{A}	50.5 ± 0.19^{A}	11.2 ± 0.17^{C}	1013.0 ± 77.3^{A}
Chambourcin	88.5 ± 2.4^{C}	36.3 ± 1.52^{B}	28.2 ± 0.02^{D}	378.0 ± 11.8^{C}

GAE, gallic acid equivalent; TE, trolox equivalent.
Data are expressed as mean ± standard deviation ($n = 3$).
Means followed by the same letter within a column indicate no significant ($P > 0.05$) difference among samples.

Table 5.8 Major individual flavonoids in extracts from four grape pomaces. *Source:* Xu et al. (2016). Reproduced with permission of John Wiley & Sons.

Flavonoid compounds (mg/g extract)	Viognier	Vidal blanc	Cabernet Franc	Chambourcin
Catechin	910.0 ± 10.5^{A}	631.0 ± 13.4^{B}	560.0 ± 45.37^{B}	214.0 ± 4.8^{C}
Epicatechin	625.0 ± 9.2^{A}	451.0 ± 22.2^{B}	215.0 ± 4.67^{C}	109.0 ± 4.17^{D}
Epigallocatechin gallate	96.1 ± 3.47^{B}	62.8 ± 0.78^{B}	171.0 ± 7.26^{A}	–
Gallocatechin gallate	99.1 ± 1.29^{C}	–	232.0 ± 3.19^{A}	146.0 ± 3.37^{B}
Epicatechingallate	427.0 ± 11.7^{A}	–	122.0 ± 2.995^{B}	56.9 ± 5.36^{C}
Quercetin-3-rhamnoside	27.1 ± 2.59^{A}	33.5 ± 1.57^{A}	–	–
Quercetin	17.3 ± 0.38^{C}	20.7 ± 0.01^{C}	56.5 ± 1.95^{A}	31.2 ± 2.26^{B}
Kaempferol	–	–	8.69 ± 0.38^{A}	3.28 ± 0.14^{B}
Rutin	255.0 ± 16.7^{C}	435.0 ± 14.0^{A}	343.0 ± 11.0^{B}	99.5 ± 0.39^{D}

Data are expressed as mean ± standard deviation ($n = 2$).
Means followed by the same letter within a row indicate no significant ($P > 0.05$) difference among samples.
-, not detectable.

Fontana et al. (2013) reported on the extraction, characterisation and biotechnological applications of phenolic compounds from grape pomace as a sustainable source. Many phenolics have been identified in grape pomace, with the most abundant being anthocyanins, hydroxybenzoic and hydroxycinnamic acids, flavan-3-ols, flavonols and stilbenes (Figure 5.4). Table 5.9 shows different techniques for sample pretreatment, as well as various solvents, extraction methods and conditions (temperature and extraction time), which can be used to obtain different total phenolic yields. The highest phenolic contents are achieved from grape pomace (475.4 or 344–618 mg/g dried grape pomace extract), using freeze drying and shaking with 80% ethanol overnight at a solvent to sample ratio of 10 : 1 (475.4 mg/g dried grape pomace extract) or freeze drying and UAE with a mixture of methanol/ethanol/water (2 : 1 : 1) for 0.3 hours at a ratio of solvent to sample of 10 : 1 (344–618 mg/g dried grape pomace extract). The next highest are obtained using freeze drying and subsequent extraction with 50% acidified ethanol (pH 1, HCl) at 60 °C for 2 hours (139 mg/g dried grape pomace extract) or oven drying at 60 °C and shaking with 50% acetone at a solvent to sample ratio of 5 : 1 and 60 °C for 0.7 hours (77.5–148.3 mg/g dried grape pomace extract).

Barba et al. (2016) summarised the green alternative methods for the extraction of antioxidant bioactive compounds from winery wastes and byproducts (Table 5.10) and indicated that conventional extraction methods are the most widespread techniques used to extract bioactive compounds from plant matrices among different phases (i.e. solid, liquid, and vapour) at industrial scale. These methods are mainly based on the heating process, facilitating the mass transfer between the different phases of the system. They consume a lot of energy and may result in degradation of the thermolabile compounds. Therefore, many emerging technologies (e.g. ultrasounds, microwave, pulsed electric fields, SFE, high-pressure processing and so on) have been applied and are attracting increasing interest due to their potential in the effective and sustainable recovery of high added-value compounds – reducing the use of organic solvents, shortening the extraction time, decreasing the processing temperature, intensifying the mass transfer process, significantly increasing extraction yields, preserving extract quality and reducing energy consumption (Nguyen et al., 2015, 2016a, 2016b, 2016c). Among the different innovative technologies available, some nonconventional technologies can selectively extract the intracellular molecules without fragmenting the treated tissue, which is highly desirable in reducing the subsequent purification steps. Selection of the most appropriate extraction technique depends on the type of waste and the compound to be extracted. However, although their extraction efficiencies have been compared to conventional extraction methods in the recovery of high added-value compounds from wine wastes and byproducts at laboratory and pilot scales (Figure 5.5), the nonconventional technologies are not yet ready for scaling up in the wine industry due to reasons relating to the equipment and installation costs. In addition, most of the nonconventional technologies need to be tested on wine wastes and byproducts at pilot scale before they can be established at industrial scale because of some differences –especially in energy consumption – which have been noticed after scaling up.

5.2.2 Other Bioactive Compounds from Wine Wastes

Li et al. (2006) evaluated extractable amounts of *trans*-resveratrol (3,5,4′-trihydroxy stilbene) in grape (*Vitis*) seeds and skins from different germplasm cultivars during

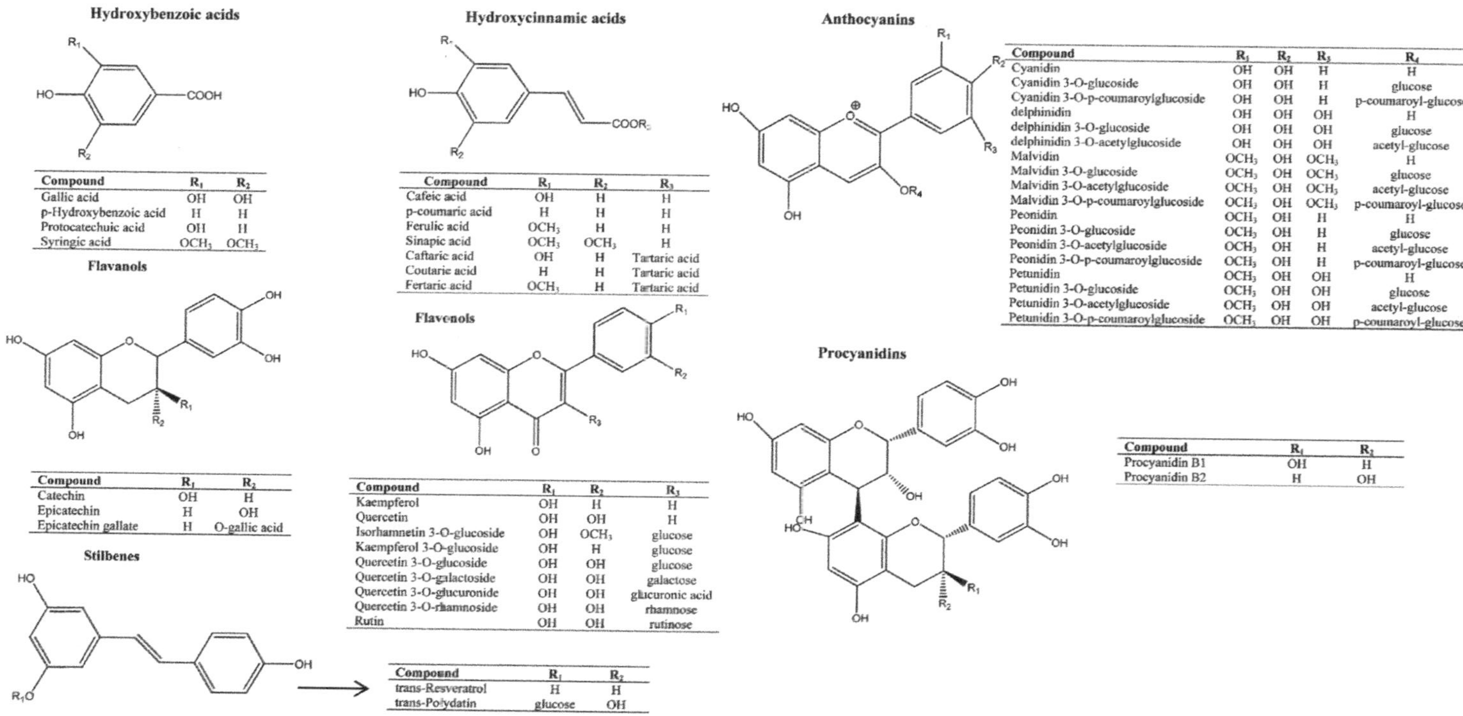

Figure 5.4 Chemical structures of the common classes of phenolics identified in grape pomace. *Source:* Fontana et al. (2013). Reproduced with permission of American Chemical Society.

Table 5.9 Solid–liquid extraction (SLE) techniques for phenolics from grape pomace. *Source*: Fontana et al. (2013). Reproduced with permission of American Chemical Society.

Sample pretreatment	Solvent/solvent mixture	Extraction mode, time (h)	Temperature (°C)	TPC (mg GAE/g GPE)
Freeze-dried and ground to a fine powder	80% ethanol (ratio 10 : 1)	Overnight shaking	ni	475.4
Freeze-dried	50% acidified ethanol (pH 1 HCl)	2	60	139
Oven-dried	50 or 75% acetone (ratio 10 : 1)	Sonication, 0.25		
Shaking, 0.5	ni	53		
Chopped fresh tissue	Methanol/acetone/water (6 : 3 : 1) (acidified 0.1% HCl)	Stirring, 0.25	ni	54
Chopped fresh tissue	Methanol/acetone/water (6 : 3 : 1) (acidified 0.1% HCl) (ratio 10 : 1)	Maceration, 0.25		
Stirring, 0.5	ni	48–54		
Air-dried at room temperature and ground to a fine powder	Water (ratio 10 : 1)	Continuous refluxing, 1.7	50	12.0–15.3
Freeze-dried and ground to a fine powder	Methanol (0.1% HCl) (ratio 20 : 1)	Stirring, 2	ni	ni
Freeze-dried	Methanol/ethanol (8 : 2)	Ultrasound, 0.25		
Shaking, 12.0	ni	ni		
ni	Methanol (ratio 10 : 1)	Shaking, 0.5	ni	2.5–4.8 (fresh weight)
Freeze-dried and ground to a fine powder	Methanol (1% 1 N HCl) (ratio 6 : 1)	Continuous stirring, 4	25	6.91–49.33
Vacuum-dried at 50 °C	50% acetone (ratio 20 : 1)	Ni	60	16.7
Oven-dried at 60 °C	60% ethanol (ratio 4 : 1)	Continuous stirring, 1	60	261.1
Oven-dried at 50 °C and crushed	80% ethanol (0.5% 0.1 N HCl) (ratio 30 : 1)	Ni	ni	41.9

Oven-dried at 60 °C and ground to a fine powder	Ethanol (ratio 4 : 1)	5	60	ni
Air-dried at room temperature	Methanol (ratio 5 : 1)	Shaking, 1.5	50	10.9
Freeze-driex and powdered in liquid nitrogen	Methanol (0.1% HCl) (ratio 50 : 1)	Shaking, 1	4	32.62–74.75
Hot air-dried at 60 °C	50% methanol and 70% acetone	1	room	ni
Freeze-dried and ground to a fine powder	Methanol (1% 1 N HCl) (ratio 4 : 1)	Stirring, 12	ni	ni
Oven-dried at 60 °C and ground to a fine powder	50% acetone (ratio 5 : 1)	Shaking, 0.7	25	77.5–148.3
Freeze-dried and ground to a fine powder	70% ethanol (ratio 4 : 1)	Shaking, 5	60	≈20
ni	Ethyl acetate (ratio 10 : 1)	0.3	ni	ni
Oven-dried at 50 °C and ground to a fine powder	53% ethanol (ratio 50 : 1)	Ultrasound, 0.5	56	5.44 (100 ml extract)
Oven-dried at 50 °C and ground to a fine powder	53% ethanol (ratio 50 : 1)	Ultrasound, 0.5	56	5.44 (100 ml extract)
Freeze-dried	Methanol/ethanol/water (2 : 1 : 1) (ratio 10:1)	Ultrasound, 0.3	ni	344–618
Freeze-dried	Methanol (ratio 10 : 1)	Ultrasound, 0.25	ni	1.9–4.5 (100 ml extract)
Freeze-dried	74% methanol	Ultrasound, 0.25	ni	ni

TPC, total phenolic content; GAE, gallic acid equivalent; GPE, grape pomace extract; ni, no information.

Table 5.10 Application of nonconventional technologies to improve the extraction of valuable compounds from winery wastes and byproducts. *Source:* Barba et al. (2016). Reproduced with permission of Elsevier.

Matrix	Extracted compounds	Treatment conditions	Extraction improvement
Pulsed electric field (PEF)			
Wine grape	Polyphenols	(1) 0.5 kV/cm, 50 pulses, 0.1 kJ/kg (2) 2.4 kV/cm, 50 pulses 2.3 kJ/kg	(1) 13% (2) 28%
Wine grape	Anthocyanins	3 kV/cm, 50 pulses	3-fold increase
Cabernet Franc grapes	Anthocyanins and tannins	0.8–5.0 kV/cm, 1–100 ms, 42–53 kJ/kg	51–62% (PEF) versus 20% (conventional)
Cabernet Franc and cabernet sauvignon grapes	Anthocyanins and total polyphenols	5 kV/cm, 1 ms, 48 kJ/kg	Increased anthocyanin content (from 87 to 172 mg/l for CF and from 168 to 269 mg/l for CS grapes)
Grape pomace	Polyphenols	1.2 kV/cm, 18 kJ/kg	Increased extraction yields depending on the temperature used
Cabernet sauvignon grape	Polyphenols	(1) 0.7 kV/cm, 200 ms, 31 Wh/kg (2) 4 kV/cm, 1 ms, 4 Wh/kg	(1) 34% more tannins (2) 19% more anthocyanin
Merlot grape	Polyphenols	500–700 V/cm, and 40–100 ms	Increased extraction yields
Grape pomace	Anthocyanins	13.3 kV/cm, 0–564 kJ/kg	22 and 55% more than USN and HVED, respectively
Vine shoots	Polyphenols and proteins	13.3 kV/cm, 0–1500 pulses, 50 °C, 50–762 kJ/kg/3-hour diffusion	Up to twofold increase in total polyphenols compared to untreated. Kaempferol (0.156 mg/g), epicatechin (1.747 mg/g), resveratrol (0.032 mg/g)
High-voltage electrical discharge (HVED)			
Grape pomace	Polyphenols		Sevenfold increase
	Polyphenols and anthocyanins	40 kV	HVED more energy efficient to recover polyphenols compared to PEF and USN
Grape seeds	Polyphenols	40 kV	HVED + supplementary ethanol extraction yielded 9 g GAE/100 g DM
Vine shoots	Polyphenols	40 kV/50 °C/10–240 kJ/kg/3-hour diffusion	Up to threefold increase in total polyphenols compared to untreated; kaempferol (0.213 mg/g), epicatechin (2.459 mg/g), resveratrol (0.414 mg/g); more energy efficient extraction compared to USN and PEF

Pulsed ohmic heating (POH)			
Grape pomace	Polyphenols	20–800 V/cm + supplementary hydroethanolic (0–50%, v/v) diffusion 60 min/20–50 °C	36% yield increase compared to 400 V/cm + supplementary hydroethanolic (30%, v/v) diffusion 60 min/50 °C compared to untreated samples
Ultrasound (USN)			
Grapes (Campbell and Gerbong varieties)	Resveratrol	47 kHz, 14 W/I, 60 °C, 30 minutes	24–28% compared to conventional solvent extraction
Cabernet Franc grapes	Polyphenols (anthocyanins and tannins)	24 kHz, 5–15 minutes, 121–363 kJ/kg	7%
Grape byproducts	Polyphenols	35 kHz, 70 °C, 1 minutes	50%
Red grape marc	Polyphenols	24 kHz, 20–75 W/ml	11–35%
Grape seeds	Polyphenols	20 kHz, 50–150 W, 30 °C, 30 minutes	Similar extraction (14 g/100 g) compared to solvent Soxhlet extraction hut reduced time
Vine shoots	Polyphenols and proteins	24 kHz, 400 W, 50 °C, 3-hour diffusion (1010–3428 kJ/kg)	Up to 45% yield increase in total polyphenols; kaempferol (0.097 mg/g), epicatechin (0.671 mg/g), resveratrol (0.024 mg/g)
Microwave-assisted extraction (MAE)			
Grape skins	Anthocyanins	100–500 W, 50–100 °C, 20–50 minutes, solvent (50–80% methanol in water)	Reduced extraction time from 5 hours to 5 minutes under the optimum conditions (2 g of sample, 500 W, 100 °C, 40% methanol in water as extraction solvent) compared to conventional extraction, And extraction of three additional acyl derivatives
Grape peel	Polyphenols	100–540 W, 3–10 minutes, 0–50 minutes, solvent (0–50% ethanol in water)	Optimum extraction of polyphenols after MW (540 W, 3 minutes), 50% ethanol and 1 g grape skin
Grape juice	Polyphenols	2.45 GHz, 900 W, 20 minutes	Improvement in the extraction of polyphenols in the extracted juice

(Continued)

Table 5.10 (Continued)

Matrix	Extracted compounds	Treatment conditions	Extraction improvement
Subcritical fluid extraction (SbFE)			
Grape skins	Anthocyanins	100–130 bar, 30–40 °C, pH 2–4, 25–30% ethanol, 25–50 ml/min CO_2 flow, 3–10% extract flow ratio	85%
White grape seeds	Polyphenols	1 ml/min CO_2 flow rate, 20-minute extraction, 35–55 °C, organic modifier density (0.85–0.95 g/ml), modifier (ethanol–methanol:10–40).	Improved recovery of gallic acid, catechin and epicatechin
Supercritical fluid extraction (SFE)			
Grape skin *Vitis vinifera*	Resveratrol	80–110 bar, 40 °C, ethanol concentration 5–15%, extraction time 5–25 minutes	Resveratrol completely recovered (100%) under optimum extraction conditions: 110 bar, 40 °C, 7.5% ethanol, 15-minute extraction time
Grape seeds, stems, skin and pomace from Palomino fino variety	Resveratrol	100–400 bar, 35–55 °C, 5% (v/v) ethanol modifier	Recovery of 49.1 mg/100 g dry sample after 400 bar/35 °C, 5% (v/v) ethanol as co-solvent
Grape seeds	Polyphenols	Pure CO_2 or CO_2 modified with 2, 5, 10 or 15% volume co-solvent	Increased extraction yield, depending on the CO_2 pressure and the percentage of ethanol
High-pressure processing (HP)			
Grape byproducts	Anthocyanin	600 MPa, 70 °C, 60 minutes, 50 : 50 ethanol–water solvent	41%
	Anthocyanin	200–600 MPa, 20–70 °C, 30–90 minutes, 20 : 80/100 : 0 ethanol–water solvent	22–83%

Grape seeds and skins	Polyphenols		Yields of total polyphenols, o-diphenols and flavonoids, for both seeds (108.3, 47 mg GAE/g dry weight, 47.2 mg CE/g dry weight) and skins (34.2, 10.1 mg GAE/g dry weight, 21.6 mg CE/g dry weight)
Grape skins	Polyphenols	30–150 °C, 15–330 minutes	Highest total phenolic compound (60.7 mg GAE) and total flavonoid (15.1 mg CE) yields obtained at 150 °C over 270 and 15 minutes, respectively; gallic acid (163.2 mg/100 g), protocatechuic acid (69.9 mg/100 g), catechin (420 mg/100 g), vanillic acid (20.6 mg/100 g), syringic acid (603 mg/100 g), coumaric acid (20.1 mg/100 g), *trans*-resveratrol (42.4 mg/100 g) and quercetin (117.1 mg/100 g) were the predominant polyphenols present in the extracts
Accelerated solvent extraction (ASE)			
Grape skins (St Laurent and Alibernet varieties)	Polyphenols, anthocyanins	150 bar, 40–120 °C	Improved recovery
Grape pomace	Polyphenols, anthocyanins, and tannins	103 bar, 50–200 °C, 5–30 minutes	Polyphenol extraction yield significantly decreased when temperatures higher than 100 °C were used at longer extraction times
Grape pomace	Polyphenols	60–140 °C, ethanol–water (30 : 70; 70 : 30, v/v) solvent	Significant increase in polyphenol recovery yield for both wet (16.2 g GAE/100 g dry matter) and dry (7.28 g GAE/100 g dry matter) grape pomace extracts when a mixture of ethanol–water at 70% was used at 140 °C
Extraction assisted by hydrotropic solvents			
Vine shoots	Polyphenols	β-cyclodextrin (β-CD) concentration (13.82–40.0 mg/ml), 33.18–66.82 °C, 0.30–6.19 hours	Polyphenol recovery yield 5.8 mg GAE/g dry matter under optimal β-CD-assisted extraction conditions (37.7 mg/ml β-CD at 66.6 °C over 4.8 hours)

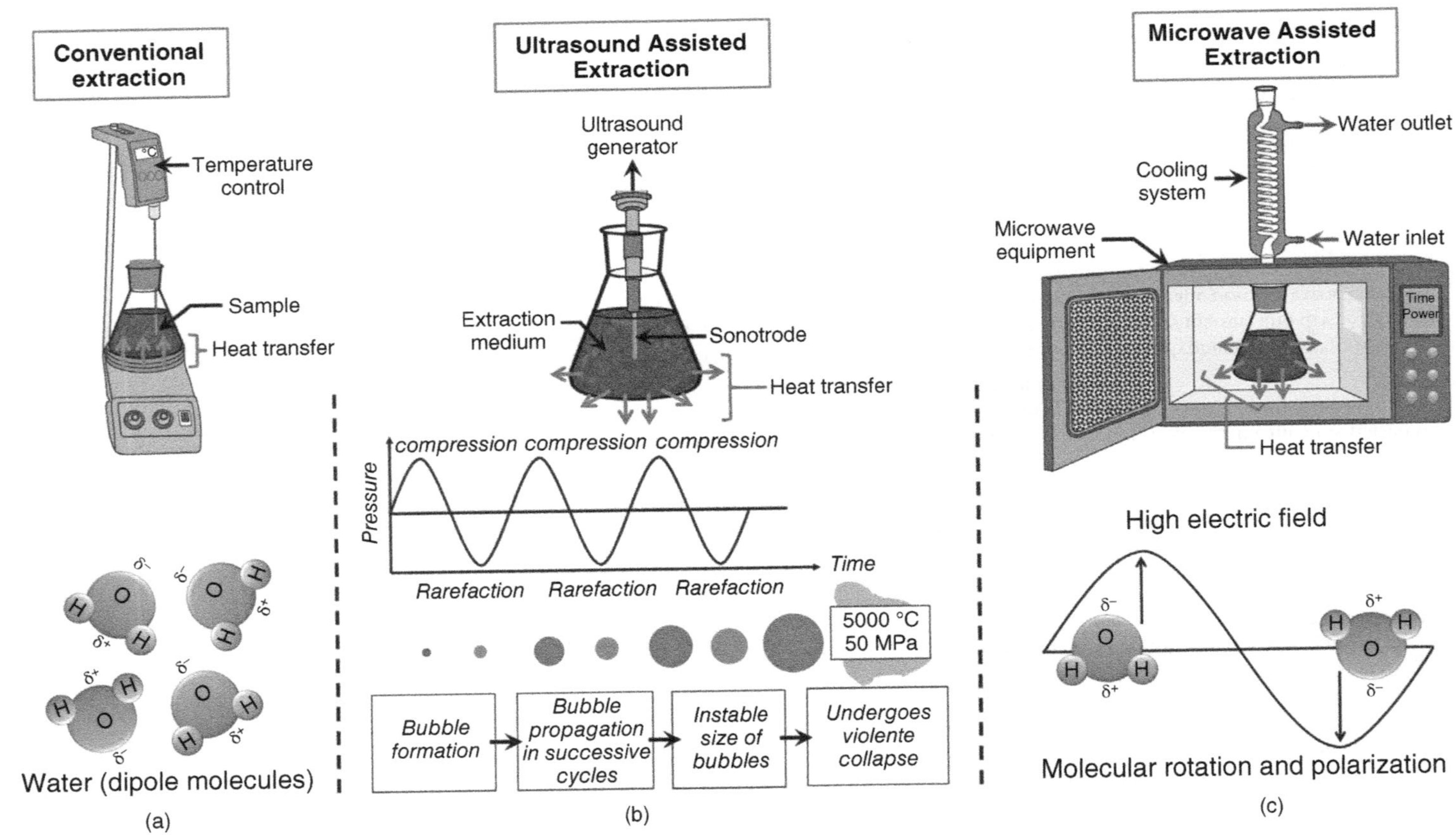

Figure 5.5 (a) Experimental set-up for conventional extraction of high added-value molecules from plant matrices used at laboratory scale. (b) Ultrasound-assisted extraction (UAE) principle and cavitational phenomenon. (c) Microwave-assisted extraction (MAE) equipment used at laboratory scale, showing the molecular rotation mechanism. *Source:* Barba et al. (2016). Reproduced with permission of Elsevier.

Table 5.11 Comparison of extractable amounts of resveratrol in grape skins and seeds (μg/g fresh weight) among different genotype groups and different intended usages. *Source:* Li et al. (2006). Reproduced with permission of American Chemical Society.

Genotype group or purpose of use	Avg. extractable resveratrol in 2003		Avg. extractable resveratrol in 2004	
	Skin	Seed	Skin	Seed
All cultivated cultivars of V and LV	0.61	1.30	1.24	1.36
All rootstocks of interspecies	69.80***	17.03***	145.11***	16.01***
All seeded cultivars of LV and V	0.61^{B}	1.37^{A}	1.33***	1.36
All seedless cultivars of LV and V	0.59		0.85	
All red-berry cultivars of LV and V	0.66***	1.34*	1.44***	1.40***
All green-berry cultivars of LV and V	0.44	1.22	0.73	1.23
All culviars of LV	0.41	1.35	0.88	1.37
All cultivars of V	0.81***	1.239	1.604***	1.341
V. vinifera wine grapes	1.03***	1.43***	2.50***	1.43***
V. vinifera table grapes	0.67	1.06	1.02	1.25

LV, hybrids between *V. labrusca* and *V. vinifera*; V, *V. vinifera*.
* and *** indicate significant differences in the same berry organ (skin or seeds) in the same year between different genotype groups or different purposes at $P < 0.05$ and $P < 0.01$ levels, respectively.
Means followed by different capital letters indicate significant difference between berry skins and seeds in the same year at $P < 0.001$ level.

two consecutive years (2003 and 2004) to find the distribution of resveratrol, one of the major stilbene phytoalexins and an active component of folk medicines used for the treatment of a wide range of diseases, which has anti-inflammatory, cardioprotective, cancer chemopreventive and antioxidant properties and provides inhibition of platelet aggregation. The results showed that the extractable resveratrol contents (ERCs) in skins and seeds (69.80 and 141.11 μg/g fresh weight for skins, and 17.03 and 16.05 μg/g fresh weight for seeds in 2003 and 2004, respectively) of interspecific rootstocks were much higher than those of all other genotype groups in both years (Table 5.11), while the ERCs in grape skins of seeded cultivars were significantly higher than those of seedless ones. Significantly greater ERCs in both skins and seeds were found in winemaking grapes compared to table grapes, and the ERCs in red grapes were significantly higher than those in green ones. The data indicate that the ERCs in grape skins and seeds are greatly varied and depend on the cultivar, genotype and purpose of use.

Mildner-Szkudlarz et al. (2011) extracted and used dietary fibre and phenolic compounds from grape pomace (skins and seeds) in sourdough mixed rye bread and found that total dietary fibre, soluble dietary fibre, insoluble dietary fibre and total phenol contents in grape pomace were 50.93, 4.53 and 46.39 g/100 g dry weight and 58.95 mg GAE/g dry weight, respectively, while ferric-reducing antioxidant power (FRAP) and DPPH radical scavenging capacity (DRSC) were 548.97 mM Fe and 471.23 mM trolox equivalents/g dry weight, respectively.

Deng et al. (2012) determined the dietary fibre content in terms of insoluble dietary fibre (IDF) and soluble dietary fibre (SDF) in five varieties of wine grape pomace skins

Table 5.12 Dietary fibre content of five varieties of wine grape pomace skins. NS, neutral sugar; UA, uronic acid. *Source:* Deng et al. (2011). Reproduced with permission of Elsevier.

Dietary fibre		Muller thurgau	Morio muscat	Cabernet sauvignon	Merlot	Pinot noir
IDF% dry matter	Glucose	4.59 ± 0.57^{C}	3.16 ± 0.07^{D}	7.33 ± 0.51^{B}	9.16 ± 0.46^{A}	8.43 ± 0.39^{A}
	Xylose	0.91 ± 0.09^{C}	0.45 ± 0.04^{D}	1.94 ± 0.11^{B}	2.13 ± 0.11^{A}	2.09 ± 0.08^{A}
	Galactose	0.67 ± 0.05^{C}	0.54 ± 0.03^{D}	0.97 ± 0.02^{B}	1.02 ± 0.03^{B}	1.17 ± 0.08^{A}
	Arabinose	0.43 ± 0.04^{C}	0.34 ± 0.02^{D}	0.76 ± 0.05^{A}	0.75 ± 0.05^{A}	0.58 ± 0.06^{B}
	Mannose	0.49 ± 0.05^{C}	0.36 ± 0.01^{D}	1.48 ± 0.04^{A}	1.54 ± 0.05^{A}	1.12 ± 0.07^{B}
	NS_{IDF}	7.08 ± 0.80^{D}	4.85 ± 0.04^{E}	12.25 ± 0.39^{C}	14.59 ± 0.39^{A}	13.39 ± 0.37^{B}
	UA_{IDF}	4.81 ± 0.66^{CD}	3.64 ± 0.21^{D}	8.53 ± 0.82^{A}	6.62 ± 0.25^{B}	5.14 ± 0.70^{C}
	$NS_{IDF} + UA_{IDF}$	11.89 ± 1.20^{C}	8.49 ± 0.17^{D}	20.78 ± 0.79^{A}	21.20 ± 0.61^{A}	18.50 ± 0.94^{B}
	KL	15.40 ± 0.27^{D}	7.94 ± 0.06^{E}	31.62 ± 0.48^{B}	28.38 ± 0.67^{C}	36.06 ± 0.75^{A}
	IDF	27.29 ± 1.46^{D}	16.44 ± 0.24^{E}	52.40 ± 0.40^{B}	49.59 ± 0.34^{C}	54.59 ± 1.57^{A}
SDF% dry matter	Glucose	0.20 ± 0.07^{C}	0.15 ± 0.04^{C}	0.33 ± 0.03^{B}	0.38 ± 0.1^{AB}	0.43 ± 0.39^{A}
	Xylose	0.02 ± 0.0^{C}	0.02 ± 0.01^{C}	0.03 ± 0.01^{AB}	0.04 ± 0.01^{A}	0.02 ± 0.0^{C}
	Galactose	0.12 ± 0.04^{AB}	0.13 ± 0.04^{A}	0.07 ± 0.03^{B}	0.15 ± 0.06^{A}	0.18 ± 0.02^{A}
	Arabinose	0.07 ± 0.02^{C}	0.11 ± 0.01^{B}	0.07 ± 0.02^{B}	0.13 ± 0.02^{AB}	0.16 ± 0.01^{A}
	Mannose	0.05 ± 0.01^{BC}	0.01 ± 0.0^{C}	0.05 ± 0.06^{BC}	0.09 ± 0.03^{B}	0.22 ± 0.08^{A}
	NS_{SDF}	0.46 ± 0.13^{C}	0.42 ± 0.10^{C}	0.54 ± 0.08^{C}	0.78 ± 0.22^{B}	1.0 ± 0.1^{A}
	UA_{SDF}	0.43 ± 0.06^{B}	0.26 ± 0.03^{C}	0.27 ± 0.04^{C}	0.73 ± 0.07^{A}	0.72 ± 0.06^{A}
	SDF	0.72 ± 0.14^{C}	0.84 ± 0.07^{C}	0.81 ± 0.06^{C}	1.51 ± 0.14^{B}	1.72 ± 0.15^{A}
$(NS_{IDF} + NS_{SDF})$% dry matter		7.54 ± 0.74^{D}	5.27 ± 0.11^{E}	12.79 ± 0.34^{C}	15.37 ± 0.48^{A}	14.39 ± 0.29^{B}
$(NA_{IDF} + NA_{SDF})$% dry matter		5.07 ± 0.64^{CD}	4.07 ± 0.25^{D}	8.80 ± 0.79^{A}	7.34 ± 0.2^{B}	5.72 ± 0.72^{C}

KL, Klason lignin; NS, neutral sugar; UA, uronic acid, expressed as mg galacturonic acid equivalents (GUAE)/g dry matter; IDF, insoluble dietary fibre; SDF, soluble dietary fibre.

$IDF = NS_{IDF} + UA_{IDF} + KL$; $SDF = NS_{SDF} + UA_{SDF}$.

Data expressed as mean ± standard deviation ($n = 3$).

Different superscript letters in the same row indicate significantly different values ($P < 0.05$).

and indicated that the IDF composed of Klason lignin (7.9–36.1% dry weight), neutral sugars (4.9–14.6% dry weight) and uronic acid (3.6–8.5% dry weight) accounted for more than 95.5% of total dietary fibre (TDF) in all five wine grape pomace varieties. The dietary fibre of white wine grape pomace (17.3–28.0% dry weight) was significantly lower than that of red wine grape pomace (51.1–56.3% dry weight). In contrast, the soluble sugar of white wine grape pomace was much higher than that of red wine grape pomace (55.8–77.5% dry weight *versus* 1.3–1.7% dry weight) (Table 5.12).

Recently, Sousa et al. (2014) determined the total dietary fibre, soluble fibre, insoluble fibre, vitamin C and anthocyanins of grape pomace (*Vitis vinifera* L.) and found that grape pomace flour consisted of 46.17 g total dietary fibre, 9.75 g soluble fibre, 36.4 g insoluble fibre/100 g, 26.25 mg ascorbic acid and 131 mg anthocyanins/100 g dry weight.

5.3 Conclusion

With increasing grape crop production and expansion of winemaking processes, a huge amount of wine residues/byproducts/wastes, such as grape bagasse/pomace (seeds and skins/peels), leaves, stems/stalks, rachis and lees, are discarded to the environment. Recovery of bioactive compounds from wine wastes, such as polyphenols, *trans*-resveratrol, dietary fibre and lignin, is highly attractive to researchers. This not only increases the economic value of grape crops and of the winemaking industry as a whole, but also reduces environmental pollution and points the way towards sustainable development. The bioactive compounds recovered from wine wastes are valuable sources for potential application in the nutraceutical, medical, pharmaceutical and cosmetic industries.

References

Adámez, J. D., Samino, E. G., Sánchez, E. V. & González-Gómez, D. 2012. *In vitro* estimation of the antibacterial activity and antioxidant capacity of aqueous extracts from grape-seeds (*Vitis vinifera* L.). *Food Control*, **24**, 136–141.

Barba, F. J., Zhu, Z., Koubaa, M., De Souza Sant'ana, A. & Orlien, V. 2016. Green alternative methods for the extraction of antioxidant bioactive compounds from winery wastes and by-products: a review. *Trends in Food Science & Technology*, **49**, 96–109.

Baydar, N. G., Özkan, G. & Yaşar, S. 2007. Evaluation of the antiradical and antioxidant potential of grape extracts. *Food control*, **18**, 1131–1136.

Bertran, E., Sort, X., Soliva, M. & Trillas, I. 2004. Composting winery waste: sludges and grape stalks. *Bioresource technology*, **95**, 203–208.

Bosso, A., Guaita, M. & Petrozziello, M. 2016. Influence of solvents on the composition of condensed tannins in grape pomace seed extracts. *Food Chemistry*, **207**, 162–169.

Bustamante, M. A., Moral, R., Paredes, C., Pérez-Espinosa, A., Moreno-Caselles, J. & Pérez-Murcia, M. D. 2008. Agrochemical characterisation of the solid by-products and residues from the winery and distillery industry. *Waste Management*, **28**, 372–380.

Cabrera, C., Artacho, R. & Giménez, R. 2006. Beneficial effects of green tea – a review. *Journal of the American College of Nutrition*, **25**, 79–99.

DENG, Q., PENNER, M. H. & ZHAO, Y. 2011. Chemical composition of dietary fiber and polyphenols of five different varieties of wine grape pomace skins. *Food Research International*, **44**, 2712–2720.

Deng, G. F., Shen, C., Xu, X. R., Kuang, R. D., Guo, Y. J., Zeng, L. S., Gao, L. L., Lin, X., Xie, J. F. & Xia, E. Q. 2012. Potential of fruit wastes as natural resources of bioactive compounds. *International Journal of Molecular Sciences*, **13**, 8308–8323.

Di Lecce, G., Arranz, S., Jáuregui, O., Tresserra-Rimbau, A., Quifer-Rada, P. & Lamuela-Raventós, R. M. 2014. Phenolic profiling of the skin, pulp and seeds of Albariño grapes using hybrid quadrupole time-of-flight and triple-quadrupole mass spectrometry. *Food Chemistry*, **145**, 874–882.

FAOSTAT 2015. FAO Statistic Division. http://faostat3.fao.org/home/E (last accessed March 20, 2017).

Farías-Campomanes, A. M., Rostagno, M. A. & Meireles, M. A. A. 2013. Production of polyphenol extracts from grape bagasse using supercritical fluids: yield, extract composition and economic evaluation. *The Journal of Supercritical Fluids*, **77**, 70–78.

Fiori, L. 2010. Supercritical extraction of grape seed oil at industrial-scale: Plant and process design, modeling, economic feasibility. *Chemical Engineering and Processing: Process Intensification*, **49**, 866–872.

Fontana, A. R., Antoniolli, A. & Bottini, R. 2013. Grape pomace as a sustainable source of bioactive compounds: extraction, characterization, and biotechnological applications of phenolics. *Journal of Agricultural and Food Chemistry*, **61**, 8987–9003.

García-Lafuente, A., Guillamón, E., Villares, A., Rostagno, M. A. & Martínez, J. A. 2009. Flavonoids as anti-inflammatory agents: implications in cancer and cardiovascular disease. *Inflammation Research*, **58**, 537–552.

González-Centeno, M. R., Knoerzer, K., Sabarez, H., Simal, S., Rosselló, C. & Femenia, A. 2014. Effect of acoustic frequency and power density on the aqueous ultrasonic-assisted extraction of grape pomace (*Vitis vinifera* L.) – a response surface approach. *Ultrasonics Sonochemistry*, **21**, 2176–2184.

Jin, B. & Kelly, J. M. 2009. Wine industry residues. In: Nigam, P. S. N. & Pandey, A. (eds).*Biotechnology for Agro-industrial Residues Utilisation: Utilisation of Agro-residues.* New York: Springer Science & Business Media.

Kammerer, D. R. & Carle, R. 2008. Process strategies for the recovery and isolation of phenolic compounds from winery by-products. *Electronic Journal of Environmental, Agricultural and Food Chemistry*, **7**, 3226–3230.

Li, X., Wu, B., Wang, L. & Li, S. 2006. Extractable amounts of *trans*-resveratrol in seed and berry skin in Vitis evaluated at the germplasm level. *Journal of Agricultural and Food Chemistry*, **54**, 8804–8811.

Melo, P. S., Massarioli, A. P., Denny, C., Dos Santos, L. F., Franchin, M., Pereira, G. E., De Souza Vieira, T. M. F., Rosalen, P. L. & De Alencar, S. M. 2015. Winery by-products: extraction optimization, phenolic composition and cytotoxic evaluation to act as a new source of scavenging of reactive oxygen species. *Food Chemistry*, **181**, 160–169.

Mildner-Szkudlarz, S., Zawirska-Wojtasiak, R., Szwengiel, A. & Pacyński, M. 2011. Use of grape by-product as a source of dietary fibre and phenolic compounds in sourdough mixed rye bread. *International Journal of Food Science & Technology*, **46**, 1485–1493.

Nguyen, V. T., Ueng, J. P. & Tsai, G. J. 2011. Proximate composition, total phenolic content, and antioxidant activity of Seagrape (*Caulerpa lentillifera*). *Journal of Food Science*, **76**, C950–C958.

Nguyen, V. T., Bowyer, M. C., Vuong, Q. V., Van Altena, I. A. & Scarlett, C. J. 2015. Phytochemicals and antioxidant capacity of Xao tam phan (*Paramignya trimera*) root as affected by various solvents and extraction methods. *Industrial Crops and Products*, **67**, 192–200.

Nguyen, V. T., Vuong, Q. V., Bowyer, M. C., Van Altena, I. A. & Scarlett, C. J. 2016a. Microwave-assisted extraction for saponins and antioxidant capacity from Xao tam phan (*Paramignya trimera*) root. *Journal of Food Processing and Preservation*, **41**, 2, e12851.

Nguyen, V. T., Pham, H. N. T., Bowyer, M. C., Van Altena, I. A. & Scarlett, C. J. 2016b. Influence of solvents and novel extraction methods on bioactive compounds and antioxidant capacity of *Phyllanthus amarus*. *Chemical Papers*, **70**, 556–566.

Nguyen, V. T., Bowyer, M. C., Van Altena, I. A. & Scarlett, C. J. 2016c. Optimisation of microwave-assisted extraction from *Phyllanthus amarus* for phenolic compounds – enriched extracts and antioxidant capacity. *Chemical Papers*, **70**, 713–725.

Nuhu, A. A. 2014. Bioactive micronutrients in coffee: recent analytical approaches for characterization and quantification. *ISRN Nutrition*, Article ID 384230.

Pedroza, M. A., Amendola, D., Maggi, L., Zalacain, A., De Faveri, D. M. & Spigno, G. 2015. Microwave-assisted extraction of phenolic compounds from dried waste grape skins. *International Journal of Food Engineering*, **11**, 359–370.

Prado, J. M., Dalmolin, I., Carareto, N. D. D., Basso, R. C., Meirelles, A. J. A., Oliveira, J. V., Batista, E. A. C. & Meireles, M. A. A. 2012. Supercritical fluid extraction of grape seed: process scale-up, extract chemical composition and economic evaluation. *Journal of Food Engineering*, **109**, 249–257.

Prasain, J. K., Peng, N., Dai, Y., Moore, R., Arabshahi, A., Wilson, L., Barnes, S., Wyss, J. M., Kim, H. & Watts, R. L. 2009. Liquid chromatography tandem mass spectrometry identification of proanthocyanidins in rat plasma after oral administration of grape seed extract. *Phytomedicine*, **16**, 233–243.

Prozil, S. O., Evtuguin, D. V., Silva, A. M. S. & Lopes, L. P. C. 2014. Structural characterization of lignin from grape stalks (*Vitis vinifera* L.). *Journal of Agricultural and Food Chemistry*, **62**, 5420–5428.

Putnik, P., Bursać Kovačević, D. & Dragović-Uzelac, V. 2016. Optimizing acidity and extraction time for polyphenolic recovery and antioxidant capacity in grape pomace skin extracts with response surface methodology approach. *Journal of Food Processing and Preservation*, in press.

Ruberto, G., Renda, A., Daquino, C., Amico, V., Spatafora, C., Tringali, C. & De Tommasi, N. 2007. Polyphenol constituents and antioxidant activity of grape pomace extracts from five Sicilian red grape cultivars. *Food Chemistry*, **100**, 203–210.

Schieber, A., Stintzing, F. C. & Carle, R. 2001. By-products of plant food processing as a source of functional compounds-recent developments. *Trends in Food Science & Technology*, **12**, 401–413.

Sousa, E. C., Uchôa-Thomaz, A. M. A., Carioca, J. O. B., De Morais, S. M. D., De Lima, A., Martins, C. G., Alexandrino, C. D., Ferreira, P. A. T., Rodrigues, A. L. M. & Rodrigues, S. P. 2014. Chemical composition and bioactive compounds of grape pomace (*Vitis vinifera* L.), Benitaka variety, grown in the semiarid region of Northeast Brazil. *Food Science and Technology (Campinas)*, **34**, 135–142.

Teixeira, A., Baenas, N., Dominguez-Perles, R., Barros, A., Rosa, E., Moreno, D. A. & Garcia-Viguera, C. 2014. Natural bioactive compounds from winery by-products as health promoters: a review. *International Journal of Molecular Sciences*, **15**, 15 638–15 678.

Vuong, Q. V., Golding, J. B., Nguyen, M. & Roach, P. D. 2010. Extraction and isolation of catechins from tea. *Journal of Separation Science*, **33**, 3415–28.

Xiang, Y., Zhang, T., Zhang, X. & Ma, L. 2003. Quantitative analysis of resveratrol from grape seeds and grape skins by high performance liquid chromatography method. *Journal of Hygiene Research*, **32**, 490–492.

Xu, Y., Burton, S., Kim, C. & Sismour, E. 2016. Phenolic compounds, antioxidant, and antibacterial properties of pomace extracts from four Virginia-grown grape varieties. *Food Science & Nutrition*, **4**, 125–133.

Yilmaz, Y. & Toledo, R. T. 2006. Oxygen radical absorbance capacities of grape/wine industry byproducts and effect of solvent type on extraction of grape seed polyphenols. *Journal of Food Composition and Analysis*, **19**, 41–48.

6

Recovering Bioactive Compounds from Edible Oil Wastes

Hoang Quoc Tuan and Nguyen Thi Thao

Department of Quality Management, School of Biotechnology and Food Technology, Hanoi University of Science and Technology, Hanoi, Vietnam

6.1 Introduction

Edible vegetable oils play an important role in human diet and food preparation habits. The variety of quality properties of edible oils has been confirmed in many studies (Savva & Kafatos, 2016). Recently, consumers have become more interested in the nutritional properties of edible oils. According to the Statistics Portal, the production of vegetable oils from palm, soybean, canola and sunflower was 173 million tonnes worldwide in 2014/15 (Statistics Portal, 2015). Production of soybean, rapeseed, palm, olive and sunflower oils has increased in recent years, and rice brand oil is becoming popular in the United States and Europe, where it is considered a nutritious oil (Table 6.1) (USDA, 2015).

Although vegetable oils are considered a nutritious resource for human diet, during vegetable oil processing, a large amount of nutritional compounds are lost along with the byproducts and wastes (Dijkstra & van Duijn, 2016). In the edible vegetable oil industry, wastes derive from the processing of raw vegetable materials into oils, which generally consists in the extraction or separation of the nutritional portion from the remains (Muktadirul Bari Chowdhury, 2013). Huge amounts of wastes are generated from the production of vegetable and animal oils and fats worldwide (Dijkstra & van Duijn, 2016). In the case of crude palm oil processing, in the production of 1 tonne of crude palm oil, more than 1 tonne of waste product is produced, consisting of 16% lignin, 50% cellulose, 23% hemicelluloses, 3.6% oil and 8% ash (Arienzo & Violante, 2007; Godswill et al., 2016). The discarded products are usually treated as refuse or used to produce animal feed and biofuels (Juan Fernández-Bolaños et al., 2006). Several decades ago, the wastes of edible oils processing were considered neither a cost nor a benefit, so they were used only as animal feed or sent to landfills or compost. However, this attitude has changed recently, for several reasons, including growing environmental concern in many countries (especially developed countries such as the United States and members of the European Union) (Laufenberg et al., 2003). For instance, the large amount of waste from olive production (around 30 million cubic metres of olive mill wastewaters (OMWs) per year) is currently one of the most serious environmental concerns in the Mediterranean basin, due to its recalcitrant characteristics and toxic

Recovering Bioactive Compounds from Agricultural Wastes, First Edition. Edited by Van Tang Nguyen.

Table 6.1 Annual production of major vegetable oils (million tonnes) from 2010/11 to 2015/16 (forecast). *Source:* USDA figures (May 2015).

	2010/11	2011/12	2012/13	2013/14	2014/15	2015/16 (f)
Coconut	3.68	3.43	3.65	3.38	3.35	3.39
Cottonseed	4.99	5.27	5.25	5.18	5.20	4.96
Olive	3.27	3.45	2.45	3.14	2.33	2.86
Palm	49.24	52.58	56.59	59.54	61.65	65.17
Palm kernel	5.75	6.17	6.58	6.98	7.27	7.55
Peanut	5.31	5.30	5.51	5.64	5.59	5.69
Rapeseed	23.04	24.10	24.90	26.60	27.21	26.65
Soybean	41.37	42.73	43.08	44.89	47.39	49.64
Sunflower seed	12.21	14.74	13.26	15.96	15.10	15.22
Total	148.85	157.77	161.27	171.31	175.09	181.12

effect on the environment (Arienzo & Violante, 2007; Dermeche et al., 2013). However, there is now growing awareness of the benefits deriving from potentially marketable components present in edible oil wastes (Laufenberg et al., 2003). Edible waste can be considered an important source of valuable products including carbohydrates, polyphenols, vitamins, sterols, lecithin and protein (da Silva & Jorge, 2014; Dermeche et al., 2013; Muktadirul Bari Chowdhury, 2013; Waldron, 2007).

Regarding the potentially marketable components present in edible oil wastes, the aim is to extract and purify bioactive compounds as nutritionally and pharmacologically functional ingredients. The bioactive compounds that have been recovered from edible oil wastes present no microbiological or toxicity hazard, so the final products will comply with existing food regulations and meet consumer preference (Council of the European Parliament, 2009). The bioactive compounds present in edible oil wastes need to be extracted from the matrix using a combined approach, including biochemical, chemical and physical aspects, in order to ensure selective extraction and modification of the targeted compounds. They can then be changed into higher-value foods, functional foods and cosmetic and pharmaceutical products. Optimisation of the extraction conditions using conventional or modern techniques and the use of recovered biomolecules have been the subject of much research (Azmir et al., 2013; Baiano, 2014; Fernández-Mar et al., 2012; Muktadirul Bari Chowdhury, 2013; Ren et al., 2013). Recovery of bioactive compounds from edible oil wastes must be economically feasible. This could be achieved by extracting and separating the compounds of interests through individual and/or combined physical and biochemical approaches in order to provide a range of components, from high to low value, all of which would contribute to achieving whole-waste exploitation (Baiano, 2014; Ren et al., 2013).

6.2 Edible Oil Processing and Byproducts

The edible oil processing industry involves the extraction and processing of oils and fats from vegetable and oilseed sources. Edible oils are used not only for human consumption but also as animal feeds and in medicinal and certain technical applications. The starting point of oil processing is the preparation step, which includes husking, cleaning, crushing and conditioning, in order to produce crude oil. The extraction processes are generally mechanical, in that boiling is applied for fruits and pressing is used for seeds and nuts. After boiling, the liquid oil is skimmed, pressed and filtered. After solvent extraction, the crude oil is separated and the solvent is evaporated and recovered. Residues are conditioned (e.g. dried) and reprocessed to yield byproducts for animal feed, biodiesel production, recovery of valuable compounds and so on. In many countries today, processors use solvent extraction as the primary method of producing edible oils (Dijkstra & van Duijn, 2016; Hamm et al., 2013).

6.2.1 Solvent Extraction Process and Wastes

The same basic process used for extraction of oils is used for edible oils, but some differences should be considered. Normally, the process uses hexane as solvent to dissolve the oil present in seeds or fruits. The mixture of solvent and oil is called 'miscella'. In general, there are three different types of extractors – horizontal belt, rotary and continuous loop – with the latter the most popular. After oils are extracted, the solvent recovery process must be applied in order to separate the solvent from both crude oil and meal. Meal or cake is the residue left after the crude oil is extracted from the raw material. Most of the cake is used for the production of animal feed with a high protein content. In the case of soybean oil processing, a small amount of cake can be used for human consumption (Dijkstra & van Duijn, 2016; Silva & Perrone, 2015; Hamm et al., 2013).

6.2.2 Refining Process and Wastes

The oil obtained from the extraction process is called 'crude oil'. Crude oil contains some undesirable compounds which might affect the quality of finished oil products, such as proteinaceous material, free fatty acids (FFAs), phosphatides and other impurities that must be removed before use. There are two different crude oil refining technologies which can be used for the refining of edible oils: chemical and physical processes (Figure 6.1). In general, in order to remove impure compounds, oil must undergo the following steps: degumming (water or acid degumming), refining, bleaching and deodorising (Hamm et al., 2013).

6.2.2.1 Deguming

This process prepares the edible oil for long-term storage, transport or further refining. It also produces lecithin, a food additive. Processors degum crude oil by mixing it with water and acid in a chemical and a physical refining process, respectively, then put it through settling or centrifugation to separate the gum (Hamm et al., 2013). The acid degumming process can be considered an alternative to water degumming in that it uses a degumming acid combined with water. Crude oil, either water-degummed or not, is treated by an acid, usually phosphoric acid, citric acid or malic acid in the presence of water (Deffense, 2011). One of the valuable residues produced by the degumming process is lecithin, a mixed phosphatide product which is used as a wetting and dispersing agent, an emulsifier and an antioxidant (Dijkstra, 2014; Hamm et al., 2013).

6.2.2.2 Alkali Neutralising (Caustic Refining)

Normally, the process used to refine edible oils is alkali neutralisation or caustic refining. Alkali neutralisation involves mixing the oil with an aqueous alkaline solution. In

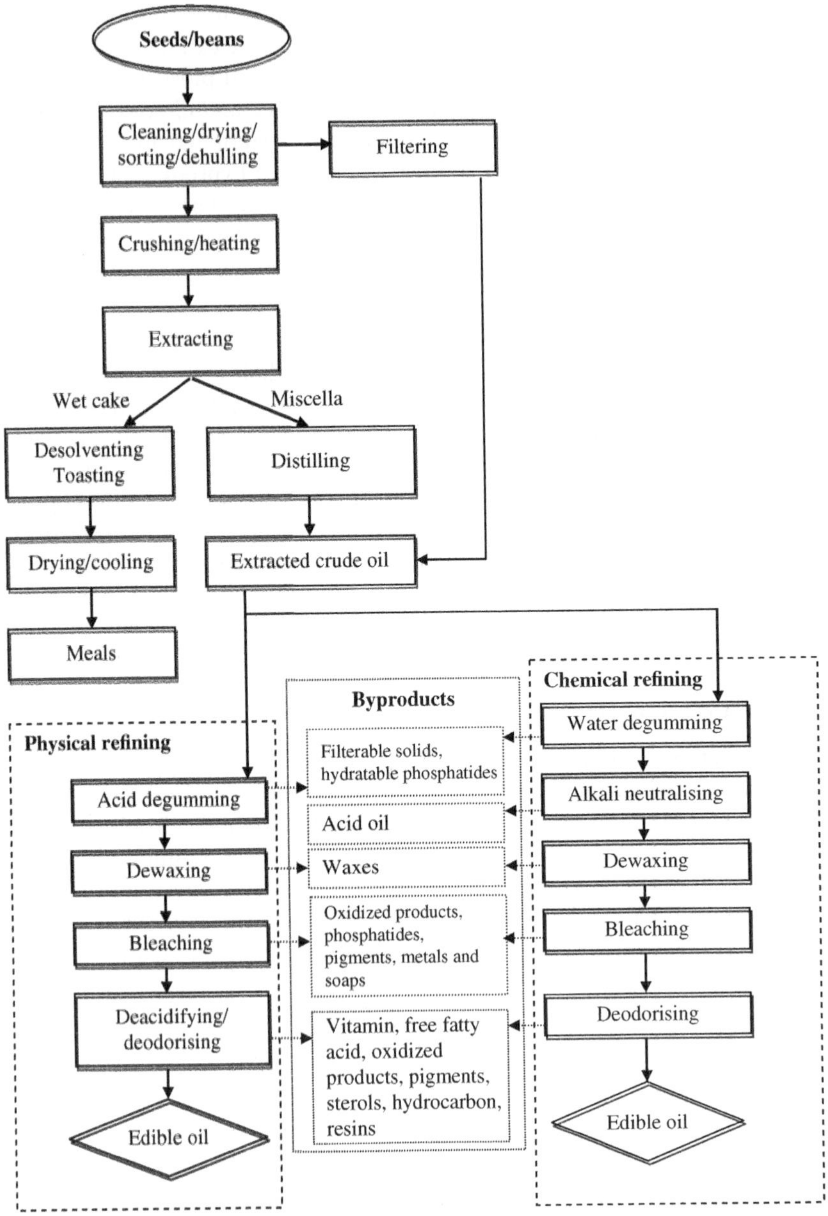

Figure 6.1 Full processing flow chart for a general edible oil process.

the edible oil industry, caustic soda is almost the only substance used for this reaction, although potassium hydroxide is used in a few factories. The alkali neutralises the FFAs in oil to create 'soap stock'. The soap stock then adsorbs colour compounds and precipitates gums and proteinaceous materials. The soap stock is removed by centrifugation or settling and the oil is washed with water to remove residual soap (Dijkstra, 2014; Hamm et al., 2013). Soap stock is a byproduct of the alkali neutralisation process which has commercial value in animal feed and in the manufacture of soap and other chemicals, for both economical and nutritional reasons (Dumont & Narine, 2007). It is a heterogeneous mixture of soap, neutral oil, gums, minor nonpolar compounds and about 50% water. Although the quantity and composition of this byproduct depend on the type and quality of crude oil and the refining conditions selected, it is always generated at a rate of about 10% (v/v) of the crude oil refined (Hamm et al., 2013).

6.2.2.3 Bleaching

In this step, as well as removing pigments, some oxidised products, residual phosphatides, soaps and trace metals, bleaching is applied to improve the flavour of the edible oil. The bleaching agents normally used in edible oil processing are neutral earth, acid-activated earth, activated carbon and silicates. However, diatomaceous earth and other inert materials can also be used to aid filtering. This process involves mixing the oil with bleaching agents, heating the mixture and then filtering (Hamm et al., 2013). Exhausted bleaching earth is a solid byproduct generated during the bleaching of edible oil from both chemical and physical processes. This waste contains about 20–50% oil, depending on the technology used and the efficiency of the equipment. The oil residues in this waste are regarded as unsuitable substances for food and feed applications for a number of different reasons, including the concentration of oxidised products and their effect on the environment (Waldron, 2007).

6.2.2.4 Deodorising

The final step in the refining process is deodorisation. In this step, low-pressure and high-temperature steam is injected through the oil in order to remove any off-flavour or odours. Deodoriser distillates are byproducts of this process with rather complex structures. Any undesirable substances that would affect the taste and flavour of the edible oil are eliminated at this step. Depending on the type of refining used (chemical or physical), the steps taken and the oil source being processed (e.g. soybean, palm, sunflower, olive, rapeseed, etc.), components such as fatty acids (30–35%), fatty esters, fatty alcohols, mono-, di- and triglycerides and bioactive compounds involving vitamin E (2–10%), sterols (3–15% – comprising about 50% sitosterol, 40% stigmasterol and campesterol and 10% other minor sterols) and hydrocarbons (2–5%) will be lost in significant amounts. The deodoriser distillates will also contain lecithin and breakdown products of fatty acids, unsaturated high-molecular-weight aldehydes, ketones and acyglycerols. The deodoriser distillate by itself has a certain commercial value – as much as the edible oil. It has therefore become an interesting material to many researchers and producers (Dermeche et al., 2013; Juan Fernández-Bolaños et al., 2006; Waldron, 2007).

6.3 Recovering Bioactive Compounds from Edible Oil Wastes

6.3.1 Phenolic Compounds

Edible oils are a valuable source of antioxidants for the human diet. As was mentioned in the previous section, however, during edible oil production, a large amount of phenolic compounds are lost in the edible oil byproducts; for example, in the case of olive oils, nearly 98% of total phenols in the olive fruit remain in its byproducts (Araújo et al., 2015;

García et al., 2016; Lafka et al., 2011; Rodis et al., 2002). Olive is known to contain an appreciable amount of phenols with good antioxidant properties. Therefore, the recovery of phenolic compounds from these wastes is of considerable interest for researchers and producers (Araújo et al., 2015; García et al., 2016; Lafka et al., 2011).

The byproducts profile of edible oils depends on several factors, including cultivars, geographic origin, maturation of seeds/fruits, climatic conditions and the process used (Dermeche et al., 2013; Lafka et al., 2011; Muktadirul Bari Chowdhury, 2013). In the case of olive byproducts, hydroxytyrosol has been found to be the most abundant phenolic compound, at about 70% of total phenolic content (El-Abbassi et al., 2012; Leouifoudi et al., 2014). The results of some studies show the concentration of these phenolic compounds in olive pomace ranges from 1624 to 2872 mg/kg (Japon-Lujan & Luque de Castro, 2007; Leouifoudi et al., 2014; Muktadirul Bari Chowdhury, 2013), but other authors report different major compounds, mainly oleuropein derivatives, oleuropein, 3,4-DPHEA-EDA and protocatechuic acid (Araújo et al., 2015; Cardoso et al., 2011; Guinda et al., 2015). Other phenolic compounds have also been identified, but in lower concentrations, such as rutin, luteolin, apigenin, vanillic acid, caffeic acid, tyrosol and verbascoside (Araújo et al., 2015). During olive oil processing, oleuropein is converted by β-glucosidase into hydroxytyrosol and other secoiridoid derivatives, such as the dialdehydic form of decarboxymethylelenolic acid linked to hydroxytyrosol, oleuropeinalgycone di-aldehyde (3,4-DHPEA-EAD), increasing their abundance in byproducts (Figure 6.2) (Obied et al., 2008). Therefore, olive byproducts could be of great economic and pharmaceutical interest, as hydroxytyrosol displays antioxidant, antimicrobial and antitumoral properties (Bulotta, 2013).

At industrial scale, many techniques have been developed in order to provide a better method for the identification, quantification and recovery of phenolic compounds, not only in olive wastes but also in other seeds/fruit oils, including enzymatic preparation,

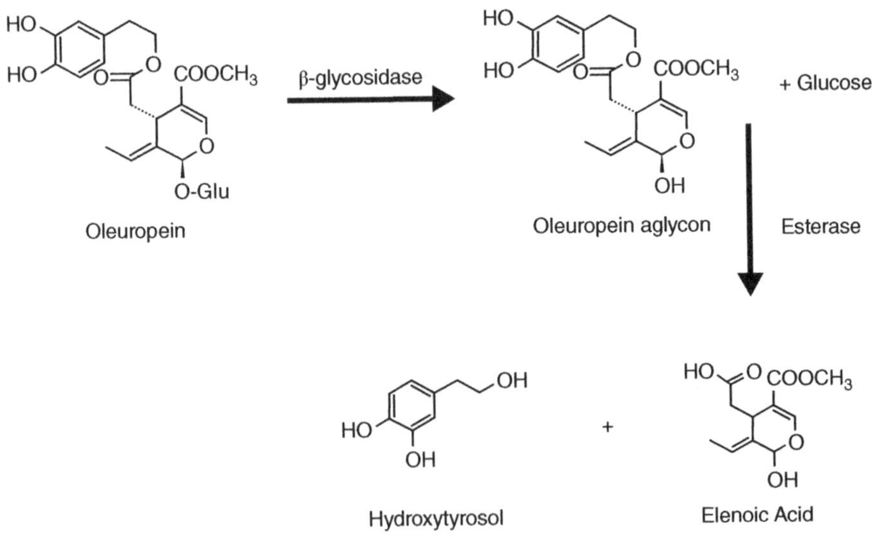

Figure 6.2 Oleuropein and its derivatives after enzymatic hydrolysis by β-glycosidase and esterase. *Source:* Khoufi et al. (2011). Reproduced with permission of Elsevier.

solvent extraction, membrane separation, centrifugation and chromatographic procedures (Baiano, 2014).

6.3.2 Hydrolysis

The release of bioactive compounds from plant cell disruption and extraction can be optimised using enzymatic preparations, either alone or in mixture. To break ester or glycosidic bonds and achieve higher recoveries, a basic, acidic and/or enzymatic hydrolysis is usually applied (Baiano, 2014). Some researchers report that edible oil wastes hydrolysed with a specific enzymatic preparation from *Aspergillus niger* grown on wheat bran or β-glucosidase-rich *Aspergillus niger* and *Trichoderma atroviride* are a potential source of bioactive phenolic compounds (Figure 6.3) (Khoufi et al., 2011).

In the basic hydrolysis process, a NaOH or KOH solution (pH 12) is added to a reactor containing byproduct in order to break mainly ester bonds, while an HCl solution (pH 2–3) is used to break both types of bonds (ester and glycosidic). After recovering bound phenolic compounds from byproducts by applying acidic or basic solution hydrolysis, ethyl acetate is extracted in a proportion of 1 : 1 (v/v). The ethyl acetate extract is evaporated to dryness by rotary evaporator under vacuum conditions and the residue is dried by spray drying (Araújo et al., 2015; Baiano, 2014; Khoufi et al., 2011).

Enzyme-assisted extraction (EAE) is a promising alternative to conventional solvent-based extraction methods. It is based on the ability of an enzyme to catalyse a reaction with specificity and region selectivity, under mild processing conditions and

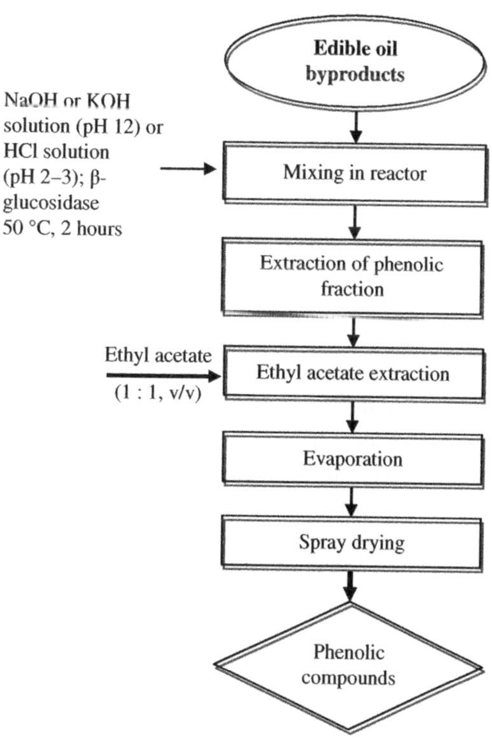

Figure 6.3 Schematic principle of phenolic extraction by enzyme.

in aqueous solutions. For these reasons, EAE is considered an environmentally friendly extraction method (Dermeche et al., 2013).

6.3.2.1 Solid–Liquid Extraction

Solid–liquid extraction (SLE) allows soluble components to be removed from solids using suitable aqueous organic solvents with optimised experimental parameters, such as temperature, time, pH, solid–liquid ratio, particle size, stirring and solvent polarity. It has the disadvantage of requiring expensive, partly toxic, inflammable, explosive and hazardous solvents and long process times (Baiano, 2014; Proestos & Komaitis, 2008). It still has a great number of industrial applications, however. It usually requires maceration or homogenisation of raw material, followed by centrifugation or sedimentation, and then separation. The organic solvent is selected according not only to its extraction efficiency but also to the application of the extracted phenolic compounds. For example, the phenolic compounds recovered from byproducts are intended to be used in the food or pharmaceutical industries; a closer control is required to avoid solvent contamination in the refined products.

Many researchers report that methanol and ethanol are the most common solvents used to extract phenolic compounds from edible oil byproducts (Alu'datt et al., 2010; Jerman Klen & Mozetič Vodopivec, 2011). For example, a mixture of water–methanol in the proportion 80 : 20 (v/v) at pH 2 is applied to extract phenols from olive pomace. Methanol concentration in food products is limited by various authorities; for example, EU Directive 2009/32/EC provides a maximum allowable concentration of 10 mg/kg food (Council of the European Parliament, 2009). Ethanol may be safer than methanol and other solvents due to its lower toxicity, but its dielectric constant is lower than that of methanol (the polarity index and dielectric constant of ethanol and methanol are 5.2 and 5.1 and 24.6 and 32.6, respectively; Nguyen et al., 2016), and therefore it is suitable only for the extraction of lipophilic and complex phenolic compounds.

Another organic solvent is ethyl acetate, which is also applied in the extraction of low- and medium-molecular-mass phenolic compounds. The application of ethyl acetate is considered safe due to its low toxicity, but it can stay inside the sample matrix and impart an unpleasant taste to foodstuffs.

An optimised extraction procedure, on a pilot scale, for the recovery of phenolic compounds from olive oil mill waste is proposed and shown in Figure 6.4.

6.3.2.2 Supercritical Fluid Extraction with CO_2

As already mentioned, the disadvantages of organic solvents, such as their flammability and toxicity, have recently been overcome by the use of supercritical fluids, notably, supercritical carbon dioxide (SC-CO_2) (Dermeche et al., 2013). However, phenolic compounds are polar compounds, so their extraction through supercritical fluid extraction (SFE) requires the addition of a polar solvent like ethanol or methanol (Araújo et al., 2015). This was confirmed by Lafka et al. (2011) in research comparing SFE using only CO_2 to liquid–solid extraction (LSE) with difference. The authors found that the extracts obtained from SFE recovered a lower phenolic content than the extracts from SLE, with the exception of extracts resulting from extraction with ethyl acetate (Lafka et al., 2011). However, the SFE extracts showed a low phenolic content and high recovery rates for nonpolar phenolic compounds, such as apigenin and luteolin, in contrast to other

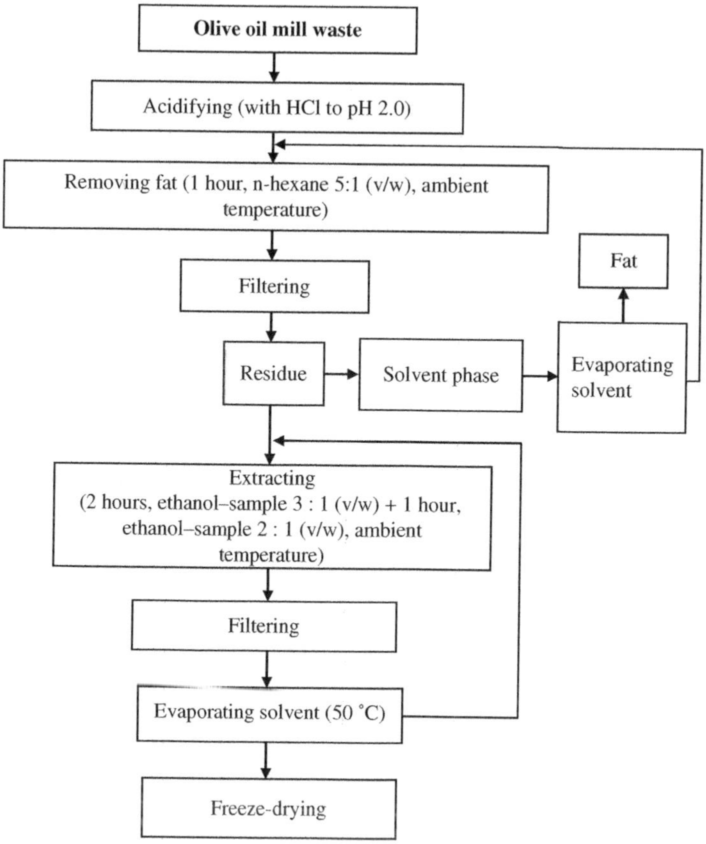

Figure 6.4 Optimised extraction procedure for the recovery of phenolic compounds from olive oil mill waste. *Source:* Lafka et al. (2011). Reproduced with permission of Elsevier.

phenol groups found in edible oil byproducts, such as hydroxytyrosol, hydroxytyrosol-glucosies and vanillin, which were not detected at all (Araújo et al., 2015). Figure 6.5 shows a simplified plan of SFE using CO_2.

6.3.2.3 Membrane Techniques

The traditional approaches applied to the concentration of bioactive compounds extracted from edible oil byproducts, such as simple steam distillation and vacuum distillation, generally require an increased temperature and high energy consumption. Therefore, these methods are inappropriate for heat-sensitive products, such as most of the samples extracted from edible oil byproducts (Bartosz Tylkowski et al., 2010; Murakami et al., 2013; Shadabi et al., 2013). They may also lead to the loss of compounds of low molecular weight, such as tyrosol and hydroxyltyrosol, which can be removed with the solvent at the evaporation step (Bartosz Tylkowski et al., 2010).

A membrane separation process can be used as an alternative to such methods, and provides reduced operating costs when used with phenolic compounds with a molecular mass <1000 Da. An extraction process based on membrane technologies has been developed by Pizzichini et al. (2011). The separation process is based on the integration

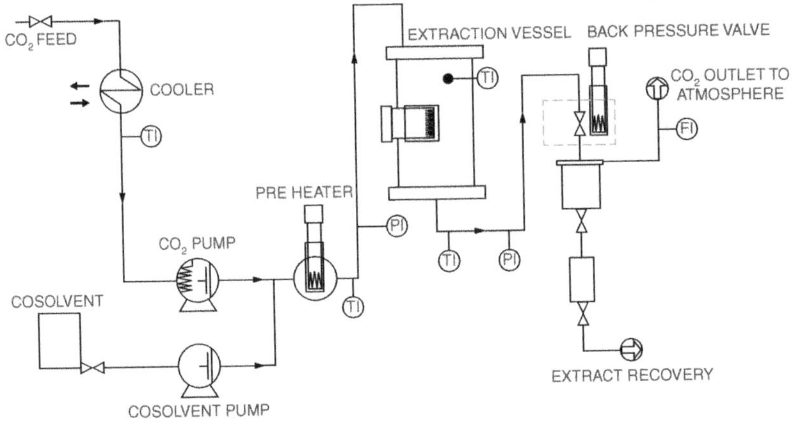

Figure 6.5 Supercritical fluid extraction (SFE) equipment. Liquid CO_2 is contained and kept at low temperature by a cooler (1) using a glycolic solution and pumped to a pre-heater by an air-pressure pump (2). The solvent is transferred to the pre-heater with CO_2 in different lines by an air-driven piston pump (3). At high pressure, the solvent and CO_2 are mixed together before being heated in a heat exchanger (4). The heated mixture passes through the extractor vessel (5), then goes to the other basket, which has a back pressure valve. Finally, liquid CO_2 expands into a tank, where the extracted substances are collected. The temperature and pressure of the extractor vessel aremonitored and controlled. *Source:* Fiori et al. (2009). Reproduced with permission of Taylor & Francis.

of suitable preliminary LSE operations, carried out on the starting matrix at room temperature or with simultaneous heating, followed by two or more consecutive tangential membrane filtration operations, carried out on the clarified liquid product coming from the extraction. If required, this can be followed by a series of adsorption and desorption operations on the active compounds contained in the various retentates of the tangential filtration operation, using specific resins. The latter step allows the titre of bioactive components of commercial interest to be increased, and urified active principles belonging to different phenolic subclasses to be obtained (Figure 6.6) (Pizzichini et al., 2011).

6.3.3 Phytosterols and Tocopherols

Phytosterols and tocopherols are natural, organic compounds that are widely distributed in all eukaryotic organisms (Piironen et al., 2000). They are naturally present in edible oils, such as sunflower, rapeseed, peanut, soybean and palm oils. In general, edible oils may contain 70–1900 mg tocopherols/kg oil (Ahsan et al., 2015; Cerretani et al., 2010) and 0.1–1.0% phytosterols. Tocopherols are useful lipid antioxidants with wide application as pharmaceutical, cosmetic and nutritional ingredients, as they help in preventing oxidation and spoilage of food. They are also valuable dietary supplements, due to their ability to reduce the risk of certain types of cancers (Husveth et al., 2000; Kamal-Eldin & Appelqvist, 1996). During the refining of vegetable oils, a large fraction of tocopherols are lost in various byproducts and waste streams, including deodorised oil distillate (DOD), stream refining distillate and acidulated soap stocks (Buczenko et al., 2003). DOD is a complex material, consisting of sterols, steryl esters, sterol hydrocarbons, tocopherol, esters, FFAs, mono-, di- and triglycerides, hydrocarbons, aldehydes, ketones and other polar and nonpolar impurities. The tocopherol content of DOD varies widely, depending on the material (e.g. oil seeds and fruits) and the conditions applied during

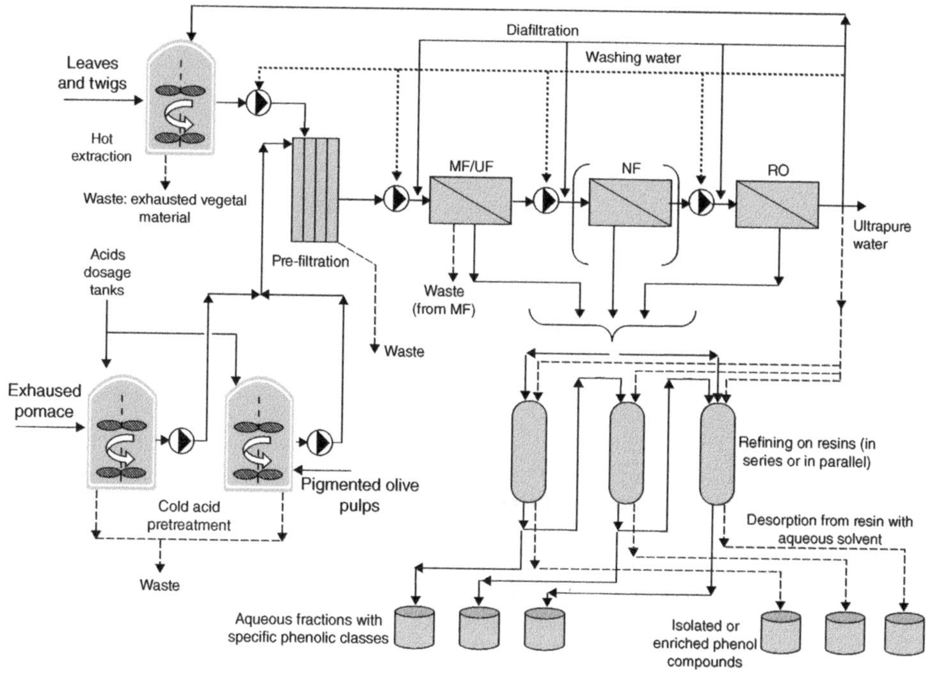

Figure 6.6 Process for producing concentrated and refined actives from tissues and byproducts of olive oils using membrane technologies. MF/UF, microfiltration/ultrafiltration; NF, nanofiltration; RO, reverse osmosis. *Source:* Adapted from Pizzichini et al. (2011).

the deodorisation process (Guthalugu et al., 2006). The concentrations of tocopherols and phytosterols in DOD range from 4 to 5% (w/w) and from 4.2 to 8.7% (w/w), respectively (Pestana-Bauer et al., 2012). Preparation of tocopherol and sterol concentrates using deodorised distillates involves a series of physical and chemical processes, including separating FFAs from the distillate in the initial step, and then separating the sterols and/or tocopherols to yield a rich tocopherol/sterol concentrate. These compounds are purified industrially from edible oil DOD by processes such as chemical methylation, shot-path distillation, solvent fractionation, ion-exchange chromatography (IEC) and SFE of CO_2.

To separate sterols and tocopherols, Smith & Winton (1964) and Smith (1967) introduced recovery procedures for these compounds in DOD (Figures 6.7 and 6.8). These procedures are suitable for use with DODs with high acid values (>40). The method developed by Smith & Winton (1964) could recover only 73% free sterols, while the method of Smith (1967) could increase the purity to 86%. In order to improve the recovery efficiency in terms of both the yield and the purity of the final product, Fizet (1996) developed a new method in which a *trans*-esterification step was applied to convert steryl esters to free form. The yield and purity of this process were 90 and 98%, respectively (Figure 6.9). The modified processes were developed in studies that could be applied to DODs with low sterol contents (5–30%) and high acid values. A second *trans*-esterification was applied in order to improve the purity of the final product. These processes allowed a product with a sterol-rich residue containing 40% free sterol

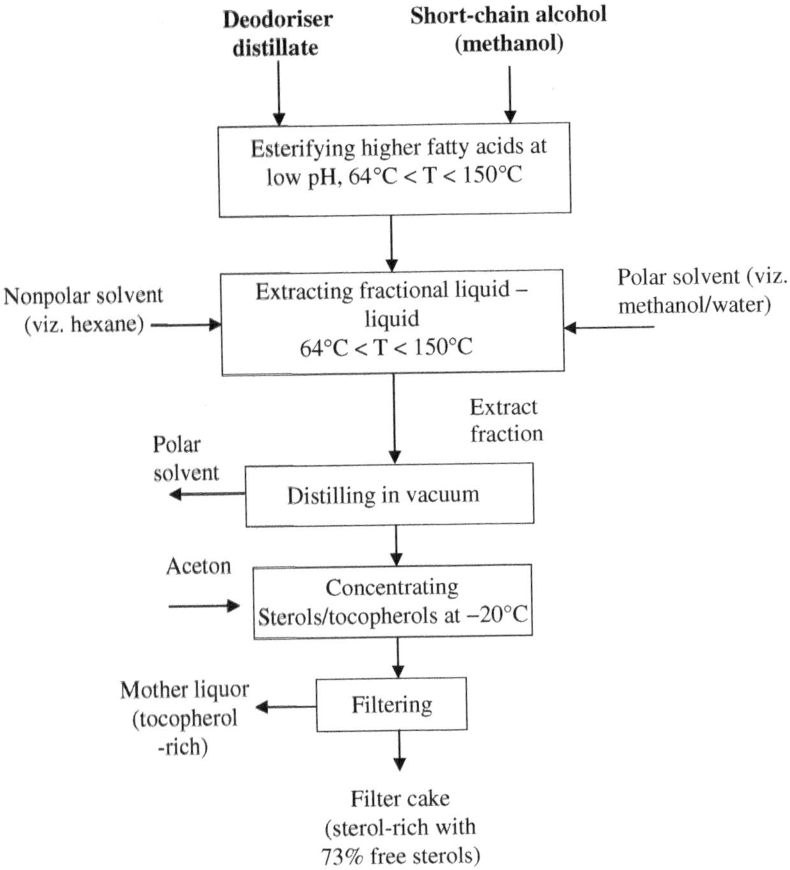

Figure 6.7 Recovery of sterols from deodorised oil distillate (DOD). *Source:* Smith & Winton (1964).

to be obtained (Jeromin et al., 1997); in fact, final products could contain 90% or even 94% sterols (Figure 6.10) (Ho, 2009; Wollmann et al., 2005).

To remove FFAs from DODs in order to reduce the volume of the material, some researchers suggest beginning with a saponification or vacuum distillation step. This can be applied to process DODs containing sterol and FFA fractions of 2–15% and 30–85%, respectively (Figure 6.11) (Czuppon et al., 2006; Copeland & Belcher, 2004). The process for the recovery of phytosterols described in Figure 6.11 can be used with DODs with an acid value <10 and/or a sterol content in the range 15–50%. In this process, the fatty acids and their esters are saponified and removed during the saponification step, and phytosterols are crystallised out of the unsaponified fraction. When the fatty acids or their esters are removed, the purity of the final phytosterol crystals is improved by the elimination of the contamination (Hattori et al., 2002; Stephane et al., 2005).

As already mentioned, both tocopherols and sterols (as steryl esters) can be purified with high yields by converting sterols to steryl esters and by converting FFAs to fatty acid methyl esters (FAMEs) (Nagao et al., 2005). These two conversions were both achieved in a single bath reaction with *Candida rugosa* lipase: (i) esterification of sterol with FFAs; (ii) esterification of FFAs with methanol (MeOH). However, the close boiling

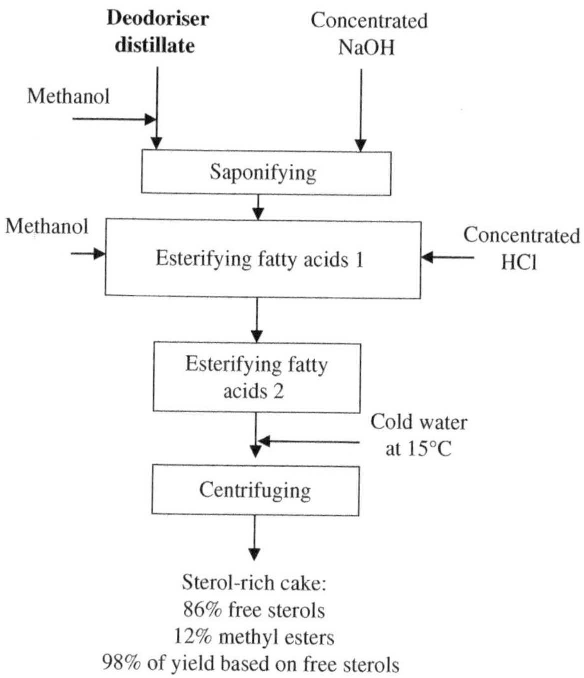

Figure 6.8 Recovery of sterols from deodorised oil distillate (DOD). *Source:* Smith (1967).

points of tocopherols and phytosterols posed a difficult problem in this recovery process (Copeland & Belcher, 2004). Recently, Nagao et al. (2005) found that the two-step *in situ* reaction converted a small amount of sterols to steryl esters and a low number of FFAs to FAMEs (<80%). These steps were not enough to provide highly efficient purification of sterols as steryl esters and tocopherols. Therefore, Nagao et al. (2005) developed a new method whereby the two-step *in situ* reaction was conducted twice to obtain higher degrees of the two reactions. The principle underlying this method is that if sterols are converted to steryl esters and FFAs are converted to FAMEs, then a relatively easy purification of tocopherols and steryl esters can be achieved because the boiling points of FAMEs, tocopherols and steryl esters are significantly different. In this process, extracted compounds were purified to 72% (yield at 88%) and 97% (yield at 97%) tocopherols and steryl esters, respectively (Figure 6.12). The oil byproducts were dehydrated in a reactor at 80 °C/5 mmHg for 30 minutes in order to recover the oil layer, then the dehydrated oil layer was filtrated to remove insoluble substances. The filtrated oil was distilled at 160 °C/0.2 mmHg to remove the FAMEs. Small amounts of FAMEs and FFAs still remained in the residue (residue 1), however, so it was continuously distilled at 175 °C/0.2 mmHg to remove the rest of these compounds (residue 2). The FAME and FFA contents in residue 2 were reduced to <0.1 and <0.8 wt %, respectively. Residue 2 was distilled at 230 °C/0.02 mmHg to purify the remaining tocopherols with a purity of 72% and a recovery of 88% (residue 3). Residue 3 still contained a steryl ester fraction, however, with a purity of 87 wt %, so it was continuously distilled at 240 °C/0.02 mmHg, and steryl esters were recovered with a purity of 97 wt % (residue 4) (Nagao et al., 2005).

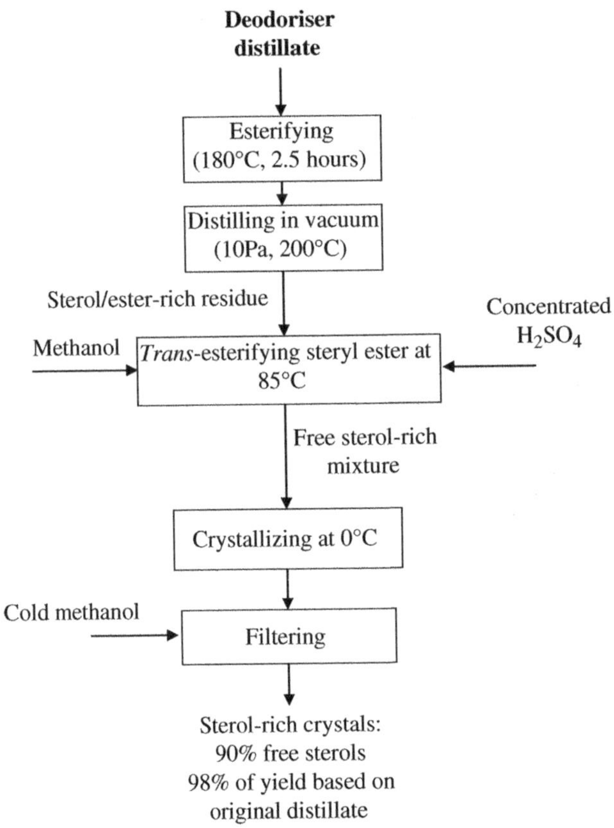

Figure 6.9 Recovery of sterols from deodorised oil distillate (DOD). *Source*: Fizet (1996).

A process for the recovery of tocopherols and phytosterols from edible oil wastes using SFE in order to improve the pre-treatment step and obtain high efficiency recoveries has been performed by some researchers. SC-CO_2 extraction has been demonstrated as a potential technology for the enrichment of these minor compounds (Lee et al., 1991; Saito & Yamauchi, 1990). In order to convert triglycerides – the major component of edible oil DOD – into FFAs, Guthalugu et al. (2006) used immobilised *Candida rogusa* lipase under SC-CO_2 medium. They found the optimal conditions were pressure 180 bar, temperature 43 °C and moisture content 40%, which provided a triglyceride hydrolysis of 94% in 1.5 hours (Figure 6.13). By contrast, the conventional method (hydrolysis in hexane) required 5 hours to achieve 88% triglyceride hydrolysis under similar conditions.

The separation and purification of some individual phytosterols, such as β-sitosterol and stigmasterol from phytosterol concentrates, have also been carried out by some researchers. These methods have been applied using various solvents, including n-pentanol and cyclohexanone, to separate and purify individual sterols. Using these processes, a final product with β-sitosterol content >85% was obtained after applying three stages of recrystallisation in cyclohexanone with recovery efficiency 39%. When the purification process was applied across five stages of fractional crystallisation using

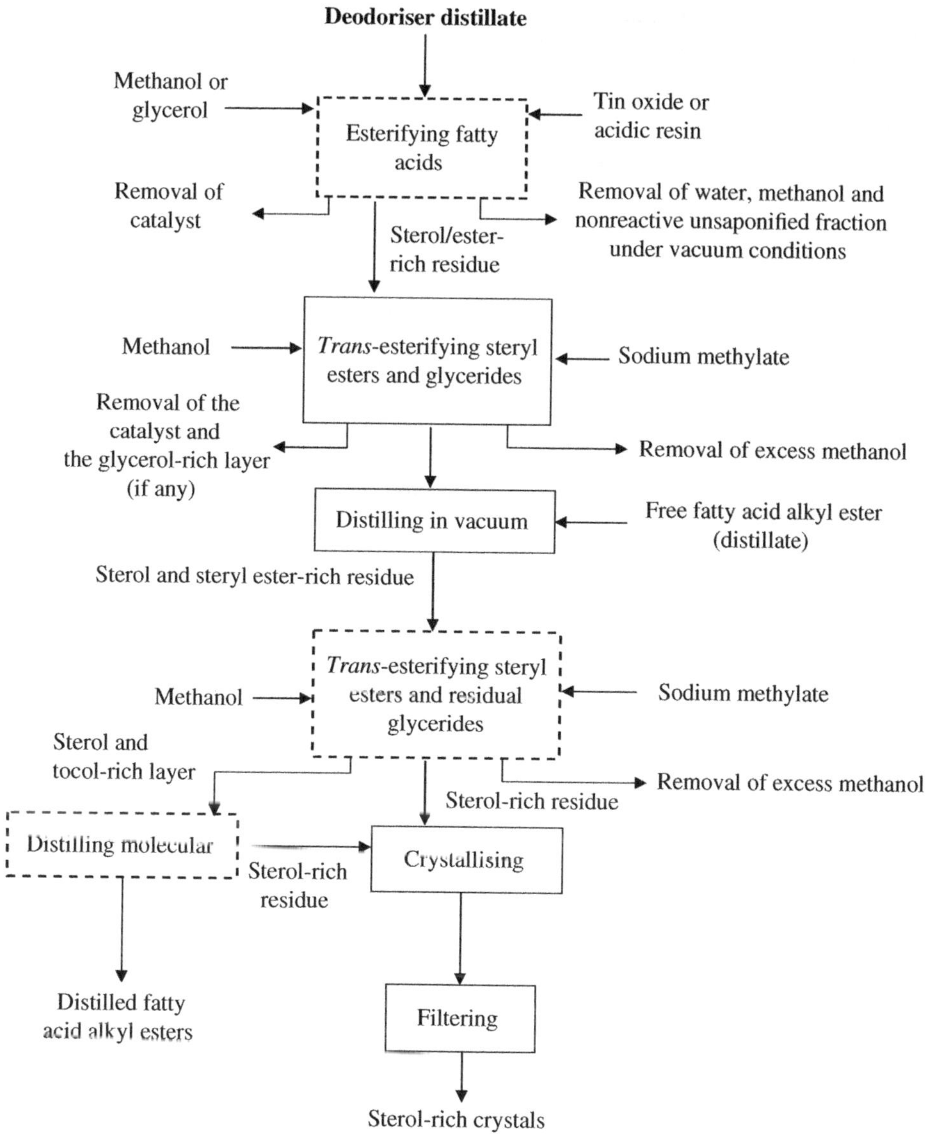

Figure 6.10 Recovery of sterols from deodorised oil distillate (DOD). *Source:* Fernandes & Cabral (2007). Reproduced with permission of Elsevier.

either n-pentanol or cyclohexanone, a final product was obtained with stigmasterol content >93% and process yield 6–8% (Arumughan et al., 2009; Koskenniska, 1982; Robinson et al., 2000).

6.3.4 γ-oryzanol

Rice bran oil is a popular edible oil in several countries, including India, Korea and China, and is known as a healthy cooking oil. Due to its high smoke point and delicate

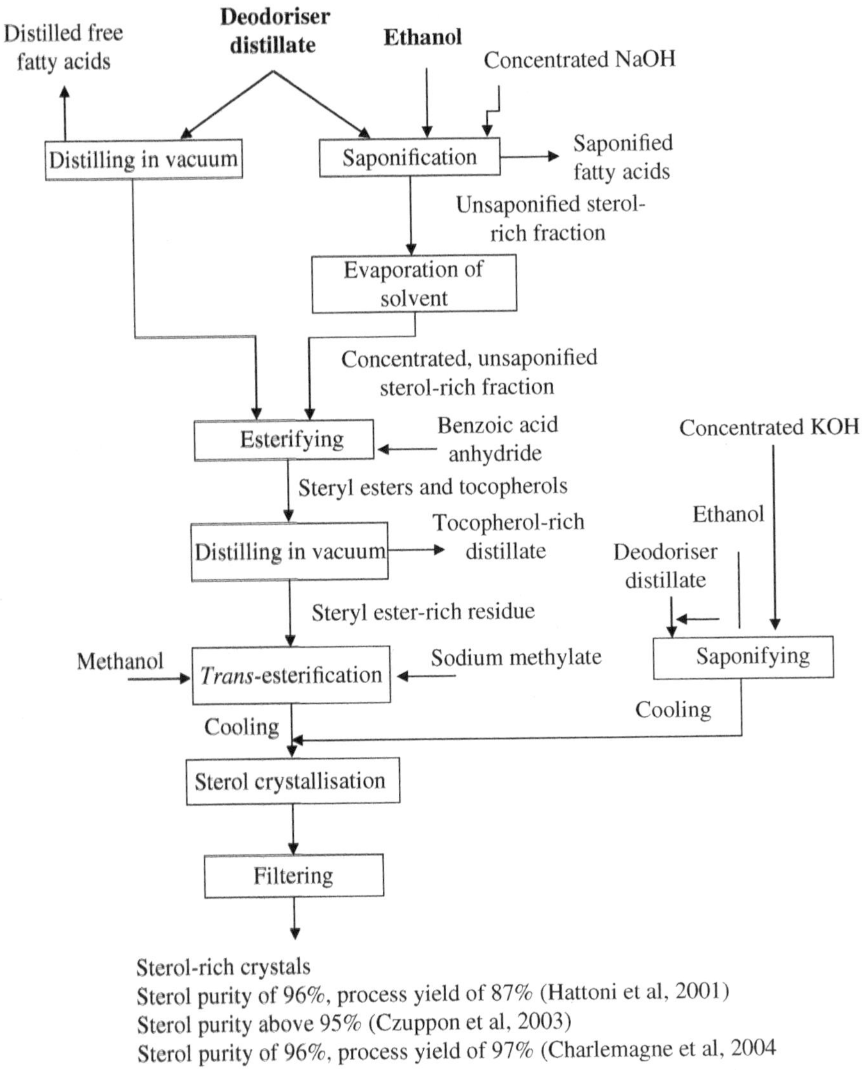

Figure 6.11 Recovery of phytosterols: (a) low sterol content and low-pH materials; (b) high sterol content and high-pH materials. *Source:* Fernandes & Cabral (2007). Reproduced with permission of Elsevier.

flavour, rice bran oil is an excellent cooking and salad oil. The nutritional qualities and health effects of rice bran oil are well established, and it is considered rich in unsaponified fraction, which contains micronutrients like vitamin E complexes, γ-oryzanol, phytosterols, polyphenols and squalene. γ-oryzanol, in particular, has the potential to be used in nutraceutical, pharmaceutical and cosmeceutical preparations. It is a mixture of ferulic acid esters of sterol and triterpene alcohols.

γ-oryzanol, which is lost during the chemical refining process, can be carried into the soap stock. The γ-oryzanol content of commercial rice bran oil, soap stock, acid oil

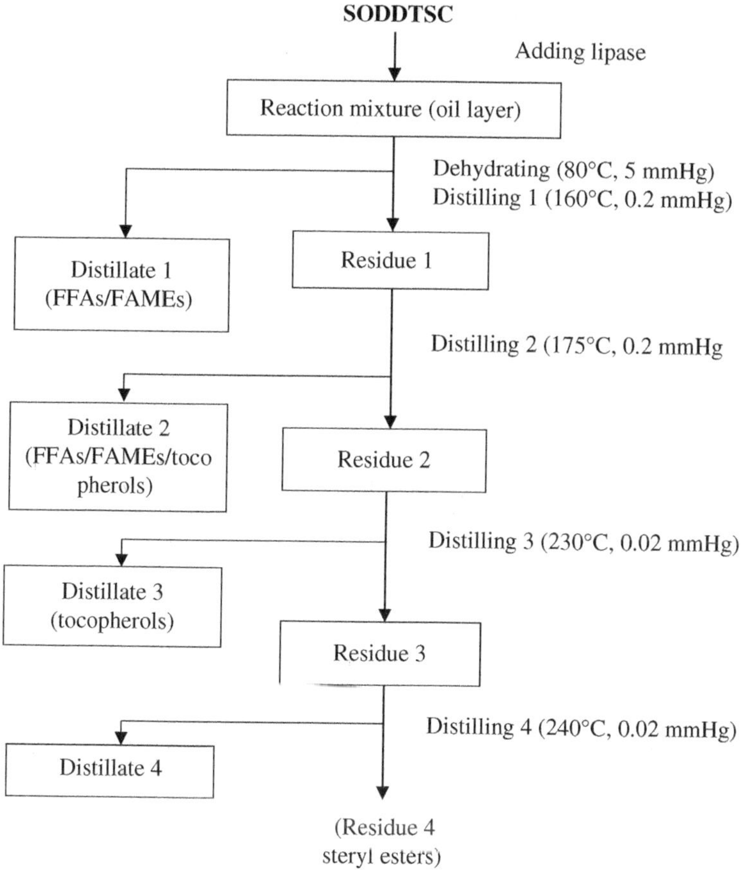

Figure 6.12 Two-step *in situ* reaction and distillation process for the purification of tocopherols and phytosterols from soybean DOD tocopherol/sterol concentrate. *Source:* Nagao et al. (2005). Reproduced with permission of Elsevier.

and DOD is in the range 1.7–2.1, 6.3–6.9, 3.3–7.4 and 0.79%, respectively. This shows that the processing steps, viz., degumming (1.1%), dewaxing (5.9%), physical refining (0%), bleaching and deodorisation, do not affect the oryzanol content appreciably, while 83–95% is lost during alkali refining. At present, the majority of both deodorised bran and soap stock is used as animal feed. Therefore, many researchers have looked to increase the value of these byproducts. Separation and purification of γ-oryzanol from rice bran oil byproducts has been under investigation for many years (Cicero & Gaddi, 2001; Jesus et al., 2010; Kaewboonnum et al., 2010).

Kaewboonnum et al. (2010) have studied conditions for the separation of γ-oryzanol from soap stock, a byproduct of the rice bran oil process. At first, they used NaOH to convert the remaining glycerides into soap, and then extracted the dried saponified material using ethyl acetate with a ratio of saponified soap stock to solvent of 15 mg/200 ml. The impurities in the extract were separated in the first 1 hour of the crystallisation step using a solvent mixture of 20% (v/v) ethyl acetate in methanol at

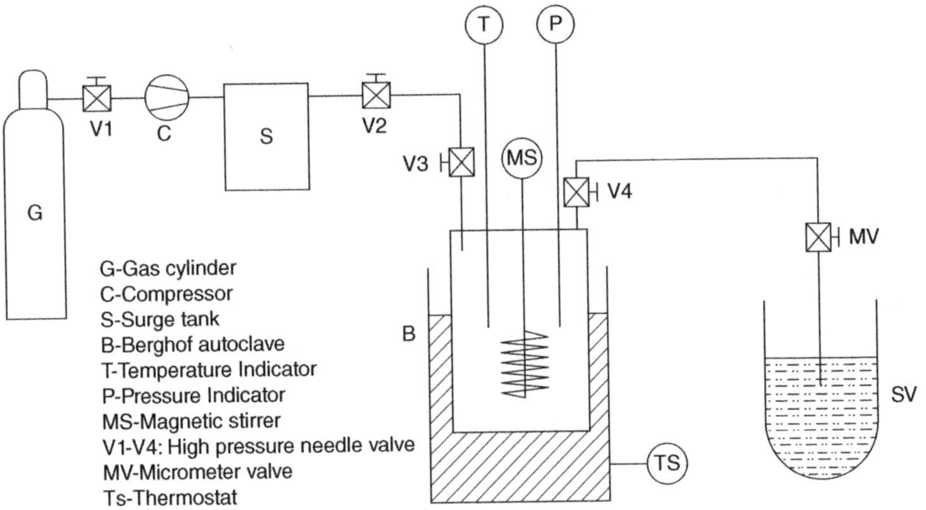

Figure 6.13 Schematic diagram of an SC-CO_2 reaction unit. *Source:* Guthalugu et al. (2006). Reproduced with permission of Elsevier.

30 °C. In the next step, the supernatant obtained was allowed to crystallise in order to recover γ-oryzanol crystals. Finally, the solvent was evaporated and recovered for reuse.

Perretti et al. (2003) demonstrated that SFE of CO_2 could be applied as a value-added scheme through the recovery of bioactive compounds from rice processing byproducts. Their results showed that the extraction conducted at 10 000 psi and 80 °C gave the highest extraction yield and the extract had a high purity, making it suitable for human consumption. While Jesus et al. (2010) found that the best operational conditions for SFE were 30 MPa/29.5 °C and 20 MPa/29.5 °C, which presented the same recovery efficiencies.

By using SFE and solvent extraction methods for the extraction of γ-oryzanol, Xu & Godbe (2000) showed that a solvent mixture of hexane and isopropanol (50 : 50, v/v) at 600 °C for 45–60 minutes produced the highest yield of γ-oryzanol (1.68 mg/g of rice bran) and demonstrated that the saponification step during solvent extraction decreased the recovery efficiency of γ-oryzanol. Meanwhile, the yield of γ-oryzanol in SFE at 680 atm and 50 °C for 25 minutes was approximately four times higher than that in solvent extraction.

For the same purposes, another oryzanol extraction and purification process was developed by Indira et al. (2005). Their process for the production of oryzanol-enriched fraction from bran oil soap stock includes the following steps: (i) stirring of bran oil byproducts (60–70% moisture, pH 10–11) in a steam-jacketed kettle at 70–90 °C for a period of 15–30 minutes to obtain a uniform dispersion; (ii) adding a suitable amount of sodium hydroxide (about 0.8%, w/w) to the hot dispersion of step (i), mixing at elevated temperature (70–90 °C) for a period of 10–15 minutes; (iii) homogenising the hot alkaline soap stock dispersion of step (ii) by passing through colloidal mixture for three to five passes over a period of 5–10 minutes in order to increase the surface area; (iv) mixing the homogenised alkaline soap stock dispersion of step (iii) in a steam-jacketed kettle at 70–90 °C for a period of 30–45 minutes in order to complete saponification;

(v) dehydrating the saponified soap stock of step (iv) by drying in an alum at 90–100 °C, a steam pressure of 1.0–1.10 kg/cm^2 for less than 1 minute; (vi) leaching the saponified and dehydrated rice bran oil soap stock of step (v) in a packed bed using an organic solvent or mixture of solvents at 25–29 °C for 2–4 hours; (vii) decanting the leached extract of step (vi); and (viii) desolvating the decanted extract of step (vii) in order to obtain the required oryzanol-enriched fraction.

Crude γ-oryzanol isolated from rice bran oil or its byproduct can be purified and crystallised by the following steps: (i) mixing the oryzanol-enriched unsaponified fraction of rice bran oil soap stock with mixtures of ketones and/or alcohols, such as acetone and methanol (1 : 3 to 4, v/v) at a particular solid–solvent ratio of 1 : 5, then refluxing at elevated temperature of 80 °C for 1 hour to facilitate to complete solubilisation of oryzanol throughout the organic solvent mixture; (ii) subjecting the hot oryzanol-enriched refluxed solvent mixture to fractional precipitation at normal temperature (around 22–29 °C) for 45–90 minutes to strip solid mucilaginous interfering impurities; (iii) following a single method or combination of two or more methods for the crystallisation of the clarified oryzanol-rich supernatant in order to improve the selectivity of oryzanol isolation (the list of crystallisation methods is as follows: temperature-controlled fractional crystallisation; anti-solvent crystallisation; temperature-controlled anti-solvent crystallisation; and high-pressure gas anti-solvent crystallisation); and (iv) recovery of the crystalline fraction on desolventisation, preferably by vacuum drying, to give oryzanol flake (Narayan et al., 2004).

6.3.5 Other Bioactive Compounds

6.3.5.1 Pectins and Oligosaccharides

The cell walls of seeds and fruits comprise a number of high-molecular-weight components, which are highly complex, and they therefore provide a range of functional components, including polysaccharides (pectins and hemicelluloses), proteins and phenolics. They contain considerable quantities of pectin polysaccharides and hemicellulosic polymers that are rich in xylans and xyloglucans (Juan Fernández-Bolaños et al., 2006). Cardoso et al. (2003) investigated the extraction of pectin from olive oil byproducts, in particular residual cell walls, which contained 45% galacturonic acid and 31% arabinose and had a methylesterification degree of 43%. They found that pectin derived from olive oil byproducts had favourable gelling properties.

In general, the pectin extraction yield from different sources varies with the processing parameters (pH, time, temperature) and sample features. Tornberg & Galanakis (2010) developed a process for isolating soluble dietary fibres and valuable polyphenols from olive oil byproducts, such as OMW (Figure 6.14). First, the products are defatted by centrifugation, then concentrated to remove water. Next, the residue (i.e. the dehydrated mixture) is extracted using ethanol and an organic acid, and the dietary fibres are separated from the polyphenols by precipitation in ethanol. The alcoholic residue is rich in polyphenols, which can be purified by filtration to obtain a mixture of polyphenols in 15–40% ethanol. The soluble dietary fibres are separated from the insoluble ones by redissolving the precipitated dietary fibres into water at 2% (w/v) and kept at 10 °C overnight; this causes the the insoluble dietary fibres to precipitate, leaving the soluble dietary fibres in solution. Soluble fibre products

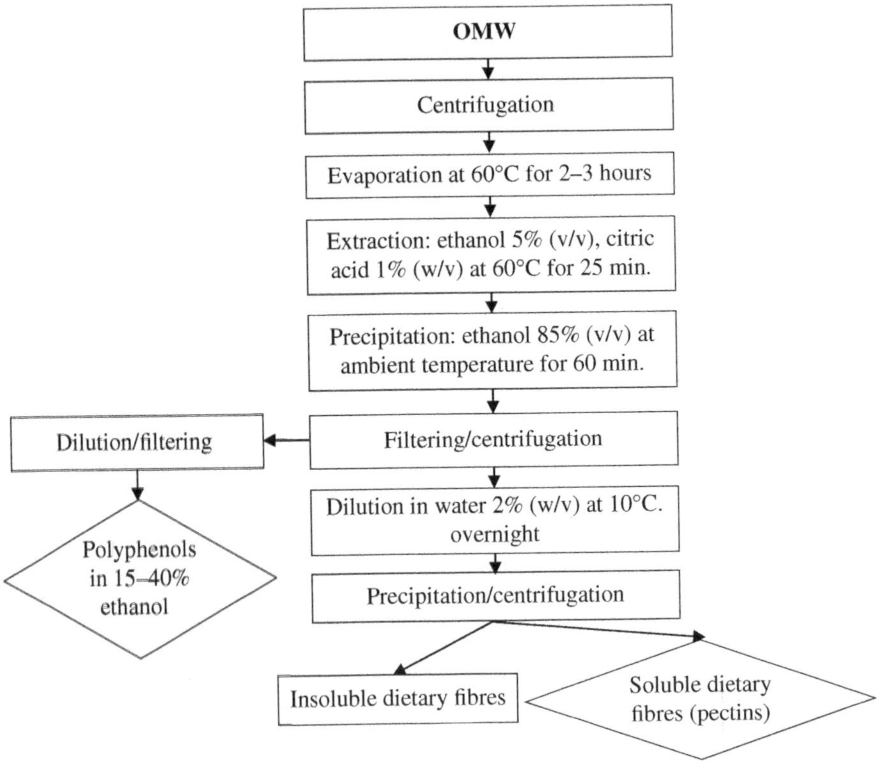

Figure 6.14 Process for isolating soluble dietary fibres and valuable polyphenols from olive mill wastewater (OMW). *Source:* Tornberg & Galanakis (2010).

include pectin from olive oil byproducts, which can be used as an additive and/or a fat replacer in other foods.

6.3.5.2 Mannitol

Saccharides are one of the valuable compounds present in oil byproducts, such as olive pulp. Simple sugar (i.e. glucose) is the main compound to be identified in pulp, following by sucrose and fructose in smaller amounts (Guillén et al., 1992). The sugar alcohol or polyol is called mannitol, which is used as an excipient in pharmacy or as a food additive (e.g. as a thickener, anti-caking agent or low-calorie sweetener). Guillén et al. (1992) developed a recovery process for the extraction of mannitol from steam-treated alperujo, an olive oil byproduct, that achieves mannitol products with high purity (Figure 6.15).

Tabera (2005) described another process for obtaining mannitol and other products derived from OMW, olive twigs and leaves. This process could be changed to work with other raw materials. In general, it comprises the following steps: (i) completed drying; (ii) extraction of mannitol by alcohols or hydroalcoholic solvents; (iii) defecation of extracts by the addition of lead (II) acetate basic in order to eliminate unexpected substances; (iv) purification and concentration of the extracts; and (v) isolation of mannitol by crystallisation with 96% alcohol. A modified process for the recovery of mannitol and

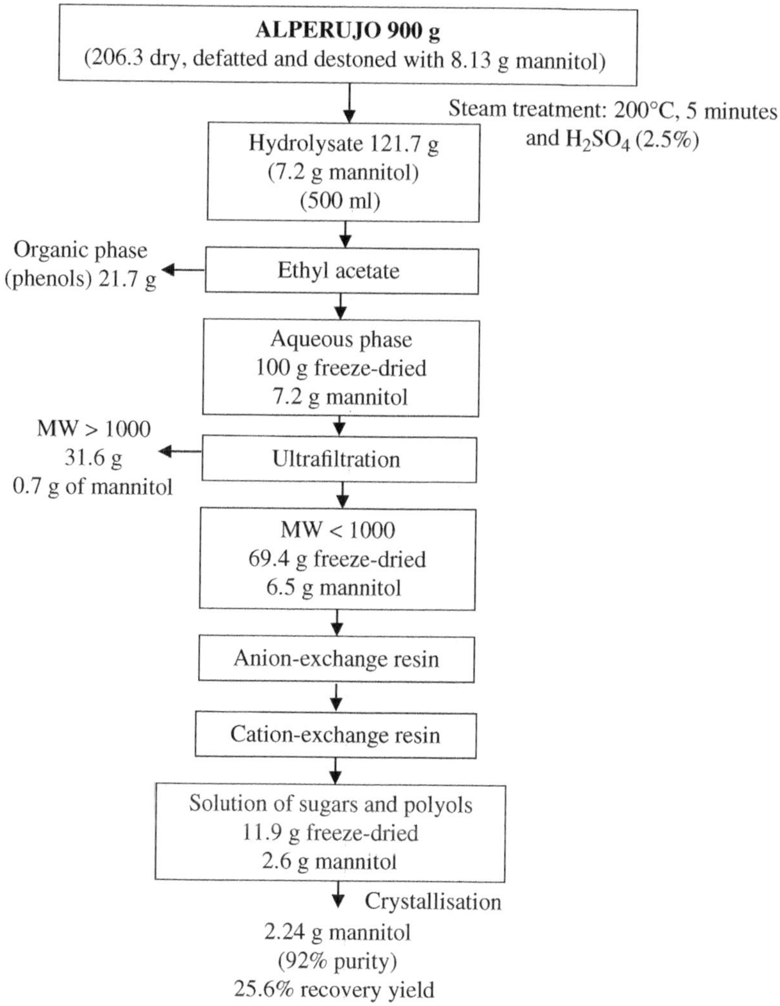

Figure 6.15 Scheme of the partial purification of mannitol in the hydrolysates obtained after hydrothermal treatment of alperujo. *Source:* Guillén et al. (1992). Reproduced with permission of Elsevier.

its derivatives from two-phase OMW (2POMW) was developed by García-Granados et al. (1994) (Figure 6.16).

6.3.6 Future Trends

The quantity, quality and profitability of edible vegetable oil byproducts are increasing annually, for a few major reasons. First, edible oils offer special nutrients and flavours not offered by animal fats, and consumption of these is increasing. Second, new regulations relating to environmental protection have been introduced, such as a Community Strategy for Waste Management implemented by the European Community, which includes the following principles:

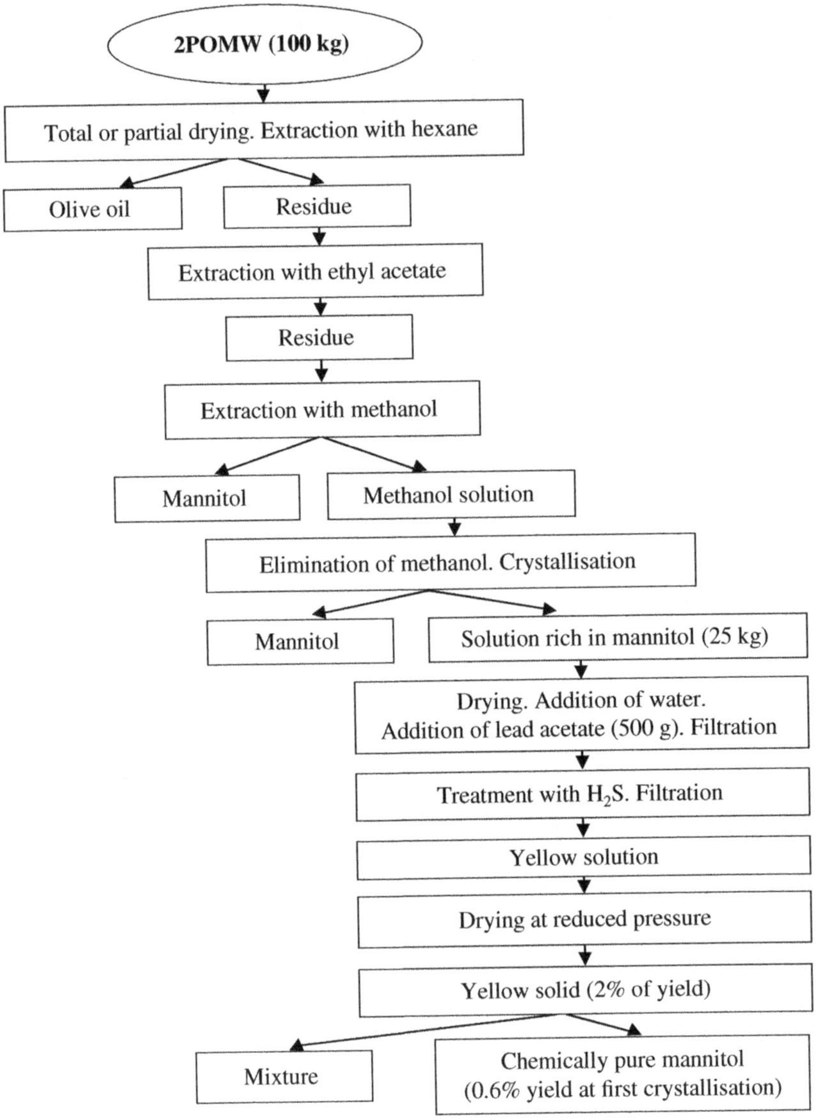

Figure 6.16 Modified process for obtaining mannitol and its derivatives from two-phase olive mill wastewater (2POMW).

(1) **the prevention principle:** waste production should be limited;
(2) **the polluter pays principle:** the polluter should pay to deal with the waste;
(3) **the precautionary principle:** waste problems should be predicted;
(4) **the proximity principle:** as far as it is possible to do so, waste problems should be addressed near or at the place of production.

Third, consumer demand for health foods and nutraceuticals is increasing. Edible oil byproducts, which have traditionally been seen as waste, are now known to be rich sources of valuable bioactive compounds, such as polyphenols and γ-oryzanol. Therefore, edible vegetable oil byproducts have become commercially viable over the last few decades, both as single products and in the form of mixtures, such as functional foods and food additives. Businesses involved in the production of edible oil byproducts have increased their market share and grown more profitable than companies producing synthetic bioactive compounds, as the natural ones are both safer and healthier. In short, the future for edible vegetable oil byproducts is positive, and it is difficult to see any limitation on the development and pricing of bioactive compounds from edible oil sources.

6.4 Conclusion

Edible oil wastes constitute an interesting source of bioactive compounds, such as phenolics, sterols and γ-oryzanol. These high-value compounds can be applied in functional food processing or in pharmaceutical preparations and nutraceuticals in order to increase their health-promoting properties. The bioactivity, bioavailability and toxicology of recovered bioactive compounds and their reactions and interactions with other food ingredients need to be carefully investigated and assessed by *in vitro* and *in vivo* studies. In this chapter, we have looked at the application of extraction and recovery methods in isolating and purifying bioactive compounds from edible oil byproducts. Each method has its advantages and disadvantages, and a proper choice is important to ensure high extraction efficiency and good bioactivity of the extracted compounds. The increasing demand for natural bioactive compounds may lead to more sophisticated and efficient extraction and purifying methods being developed in the future.

References

Ahsan, H., Ahad, A. & Siddiqui, W. A. 2015. A review of characterization of tocotrienols from plant oils and foods. *Journal of Chemical Biology*, **8**, 45–59.

Alúdatt, M. H., Alli, I., Ereifej, K., Alhamad, M., Al-Tawaha, A. R. & Rababah, T. 2010. Optimisation, characterisation and quantification of phenolic compounds in olive cake. *Food Chemistry*, **123**, 117–122.

Araújo, M., Pimentel, F. B., Alves, R. C. & Oliveira, M. B. P. P. 2015. Phenolic compounds from olive mill wastes: health effects, analytical approach and application as food antioxidants. *Trends in Food Science & Technology*, **45**, 200–211.

Arienzo, M. & Violante, P. 2007. Improving waste management and co-product recovery in vegetable oil processing. In: Waldron, K. (ed.) *Handbook of Waste Management and Co-product Recovery in Food Processing*. Abington: Woodhead.

Arumughan, C., Sobankumar, D. R., Sudaresan, A., Nair, S. S., Yohesh, K. & Rajam, L. 2009. Process for the preparation of high purity phytosterols. Patent No. US 7632530 B2.

Azmir, J., Zaidul, I. S. M., Rahman, M. M., Sharif, K. M., Mohamed, A., Sahena, F., Jahurul, M. H. A., Ghafoor, K., Norulaini, N. A. N. & Omar, A. K. M. 2013. Techniques for extraction of bioactive compounds from plant materials: a review. *Journal of Food Engineering*, **117**, 426–436.

Baiano, A. 2014. Recovery of biomolecules from food wastes – a review. *Molecules*, **19**, 14 821–14 842.

Bartosz Tylkowski, Boryana Trusheva, Vassya Bankova, Marta Giamberini, Georgi Peev & Nikolova., A. 2010. Extraction of biologically active compounds from propolis and concentration of extract by nanofiltration. *Journal of Membrane Science*, **348**, 124–130.

Buczenko, G. M., De Oliveira, J. S. & Von Meien, O. F. 2003. Extraction of tocopherols from the deodorized distillate of soybean oil with liquefied petroleum gas. *European Journal of Lipid Science and Technology*, **105**, 668–671.

Bulotta, S., Oliverio, M., Russo, D. & Procopio, A. 2013. Biological activity of oleuropein and its derivatives. In: Ramawat, K. G. & Mérillon, J.-M. (eds) *Natural Products*. Berlin: Springer.

Cardoso, S. M., Coimbra, M. A. & Lopes Da Silva, J. A. 2003. Calcium-mediated gelation of an olive pomace pectic extract. *Carbohydrate Polymers*, **52**, 125–133.

Cardoso, S. M., Falcão, S. I., Peres, A. M. & Domingues, M. R. M. 2011. Oleuropein/ligstroside isomers and their derivatives in Portuguese olive mill wastewaters. *Food Chemistry*, **129**, 291–296.

Cerretani, L., Lerma-García, M. J., Herrero-Martínez, J. M., GALLINA-TOSCHI, T. & SIMÓ-ALFONSO, E. F. 2010. Determination of tocopherols and tocotrienols in vegetable oils by nanoliquid chromatography with ultraviolet–visible detection using a silica monolithic column. *Journal of Agricultural and Food Chemistry*, **58**, 757–761.

CICERO, A. F. & GADDI, A. 2001. Rice bran oil and gamma-oryzanol in the treatment of hyperlipoproteinaemias and other conditions. *Phytotherapy Research*, **15**, 277–289.

COPELAND, D. & BELCHER, W. M. 2004. Methods for treating deodorizer distillate. Patent No. US6750359 B1.

Council of the European Parliament. 2009. Directive 2009/32/EC of the European Parliament and of the Council of 23 April 2009 on the approximation of the laws of the Member States on extraction solvents used in the production of foodstuffs and food ingredients.

Czuppon, T., Kemeny, Z., Kovari, E. & Recseg, K. 2006. Process for recovery of plant sterols from by-product of vegetable oil refining. Patent No. US 20060135794 A1.

Da Silva, A. C. & Jorge, N. 2014. Bioactive compounds of the lipid fractions of agro-industrial waste. *Food Research International*, **66**, 493–500.

Deffense, E. 2011. Chemical degumming. The AOCS Lipid Library. Available from: http://lipidlibrary.aocs.org/OilsFats/content.cfm?ItemNumber=40322 (last accessed 20 March 2017).

Dermeche, S., Nadour, M., Larroche, C., Moulti-Mati, F. & Michaud, P. 2013. Olive mill wastes: biochemical characterizations and valorization strategies. *Process Biochemistry*, **48**, 1532–1552.

Dijkstra, A. J. 2014. Edible oil processing. Available from: http://lipidlibrary.aocs.org/OilsFats/content.cfm?ItemNumber=40318&navItemNumber=19214 (last accessed 20 March 2017).

Dijkstra, A. J. & Van Duijn, G. 2016. Vegetable oils: oil production and processing A2 – Caballero, Benjamin. In: Finglas, P. M. & Toldrá, F. (eds) *Encyclopedia of Food and Health*. Oxford: Academic Press.

Dumont, M. J. & Narine, S. S. 2007. Soapstock and deodorizer distillates from North American vegetable oils: review on their characterization, extraction and utilization. *Food Research International*, **40**, 957–974.

El-Abbassi, A., Kiai, H. & Hafidi, A. 2012. Phenolic profile and antioxidant activities of olive mill wastewater. *Food Chemistry*, **132**, 406–412.

Fernandes, P. & Cabral, J. M. S. 2007. Phytosterols: applications and recovery methods. *Bioresource Technology*, **98**, 2335–2350.

Fernández-Mar, M. I., Mateos, R., García-Parrilla, M. C., Puertas, B. & Cantos-Villar, E. 2012. Bioactive compounds in wine: Resveratrol, hydroxytyrosol and melatonin: a review. *Food Chemistry*, **130**, 797–813.

Fiori, L., De Faveri, D., Casazza, A. A. & Perego, P. 2009. Grape by-products: extraction of polyphenolic compounds using supercritical CO2 and liquid organic solvent – a preliminary investigation Subproductos de la uva: extracción de compuestos polifenólicos usando CO2 supercrítico y disolventes orgánicos líquidos – una investigación preliminar. *CyTA – Journal of Food*, **7**, 163–171.

Fizet, C. 1996. Process for tocopherols and sterols from natural sources. Patent No. US 5487817 A.

García, A., Rodríguez-Juan, E., Rodríguez-Gutiérrez, G., Rios, J. J. & Fernández-Bolaños, J. 2016. Extraction of phenolic compounds from virgin olive oil by deep eutectic solvents (DESs). *Food Chemistry*, **197**(Part A), 554–561.

García-Granados López De Hierro, Andres Y Martínez Nieto & LEOPOLDO. 1994. Procedimiento de obtenci'on de manitol y productos derivados a partir de alpeorujo proce- dente del proceso de aceituna seg'un el procedimiento de dos fases. Patent No. ES2060549.

Godswill, N. N., Frank, N. E. G., Walter, A. N., Edson, M. Y. J., Kingsley, T. M., Arondel, V., Martin, B. J. & Emmanuel, Y. 2016. Oil palm A2. In: Gupta, S. K. *Breeding Oilseed Crops for Sustainable Production*. San Diego, CA: Academic Press.

Guillén, R., Heredia, A., Felizón, B., Jiménez, A., Montaño, A. & Fernández-Bolaños, J. 1992. Fibre fraction carbohydrates in *Olea europaea* (Gordal and Manzanilla var.). *Food Chemistry*, **44**, 173–178.

Guinda, Á., Castellano, J. M., Santos-Lozano, J. M., Delgado-Hervás, T., Gutiérrez-Adánez, P. & Rada, M. 2015. Determination of major bioactive compounds from olive leaf. *LWT – Food Science and Technology*, **64**, 431–438.

Guthalugu, N. K., Balaraman, M. & KADIMI, U. S. 2006. Optimization of enzymatic hydrolysis of triglycerides in soy deodorized distillate with supercritical carbon dioxide. *Biochemical Engineering Journal*, **29**, 220–226.

Hamm, W., Hamilton, R. J. & Calliauw., G. (eds) 2013. *Edible Oil Processing*. Chichester: John Wiley & Sons.

Hattori, Y. K. C. R. L., Horio, M. K. C. R. L. & Kono, J. K. C. R. L. 2002. Process for producing phytosterols by saponification in an alcohol/water solvent. Patent No. EP1226156A1.

HO, D. S. S. 2009. Recovery of phytonutrients from oils. Patent No. US7544822.

Husveth, F., Manilla, H. A., Gaal, T., Vajdovich, P., Balogh, N., Wagner, L., Loth, I. & Nemeth, K. 2000. Effects of saturated and unsaturated fats with vitamin E supplementation on the antioxidant status of broiler chicken tissues. *Acta Vet Hung*, **48**, 69–79.

Indira, T. N., Narayan, A. V., Barhate, R. S., Raghavarao, K. S. M. S., Khatoon, S., Channaiah, G., Rao, A. R. G. R. A. & Prakash, V. 2005. Process for the production of oryzanol enriched fraction from rice bran oil soapstock. Patent No. US20040192948 A1.

Japon-Lujan, R. & Luque De Castro, M. D. 2007. Static-dynamic superheated liquid extraction of hydroxytyrosol and other biophenols from alperujo (a semisolid residue of the olive oil industry). *Journal of Agricultural and Food Chemistry*, **55**, 3629–3634.

Jerman Klen, T. & Mozetič Vodopivec, B. 2011. Ultrasonic extraction of phenols from olive mill wastewater: comparison with conventional methods. *Journal of Agricultural and Food Chemistry*, **59**, 12 725–12 731.

Jeromin, L., Johannisbauer, W., Gutsche, B., Jordan, V. & Wogatzki, H. 1997. Recovery of tocopherol and sterol from tocopherol and sterol containing mixtures of fats and fat derivatives. Patent No. US 5627289.

Jesus, S. P., Grimaldi, R. & HENSE, H. 2010. Recovery of γ-oryzanol from rice bran oil byproduct using supercritical fluid extraction. *Journal of Supercritical Fluids*, **55**, 149–155.

Juan Fernández-Bolaños, Guillermo Rodríguez, Rocío Rodríguez, Rafael Guillén & Jiménez., A. 2006. Potential use of oilive by-products – extraction of interesting organic compounds from olive oil waste. *Grasas y ceites*, **57**, 95–106.

Kaewboonnum, P., Vechpanich, J., Santiwattana, P. & Shotipruk, A. 2010. γ-oryzanol recovery from rice bran oil soapstock. *Separation Science and Technology*, **45**, 1186–1195.

Kamal-Eldin, A. & Appelqvist, L. A. 1996. The chemistry and antioxidant properties of tocopherols and tocotrienols. *Lipids*, **31**, 671–701.

Khoufi, S., Hamza, M. & Sayadi, S. 2011. Enzymatic hydrolysis of olive wastewater for hydroxytyrosol enrichment. *Bioresource Technology*, **102**, 9050–9058.

Koskenniska, L. A. 1982. A process for the isolation of beta-sitosterol from a steroid mixture. Patent No. EP0057075A1.

Lafka, T. I., Lazou, A. E., Sinanoglou, V. J. & Lazos, E. S. 2011. Phenolic and antioxidant potential of olive oil mill wastes. *Food Chemistry*, **125**, 92–98.

Laufenberg, G., Kunz, B. & Nystroem, M. 2003. Transformation of vegetable waste into value added products: (A) the upgrading concept; (B) practical implementations. *Bioresource Technology*, **87**, 167–198.

LEE, H., CHUNG, B. H. & PARK, Y. H. 1991. Concentration of tocopherols from soybean sludge by supercritical carbon dioxide. *Journal of the American Oil Chemists Society*, **68**, 571–573.

Leouifoudi, I., Zyad, A., Amechrouq, A., Oukerrou, M. A., Mouse, H. A. & Mbarki, M. 2014. Identification and characterisation of phenolic compounds extracted from Moroccan olive mill wastewater. *Food Science and Technology (Campinas)*, **34**, 249–257.

Muktadirul Bari Chowdhury, A. K. M., Akratos, C. S., Vayenas, D. V. & Pavlou, S. 2013. Olive mill waste composting: a review. *International Biodeterioration & Biodegradation*, **85**, 108–119.

Murakami, A. N. N., Amboni, R. D. D. M. C., Prudêncio, E. S., Amante, E. R., Fritzen-Freire, C. B., Boaventura, B. C. B., Muñoz, I. D. B., Branco, C. D. S., Salvador, M. & Maraschin, M. 2013. Concentration of biologically active compounds extracted from *Ilex paraguariensis* St. Hil. by nanofiltration. *Food Chemistry*, **141**, 60–65.

Nagao, T., Kobayashi, T., Hirota, Y., Kitano, M., Kishimoto, N., Fujita, T., Watanabe, Y. & Shimada, Y. 2005. Improvement of a process for purification of tocopherols and sterols from soybean oil deodorizer distillate. *Journal of Molecular Catalysis B: Enzymatic*, **37**, 56–62.

Narayan, A. V., Barhate, R. S., Indira, T. N., Tiku, P. K., Raghavarao, K. S., Rao, A. R. G. R. A. & Prakash, V. 2004. Process for crystallization of oryzanol from oryzanol enriched fraction derived from rice bran oil soap stock. Patent No. WO2004055040A1.

Nguyen, V. T., Pham, H. N. T., Bowyer, M. C., Van Altena, I. A., Scarlett, C. J. 2016. Influence of solvents and novel extraction methods on bioactive compounds and antioxidant capacity of *Phyllanthus amarus*. *Chemical Papers*, **70(5)**, 556–566.

Obied, H. K., Bedgood, D., Mailer, R., Prenzler, P. D. & ROBARDS, K. 2008. Impact of cultivar, harvesting time, and seasonal variation on the content of biophenols in olive mill waste. *Journal of Agricultural and Food Chemistry*, **56**, 8851–8858.

Perretti, G., Miniati, E., Montanari, L. & Fantozzi, P. 2003. Improving the value of rice by-products by SFE. *Journal of Supercritical Fluids*, **26**, 63–71.

Pestana-Bauer, V. R., Zambiazi, R. C., Mendonça, C. R. B., Beneito-Cambra, M. & Ramis-Ramos, G. 2012. γ-oryzanol and tocopherol contents in residues of rice bran oil refining. *Food Chemistry*, **134**, 1479–1483.

Piironen, V., Lindsay, D. G., Miettinen, T. A., Toivo, J. & Lampi, A. M. 2000. Plant sterols: biosynthesis, biological function and their importance to human nutrition. *Journal of the Science of Food and Agriculture*, **80**, 939–966.

Pizzichini, D., Russo, C., Vitagliano, M., Pizzichini, M., Romani, A., Ieri, F., Pinelli, P. & Vignolini, P. 2011. Process for producing concentrated and refined actives from tissues and byproducts of *Olea europaea* with membrane technologies. Patent No. EP2338500A1

Proestos, C. & Komaitis, M. 2008. Application of microwave-assisted extraction to the fast extraction of plant phenolic compounds. *LWT – Food Science and Technology*, **41**, 652–659.

Ren, Q., Xing, H., Bao, Z., Su, B., Yang, Q., Yang, Y. & Zhang, Z. 2013. Recent advances in separation of bioactive natural products. *Chinese Journal of Chemical Engineering*, **21**, 937–952.

Robinson, P. L., Cuff, T. J., Byrne, J. F. & Yan, Z. Q. 2000. Isolation and purification of sterols from neutrals fraction of tall oil pitch by single decantation precipitation. Patent No. WO2000004039A1.

Rodis, P. S., Karathanos, V. T. & Mantzavinou, A. 2002. Partitioning of olive oil antioxidants between oil and water phases. *Journal of Agricultural and Food Chemistry*, **50**, 596–601.

Saito, M. & Yamauchi, Y. 1990. Isolation of tocopherols from wheat germ oil by recycle semi-preparative supercritical fluid chromatography. *Journal of Chromatography A*, **505**, 257–271.

Savva, S. C. & Kafatos, A. 2016. Vegetable oils: dietary importance A2 – Caballero, Benjamin. In: Finglas, P. M. & Toldrá, F. (eds) *Encyclopedia of Food and Health*. Oxford: Academic Press.

Shadabi, S., Ghiasvand, A. & Hashemi, P. 2013. Selective separation of essential phenolic compounds from olive oil mill wastewater using a bulk liquid membrane. *Chemical Papers*, **67**, 730–736.

Silva, F. D. O. & Perrone, D. 2015. Characterization and stability of bioactive compounds from soybean meal. *LWT – Food Science and Technology*, **63**, 992–1000.

Smith, F. E. 1967. Separation of tocopherols and sterols from deodorizer sludge and the like. Patent No. US 3335154.

Smith, F. E. & Winton, B. 1964. Process for separating tocopherols and sterols from deodorizer sludge and the like. Patent No. US 3153055.

Statistics Portal. 2015. http://www.statista.com/statistics/263937/vegetable-oils-global-consumption/ (last accessed 20 March 2017).

Stephane, B., Dominique, C. & David, D. 2005. Sterol recovering method. Patent No. US 2835682.

Tabera, G. J. J., Ruiz, R. A., Senorans, R. F. J., Ibanez, E. E., Reglero, R. G. J., Albi, V. T., Lanzon, R. A., Perez, C. M. C., Guinda, G. M. A. & Rada, R. M. 2005. Procedimiento para obtener compuestos de alto valor añadido a partir de hoja de olivo. Patent No. WO2005075614A1.

Tornberg, E. & Galanakis, C. 2010. Olive waste recovery. Patent No. US 20100240769.

USDA. 2015. Oilseeds: world markets and trade. Availabel from: http://www.fas.usda.gov/data/oilseeds-world-markets-and-trade (last accessed 20 March 2017).

Waldron, K. (ed.) 2007. *Handbook of Waste Management and Co-product Recovery in Food Processing*. Abington: Woodhead.

Wollmann, G., Schwarzer, J. & Gutsche, B. 2005. Processes for producing sterols from fatty acid production residues. Patent No. US 6956125.

Xu, Z. & Godbe, J. S. 2000. Comparison of supercritical fluid and solvent extraction methods in extracting γ-oryzanol from rice bran. *Journal of the American Oil Chemists' Society*, 77, 547–551.

7

Recovering Bioactive Compounds from Cane Sugar Wastes

Nguyen Thi Thao and Hoang Quoc Tuan

Department of Quality Management, School of Biotechnology and Food Technology, Hanoi University of Science and Technology, Hanoi, Vietnam

7.1 Introduction

Sugarcane (*Saccharum oficinarum* Linn.) is a well-known crop of the family Poaceae. It is the main crop used as raw material in the sugar industry, accounting for 70% of total sucrose production in the world. Sugarcane is a perennial grass, indigenous to tropical South Asia and South East Asia. Nowadays, the production of sugarcane extends widely in tropical countries from the northern to the southern hemisphere, with Brazil, China and India leading in sugarcane production. The cane has a thick longitudinal stalk, which is generally 3–5 m in height, approximately 5 cm in diameter and characterised by its sweet taste, due to its high sucrose content. The peel has different colours, from yellow to purple, depending on the variety. The yellow variety has a strong stem with long fibres which are suitable for sugar production at industrial scales. The purple one, on the other hand, is suitable for chewing and use as a dessert, due to its soft structure. Although it is produced across large areas, this purple variety is not suitable for sugar production.

Cane sugar is the main product of the sugarcane processing industry. In 2014/15, global sugar production amounted to approximately 175.1 million tonnes. During this period, Asia was the largest sugar-producing region in the world, yielding approximately 66.12 million tonnes, with India, China and Thailand the region's top producers. Sugar production in Brazil has fluctuated over the past 5 years: in 2010/11, it produced approximately 38.4 million tonnes, but this decreased to roughly 36 million tonnes in 2014/15. In 2013, Brazil was the leading global sugarcane producer, with a production of 739.27 million tonnes. The purity of cane sugar depends on the kind of product it is to be used for. Its colour varies across a wide range, from very white (sucrose purity 99.9%) to yellow or brown (lower purity).

The cane sugar industry generates a large amount of byproducts and wastes. Recently, there has been extensive exploitation of valuable compounds and value-added products produced by this industry. In this chapter, we discuss the valuable bioactive compounds in different parts of sugarcane, their availability in sugar byproducts and wastes and the principles and technologies used to recover them.

Recovering Bioactive Compounds from Agricultural Wastes, First Edition. Edited by Van Tang Nguyen.

7.2 Chemical Composition and Functional Properties of Sugarcane

Sugarcane has long been exploited, thanks to the economical and medicinal value of this high-yield crop. In many tropical countries, sugarcane juice is not only a nutritious drink but also a medicinal therapy used for the treatment of various diseases, including jaundice, haemorrhage, dysuria, anuria and other urinary diseases. Modern research has focused on the investigation of the chemical and phytochemical properties of sugarcane. Singh et al. (2015) have provided a review of the phytochemical profile and potential health benefits of sugarcane and its products. The phytochemistry of sugarcane wax (obtained from the leaves and stalks of sugarcane), leaves and juice, as well as sugarcane products, has revealed the presence of various fatty acids, alcohols, phytosterols, higher terpenoids, flavonoids, O- and C-glycosides and phenolic acids. Overall, phytopharmacological study proves that sugarcane has great potential as a valuable medicinal product due to its rich phytochemical composition.

7.2.1 Sugarcane Juice

Sugarcane juice, which is obtained by grinding and pressing sugarcane culms, is used for the production of sugar (sucrose) and various other valuable products, including raw sugar/brown sugar, jaggery and molasses. Basically, the juice consists of water (75%), sucrose (10–15%) and other nonsugar compounds (parts other than sucrose). Because of the phytochemical and chemical benefits of the nonsugar part, the juice is widely used in India as a therapeutic drink for various different illnesses. Much research has focused on understanding the chemical composition of this nonsugar part, paying particular attention to its phenolic compounds, minerals, reduced sugars (glucose and fructose) and wax.

Phenolic compounds are secondary metabolites that are derivatives of the pentose phosphate, shikimate and phenylpropanoid pathways in plants (Randhir et al., 2004). These compounds, one of the most widely occurring groups of phytochemicals, are of considerable physiological and morphological importance in plants. They play an important role in growth and reproduction, providing protection against pathogens and predators (Bravo, 1998), and contribute towards the colour and sensory characteristics of fruits and vegetables. Three well-known classes of phenolic compounds are polyphenols, flavonoids and tannins. Polyphenols and flavonoids are widely distributed in plants and have significant applications in the health and food industries. These phytochemicals have been known to reduce many chronic diseases, such as cardiovascular and heart diseases, diabetes, obesity and certain cancers, to improve endothelial functions and to reduce blood pressure. These compounds present in large quantities in both edible and non-edible plants. Sugarcane juice is believed to contain considerable amounts of nonsugar compounds or phytochemicals that have important health benefits. With the development of new analytical methods and sophisticated and sensitive instruments, many phenolic compounds has been identified and quantified in sugarcane juice.

The total phenolic content of sugarcane juice, as determined by Duarte-Almeida et al. (2006) using Folin–Ciocalteau reagent, is relatively high: 160 mg catechin equivalents/l. This means that the consumption of a glass of 250 ml sugarcane juice would provide an intake of 40 mg of phenolics. In this way, sugarcane juice represents an important source of these antioxidant compounds in our diet. The phenolic composition of sugarcane juice was determined and quantified by high-performance liquid chromatography (HPLC). Phenolic acids including hydroxycinnamic, sinapic and caffeic acids; flavones such as apigenin, luteolin and tricin; and tricin and its derivatives, mainly the O- and C-glycosides of tricin, account for the highest concentrations. The concentrations of the main phenolic compounds in sugarcane juice are as follows: apigenin (27.8 mmol/l), luteolin (9.2 mmol/l), tricin (35.7 mmol/l), caffeic acid (4.1 mmol/l), hydroxycinnamic acid (85.8 mmol/l) and sinapic acid (25.4 mmol/l). A tricin derivative was found in the highest proportion (>10% of the total). Recently, Kong et al. (2016) used HPLC to determine the gallic, chlorogenic, caffeic and ferulic acid contents in sugarcane extracts: 0.87, 1.77, 11.64 and 10.49 mg/g, respectively.

The flavonoid compounds in sugarcane juice have also been extensively studied. Four new minor flavones – swertisin, tricin-7-O-neohesperoside-4′-O-rhamnoside, tricin-7-O-methylglucuronate-4′-O-rhamnoside and tricin-7-O-methylglucuronide – were isolated and identified (Colombo et al., 2009). In addition, some novel acylated flavone glycosides were also isolated, including tricin-7-O-β-(6′-methoxycinnamic)-glucoside, luteolin-8-C-rhamnosyl glucoside and tricin-4′-O-(erthroguaicylglyceryl)-ether. Along with orientin, the flavones diosmetin-8-C-glucoside (2), vitexin (3) schaftoside (9), isoschaftoside (10) and 4′,5′-dimethyl-luteolin-8-C-glucoside (11) were the most relevant compounds (Vila et al., 2008).

Phenolics exhibit important functionalities, including antioxidant properties, due to the strong correlation between total phenolic content and antioxidant activity. Therefore, products with a high phenolic content such as green tea, olive and grape always display high antioxidant capacity. However, the antioxidant capacity also depends on the composition of the phenolic compounds. The antioxidant activity of sugarcane juice (EC_{50}) was determined using β-carotene and DPPH as the detection reagents, and the main compounds with free radical-scavenging activity in juice and leaves extracts were identified by HPLC-UV/pulsed amperometric detector (PAD) analysis combined with HPLC microfractionation monitored by thin-layer chromatography (TLC). The flavonoid content of sugarcane juice (0.241 ± 0.001 mg total flavonoids/ml juice) is comparable to that of other foods rich in flavonoids, suggesting its potential as a dietary source of natural antioxidants (Vila et al., 2008).

In addition to its antioxidant activity, the phenolic extract obtained from sugarcane juice shows a protective effect against *in vivo* MeHgCl intoxication and potent inhibition of *in vivo* lipoperoxidation of rat brain homogenates, indicating potential beneficial health effects and/or a therapeutic application. One flavone, tricin-7-O-β-(6″-methoxycinnamic)-glucoside, was isolated and identified by spectroscopic methods. It was shown to have higher antioxidant activity than trolox by DPPH assay, but lower activity by a β-carotene/linoleic acid system. It showed *in vitro* antiproliferative activity against several human cancer cell lines, with higher selectivity towards cells of the breast-resistant NIC/ADR line (Duarte-Almeida et al., 2007).

7.2.2 Sugarcane Tops and Leaves

Sugarcane tops and leaves are also important sources of phenolic compounds. A study of sugarcane leaves was carried out using extraction with 50% ethanol and fractionation by petroleum ether, ethyl acetate and n-butyl alcohol (Sun et al., 2014). Eight phenolic compounds in EtOAc extract were purified though silica gel and sephadex LH-20 column chromatography, and then identified by nuclear magnetic resonance (NMR) and electrospray ionisation mass spectrometry (ESI-MS). Eight phenolic compounds from EtOAc extract were identified: caffeic acid, cis-p-hydrocycinnamic acid, quercetin, apigenin, albanin A, australone A, moracin M and 5′-geranyl-5,7,2′,4′-tetrahydroxylflavone. The antioxidant- and nitrite-scavenging capacities of different solvent extracts were evaluated, and EtOAc extract was found to possess the highest total phenolic content and presented the strongest oxygen radical absorbance capacity (ORAC). In sugarcane leaves, luteolin-8-C-(rhamnosylglucoside) (1) was the most important compound with strong radical scavenging activity (Vila et al., 2008). The studies on sugarcane tops and leaves revealed that these wastes could be considered as a source of natural antioxidants.

7.2.3 Sugarcane Wax

Sugarcane wax is a whitish to dark yellowish powdery layer which deposits on the surfaces of sugarcane stalks and leaves. Generally, wax in plants serves as an energy store and has a protective role. The wax layer helps to prevent dehydration by producing a cuticle: a lipophilic layer coating all aerial, primary organs. As the cuticle also forms the plant–environment interface, its primary function is to block nonstomatal water loss (Kerstiens, 1996; Riederer & Schreiber, 2001). The composition of sugarcane wax and its application in the food, pharmaceutical and other industries have been well investigated.

Sugarcane wax accounts for a considerable amount of the total mass of sugarcane leaves and peel. Basically, it is composed of long-chain aliphatic fatty acids and alcohols. It makes up 0.1–0.3% of the total content of sugarcane (depending on its variety and stage of maturity). In peel, crude wax yields about 0.95% by dry weight basis. The composition of sugarcane wax has been well defined by different analytical methods, including TLC, Fourier transform–infrared spectroscopy (FT-IR), gas chromatography–mass spectrometry (GC-MS) and liquid chromatography–mass spectrometry with atmospheric pressure chemical ionisation (LC-MS/APCI). It comprises different classes of compounds, including alkanes (isobutene, 0.35%; triptane, 24.22%; cyclopentane, 2.04%; cyclohexane, 2.07%), fatty acids (propanoic acid, 0.78%; tetradecanoic acid, 0.07%; hexadecanoic acid, 0.75%) and ester as the largest part (sec-butyl isothiocyanate, 66.26%). It also contains aldehyde (pentadecanol, 0.11%) and alcohol (isotridecyl alcohol, 0.22%). The melting point and molecular weight of crude wax are of 62.1 °C and 1706 Da, respectively. GC-MS showed the compostion of crude wax as follows: alkanes (28.83%), ester (66.26%), fatty acids (4.58%), aldehyde (0.11%) and alcohol (0.22%) (Inarkar & Lele, 2012).

The compounds of most interest in sugarcane wax are policosanols, due to their pharmaceutical effects. These are a mixture of long-chain alcohols that are widely known to have beneficial effects on human health, and can therefore be used in low-density lipoprotein (LDL) control or as protectants against atherosclerosis and hepatic injury progression (Gnanaraj, 2012). Other major pharmacologically active components of sugarcane wax are long-chain aliphatic fatty acids, which present at lower concentrations. The mixture of these acids is known as D-003. Although fatty acids and fatty alcohols are reported as the major constituents in sugarcane wax, various phytosterols, steroids and higher terpenoids have also been found.

7.3 Cane Sugar Production and Its Byproducts/Wastes

Sugarcane is versatile crop, considered a source of food, fibre, fodder, fuel, chemicals and fertiliser. The main product of the cane sugar industry is sucrose, which is isolated from sugarcane following various processing stages. Sugarcane is harvested when the sugar content of the juice in the stalk reaches its highest level. After harvesting, the juice has a soluble content around 15–20% and a sugar content of 10–15%, depending on the sugarcane variety and stage of maturity. The cane is then delivered to the sugar mill, where it is crushed and pressed to collect sugarcane juice. The sugar is exhausted by consecutive pressing and then extracted. The fibre (sugarcane bagasse) is removed and the juice is clarified by lime, sulfite, carbonate or phosphate to remove impurities. The first clarification is carried out by decantation, followed by filtration. Finally, the clear juice is separated from solid or precipitated impurities. The precipitated impurities removed from the filtration process are de-watered to become filter cake or press mud.

Clear juice is evaporated to remove water in order to increase the soluble content from 15 to 60 °Brix. This is done through a multiple-effect evaporation system (normally three to five effects or evaporators), where the working pressure in the evaporators is gradually reduced from high pressure to vacuum. After evaporation, sucrose is further concentrated by evaporation at vacuum pressure in a sugar boiler system in order to reach a supersaturated syrup state, which can facilitate the crystallisation of sucrose crystals. Further centrifugation can be applied to separate the sugar crystals from the liquid phase, which will contain high levels of nonsugar fractions. The final sugar product must achieve a purity of at least 99% (depending on the technology used and the intended quality specification).

The final molasses (or blackstrap molasses) is an important byproduct of the sugar industry. It contains a high amount of reduced sugars, phytochemicals (polyphenols) and products of enzymatic and non-enzymatic browning reactions, such as caramelisation reactions and Maillard reactions between reduced sugars and the hydrolysed products of protein (amino acids). The composition and functionality of molasses and of noncentrifuged cane sugar will be discussed in the next section.

The byproducts and wastes of the cane sugar industry are vast potential reserves for human and animal consumption and are capable of providing energy as a renewable source. On average, the processing of 100 tonnes of sugarcane in a factory yields 10 tonnes of sugar, 30–40 tonnes of bagasse, 4.45 tonnes of molasses, 3 tonnes of filter mud (Solomon, 2011), 120 tonnes of flue gases and 1500 kWh of surplus electricity. Byproducts such as bagasse, molasses and press mud are rich sources of carbon compounds and minerals, providing ample opportunity for the extraction of valuable compounds, physicochemical transformation and microbial fermentation to produce value-added products such as building and structural materials, pharmaceuticals and fermentation or enzyme substrates for the production of valuable chemicals, medicines and pesticides (Solomon, 2011).

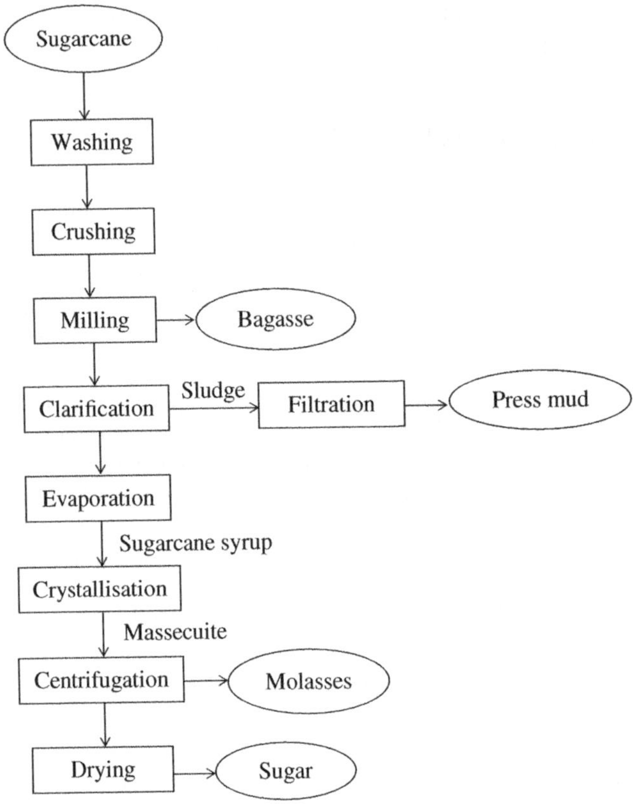

Figure 7.1 Simplified diagram of the main products and byproducts of the cane sugar processing industry.

7.4 Recovering Bioactive Compounds from Cane Sugar Byproducts/Wastes

7.4.1 Potential Bioactive Compounds in Wastes from the Cane Sugar Industry

7.4.1.1 Molasses

Blackstrap molasses (or final molasses) is a byproduct of sugar manufacture. It is a heavy and viscous liquid that must be separated from the final low-grade massecuite, in which no further sugar can be crystallised out by the usual methods. In general, molasses is not used for human consumption, due to its colour and chemical composition. The main constituents of molasses are sucrose (30–35%), glucose and fructose (10–25%), nonsugar compounds (2–3%), water and minerals (Solomon, 2011). It is therefore a rich source of carbohydrate, which can be used for fermentation in alcohol production, carbon dioxide, citric acid, baker's yeast, mono-glutamate, itaconic acid, acetone, butanol, dextran and so on. However, sugarcane molasses is also known as a wealthy source of phenolic compounds that exhibit significant biological properties, including antioxidant, antimutation and anti-inflammatory effects, which provide health benefits.

Recently, a number of researchers have begun looking at the phenolic compounds of sugarcane molasses and their functionality.

Among various products from sugar processing factories, molasses has the highest phenolic content and the highest antioxidant activity. In one study, the total phenolic content was 101.3 mg gallic acid equivalents (GAE)/g of extract, with IC_{50} values for trolox and sugarcane molasses extract of 125.33 μg/ml and 126.0 μg/ml, respectively (Yu et al., 2016). Ten phenolic constituents were identified in the fractions, including known antioxidative compounds, viz., schaftoside, isoschaftoside, ferulic acid, *p*-coumaric acid and *p*-hydroxybenzaldehyde (Asikin et al., 2013). With abundant polyphenols, sugarcane molasses is a rich source of antioxidant compounds with peroxyl radical scavenging effects. The potent antioxidant activity of sugarcane molasses against 2,2′-azobis(2-amidinopropane) dihydrochloride (AAPH)-induced peroxyl radicals has been determined using a fractionation method. Seven methanolic fractions of sugarcane molasses (F1–F7) were separated via bioassay-guided fractionation and evaluated by ORAC, cellular antioxidant activity (CAA) and oxidative DNA damage protective activity assays. The ORAC value of sugarcane molasses fractions ranged from 4399 to 6266 μmol trolox equivalents (TE)/g, whilst the EC_{50} value for CAA ranged from 3.7 to 5.9 μg/ml. Moreover, it was found that sugarcane molasses fractions, particularly F6 and F7, could protect against oxidative DNA damage caused by peroxyl radicals at an effective concentration of 100 μg/ml.

In addition to its antioxidant activity, sugarcane molasses also displays inhibitory effects on α glucosidase and α-amilase. Kong et al. (2016) assessed these effects using a hypoglycaemic model. They extracted hypoglycaemic fractions using macroporous and ion-exchange resins and found a total phenolic content of 0.179 ± 0.003 mg GAE/mg extract. Four phenolic acids were identified and quantified by HPLC: caffeic acid (11.64 mg/g), ferulic acid (10.49 mg/g), chlorogenic acid (1.77 mg/g) and gallic acid (0.87 mg/g). The inhibitory activity of sugarcane molasses was 4.693 mg/ml (K_m = 1.099 ml/mg) and 4.254 mg/ml (K_m = 0.238 ml/mg) for α-glucosidase and α-amylase, respectively, which shows that it could be a useful addition to medicinal preparations as a nutraceutical and functional food for diabetic patients.

Moreover, phenolic compounds from sugarcane molasses possess antibacterial activity against cariogenic bacteria. Colombo et al. (2006) studied the isolation and structural determination of bioactive compounds present in sugarcane molasses. They isolated dehydrodiconiferylalcohol-9′-O-beta-D-glucopyranoside (1) and isoorientin-7, 3′-O-dimethyl ether (2) as antibacterial active compounds against cariogenic bacteria, and elucidated their structures using H-NMR, ^{13}C-NMR and ESI-MS. The activity of these isolated compounds against *Streptococcus mutans* and *Streptococcus sobrinus* was assessed using a minimum inhibitory concentration (MIC) test. The MIC of compounds 1 and 2 against both *S. mutans* and *S. sobrinus* was more than 4 mg/m in both cases (Takara et al., 2007).

7.4.1.2 Bagasse

Bagasse is the fibrous residue left over after sugarcane is crushed. When it comes out of the milling plant, it is called 'mill wet bagasse,' because of its high moisture content (50%). Bagasse accounts for one-third of the cane ground. The approximate composition of mill wet bagasse is: fibre, 46–52%; water, 43–53%; sugar, 2.5–3.0%; and mineral constituents, 0.55%. Because of its high combustibility, it is used as a captive fuel in sugar factories. The appropriate composition of dry bagasse is mainly cellulose, 45% and penstosans, 28%.

Recently, much interest has focused on the phenolic content of bagasse. Xu et al. (2005) carried out a complete determination of the ferulic and *p*-coumaric acids and related compounds in the cell wall of sugarcane bagasse. They performed a mild alkaline hydrolysis of the alkali-soluble lignin preparations and acid hydrolysis of the 90% acidic dioxane-soluble lignin fractions, which released 48.8% of the total ester-linked *p*-coumaric acid and 43.8% of the total esterified ferulic acid from the alkali-soluble lignin fractions and 38.8% of the total ether-linked *p*-coumaric acid and 38.5% of the total ether-linked ferulic acid from the 90% acidic dioxane-soluble lignin preparations. Based on the dry weight, it has been found that sugarcane bagasse contains 1.76% *p*-coumaric acid and 1.29% ferulic acid, as well as minor quantities of related phenols. Further study of solubilised lignin samples through a combination of ultraviolet (UV), Fourier transform infrared (FT-IR) and ^{1}H and ^{13}C NMR spectroscopy confirmed that a predominant amount of *p*-coumaric acid (69.5–76.4%) is ester-linked to the cell wall components, mainly lignin. On the other hand, about half of the ferulic acid (44.0–55.0%) is esterified to the cell wall hemicelluloses, and the remainder is etherified through the phenolic oxygen to the lignin component of the cell wall.

Sugarcane bagasse contains natural compositions that can significantly inhibit foodborne pathogen growth. The phenolic content of sugarcane bagasse has been found to be >4 mg/g dry bagasse, with 470 mg quercetin/g polyphenol. Sugarcane bagasse extract shows bacteriostatic activity against the growth of *Staphylococcus aureus, Listeria monocytogenes, Escherichia coli* and *Salomonella typhimurium*. Additionally, sugarcane bagasse extract can increase the electric conductivity of bacterial cell suspensions, causing cellular leaking of electrolytes. The results of sodium dodecyl sulfate polyacrylamide gel electrophoresis suggest the antibacterial mechanism probably results from damage caused to cellular proteins by sugarcane bagasse extract. Scanning electron microscopy (SEM) and transmission electron microscopy (TEM) show that sugarcane bagasse extract can change its cell morphology and internal structure (Zhao et al., 2015).

The lipophilic phytochemical composition of sugarcane bagasse and straw, the two major residues of the sugarcane processing industry, was investigated in detail by GC-MS. Their lipid compositions were completely different: while the sugarcane bagasse extracts were dominated by n-aldehydes (ca. 48% of all identified lipids) and n-fatty alcohols (ca. 23%), with lower amounts of n-fatty acids (10%) and steroid ketones (14%), the straw extracts were strongly dominated by n-fatty acids (accounting for ca. 60% of all identified compounds) and significant amounts of steroid compounds, particularly sterols (10%) and steroid ketones (14%). Tocopherols and triterpenols were also found, being particularly abundant among the extracts of sugarcane straw. Sugarcane bagasse and straw can thus be considered promising feedstocks for highly valuable phytochemicals for use in the nutraceutical and pharmaceutical industries (del Río et al., 2015).

7.4.1.3 Press Mud

In the manufacture of cane sugar, the precipitated impurities present in the cane juice after removal by filtration in rotary vacuum filters or by batch-type filter presses form a cake. This is produced at a rate of around 3% of the weight of cane in sulfitation factories and 7% in carbonation factories. However, its composition varies according to the locality, cane variety, milling efficiency and method of clarification. Filter cake contains about 50–70% moisture, 5–14% crude wax and fat, 15–30% fibre, 5–15% sugar and 5–15% crude protein, with appreciable amounts of Si, Ca, P_2O_5, MgO, Fe and Mn. Filter cake produced as an impurity during sugar extraction has manifold uses. The appreciable amount of sugarcane wax in press mud makes its use meaningful in pharmaceutical and industrial applications, but in this context we are interested in the bioactive compounds with pharmaceutical application, including:

- **N-triacontanol:** A growth regulator isolated from sugarcane press mud. Sugarcane wax contains 10–15% n-triacontanol. After the wax has been extracted from press mud, it is esterified and the compound is further purified and fractionally crystalised using suitable solvent.
- **Policosanol:** A new, higher aliphatic alcohol, policosanol (an octacosanol) is isolated from sugarcane wax, which has been found to lower the synthesis of cholesterol in the liver, alleviating cardiac problems. This compound has also been reported to increase physical stamina, repair damaged nerve cells and even stimulate sex hormones.
- **Wax:** The wax content in filter mud has attracted considerable interest over the years and has been recovered for commercial use. It is used as a substitute for imported carnauba wax.

7.4.2 Recovering Bioactive Compounds from Cane Sugar Byproducts/Wastes

7.4.2.1 Production of Coumaric Acid from Sugarcane Bagasse

Sugarcane bagasse comprises 19–24% lignin, 27–32% hemicellulose, 32–44% cellulose and 4.5–9.0% ash (Solomon, 2011). Coumaric acid belongs to the phenolic acid class of phenolic compounds in plants. It is highly appreciated in the fields of functional foods and cosmetics thanks to its antioxidant and antibacterial activities. Together with ferulic and *p*-coumaric acids, coumaric acid is a significant component of sugarcane bagasse, because it is an important building block of the cell wall that helps keep the rigid structure of the cane (Xu et al., 2005). Hydrolysis of this linkage by means of alkali or acid leads to the liberation of coumaric acid. This process is illustrated in Figure 7.2.

Phenolic acids were released from sugarcane bagasse by alkali hydrolysis at 30 °C for 4 hours. The alkaline hydrolysates contained freed coumaric acid, other viscous polysaccharides and ultrafiltrated proteins. The permeates were then purified using anion-exchange resin. The phenolic acids bound by the resin were desorbed by a mixture of water–ethanol–HCl solution (36 : 60 : 4) after the resin was washed by water, ethanol and diluted HCl. The combined eluents were concentrated for crystallisation, and the crystals were then filtered and washed using 1% (v/v) HCl. After this purification process, the purity of the products reached 89.7%, based on coumaric acid. Results of HPLC/MS and HPLC analysis using coumaric and ferulic acids as the standards showed that the main component of the purified bagasse hydrolysate was *p*-coumaric acid, rather than ferulic acid. The purified products showed the same antioxidant activity, reducing power and free radical-scavenging capacity as the standard *p*-coumaric acid (Ou et al., 2009). This technology can thus be applied at industrial scale.

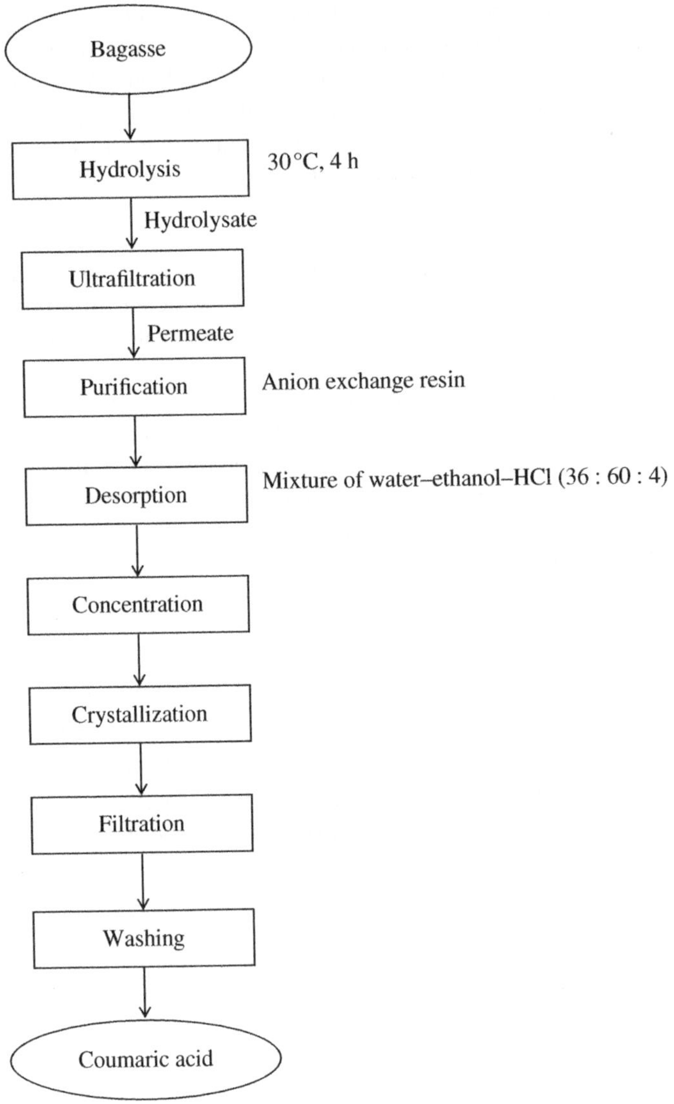

Figure 7.2 Scheme of coumaric acid production by alkaline hydrolysis from sugarcane bagasse.

7.4.2.2 Production of Hemicellulosic Furfural from Sugarcane Bagasse

Uppal & Kaur (2011) studied the chemical composition of sugarcane bagasse and its potential in the production of hemicellulosic furfural. First, the fine bagasse sample of two varieties was prepared and analysed for its chemical composition. The bagasse was then treated with various concentrations (1, 2 and 3 M) of different acids (acetic, formic, phosphoric and nitric) and autoclaved for 180, 300 and 420 minutes with a solid–liquid ratio of 1 : 15 at a temperature of 110 °C and a steam pressure of 1.05 kg/cm (to allow furfural production). Furfural was extracted and identified on the basis of TLC, boiling point, IR and NMR data. The findings indicated that the reaction of bagasse with 3.0 M nitric acid for 300 minutes gave the highest furfural yield (119.94 mg/g).

7.4.2.3 Production of Antioxidants from Sugarcane Molasses

Supercritical carbon dioxide (SC-CO_2) fluid extraction with piecewise distillation separation was used to obtain antioxidants from sugarcane molasses. Extraction pressure, time, temperature, CO_2 flow rate and ethanol concentration were selected as the independent variables. ORAC was used to evaluate the antioxidant activity of the extract. The results showed that the optimal extraction conditions in terms of obtaining the highest ORAC of sugarcane molasses extract were an extraction pressure of 33.3 MPa, a extraction temperature of 43.3 °C, an extraction time of 86.7 minutes, an ethanol concentration of 90% and a CO_2 flow rate of 20 l/h. Under these conditions, the experimental ORAC value was 2584.9 μmol TE/g. This study indicated that SC-CO_2 fluid extraction with piecewise distillation separation can effectively extract the antioxidants from sugarcane molasses (Guan et al., 2014).

7.4.2.4 Recovering Bioactive Compounds from Press Mud

Press mud contains a considerable amount of lipophic materials, including long-chain fatty alcohols and phytosterols. These compounds can be recovered from press mud to use in nutraceuticals or pharmaceuticals. Among them, policosanol is a mixture of alcohols, in which octacosanol can be used as an LDL control or protectant against atherosclerosis, with good tolerance in the human body. Therefore, the extraction of policosanols from press mud by supercritical fluid extraction (SFE) has been extensively studied (Chonde & Raut, 2012; Inarkar & Lele, 2012).

SFE is considered the most effective technique for the extraction of valuable botanicals. SFE is the separation of one component (the extractant) from another (the maxtrix) using supercritical fluids, and CO_2 is the best choice because of its near-ambient critical temperature (31 °C) – biological materials can be processed at temperature around 35 °C. The density of the SC-CO_2 at a pressure around 200 bar is close to that of hexane, and its solvation characteristics are also similar to those of hexane, so that it can act as a nonpolar solvent. CO_2 is a Generally Recognised as Safe (GRAS) solvent and does not leave residues in the exhausted raw material, which is particularly important in the food and pharmaceutical industries. It is a powerful solvent for a wide range of compounds, relatively inert, inexpensive, nonflammable, recyclable and available at high purities. Themes that have focused on the extraction of policosanols from sugarcane press mud are as follows:

- Preparation of crude wax from sugarcane press mud.
- SFE of long-chain n-alcohols from sugarcane crude wax.
- Purification of policosanols from crude sugarcane wax using SC-CO_2 extraction.
- Extraction of microcrystalline wax from press mud waste.

Preparation of crude wax is performed as follows. The filter mud is dried to a water content of about 10% and the crude wax is extracted with hot petroleum solvent. The yield of wax from dried mud, at both laboratory and pilot scales, was varied from factory to factory in a range from 5.1 to 18.0% by dry weight and from 4.5 to 16.4% by dry weight for the laboratory and pilot scale, respectively. This outcome revealed that the extraction at larger scale was less efficient than that at smaller scale but its extraction time was shorter. The colour of crude wax is generally dark brown or green, due to the presence of pigments or phytochemicals (Carter et al., 1954). In order to partly remove pigments or impurities from the crude wax, the wax can be bleached by oxidation with chromic acid. Oxidation with chromic acid is carried out in two stages: the first is completed for 5 hours, but some waxes require longer (13–14 hours). Bleached waxes are normally very pale in colour. The properties of crude waxes vary widely from mill to mill.

Recently, SFE has been applied to the extraction of valuable bioactive compounds from sugarcane crude wax. De Lucas et al. (2007) studied the SC-CO_2 extraction of long-chain alcohols from sugarcane crude wax and found the best extraction conditions consisted of an extraction pressure of 350 bar, an extraction temperature of 100 °C and a KOH concentration of 20% by dried crude wax. The results from this study were comparable to those obtained by the existing organic solvent extraction with a multiple-stage refining process. In particular, the extraction yield using SC-CO_2 extraction (78.24% by weight) was considerably higher than that using organic solvent (22.00% by weight), with a higher purity of alcohols.

The extraction of crude wax by SFE yielded a significant proportion of policosanol. However, in order to obtain pure policosanol, further purification steps should be applied. Sittiwanichai et al. (2016) investigated the purification of policosanol from crude sugarcane wax using SC-CO_2 extraction. In general, crude sugarcane wax contains many impurities, including phytosterols, chlorophyll and other chemicals; therefore, it can be extracted using solvents with different polarity indexes, including polar protic solvents (i.e. water, 9.0; methanol, 5.1; ethanol, 5.2; isopropanol, 3.9), polar aprotic solvents (i.e. acetonitrile, 5.8; acetone, 5.1; ethyl acetate, 4.4), weak polar solvents (i.e. toluene, 2.4) and nonpolar solvents (i.e. hexane, 0.0). Solvents with a wide range of dielectric constants, including polar protic solvents (such as water, 80.37; methanol, 32.63; and ethanol, 24.3), polar aprotic solvents (such as acetonitrile, 37.5; acetone, 20.7; and ethyl acetate, 6.02), weak polar solvents (such as toluene, 2.38) and nonpolar solvents (such as hexane, 1.88) are normally selected (Maryott & Smith, 1951). The yields obtained from precipitation by various solvent systems were collected from each experiment. Water extraction obtained the highest yield (79.31% by weight), but phytosterols and other impurities could not be removed from crude wax. Other polar solvents, such as ethanol, isopropanol, acetone and ethyl acetate, and nonpolar solvents, such as hexane, were capable of removing phytosterols and some impurities. Crude sugarcane wax contained three phytosterols: campesterol, stigmasterol and beta-sitosterol. Policosanol composition was mainly found to be a mixture of tetracosanol (C_{24}), isopropanol, hexacosanol (C_{26}), triacontanol (C_{23}), octacosanol (C_{28}) and some docosanol (C_{22}), according to comparison of the peak areas to the standards. Crude sugarcane waxes contained many compounds, including policosanols, phytosterols and other impurities. Most impurities were removed from the crude sugarcane wax and policosanol compositions were enriched by octacosanol. Research shows that ethyl acetate extraction has many advantages over acetone as solvent in terms of providing greater policosanol content and a higher yield of waxes. The yields of tetracosanol (C_{24}), hexacosanol (C_{26}) and octacosanol (C_{28}) extracted by acetone and isopropanol were 90.63, 75.00 and 97.63% by weight, respectively. Solvent purification can be improved by repeating the purification step with the same solvent and utilising mixed solvents of ethyl acetate and acetone; this may be more effective at removing chlorophylls and a number of impurities than the use of a single solvent.

In a study of natural higher aliphatic alcohol (HAA) from sugarcane wax, purified HAA and its higher fatty acids (HFA) byproducts were extracted and purified through a series of processes involving saponification and hydrolysis with alkali, calcification, Soxhlet extraction and crystallisation by gradient cooling (Chen et al., 2007). The extraction process was described and optimised using an orthogonal experimental design. Based on identification and quantification processes using GC-MS, an HAA purity up to 90.7% was achieved. The total content of the main bioactive components, including a considerable number of carbon atoms from C22 to C30, could reach up to 40.5%.

In order to get wax with high purity, a study of the extraction of microcrystalline wax was carried out by Phukan & Boruah (1999), who developed a method for isolating and refining microcrystalline wax from press mud waste. Physicochemical and X-ray powder diffraction (XRD) analyses were applied to characterise isolated microwax. The yield of hard wax was obtained in a range from 3.50 to 4.10%. The colour of the microwax was light cream, and it had a nominal acid value and saponification number. The degree of crystallinity (%) of the wax was found to be 31% for the refined variety and 34% for the doubly refined variety. This method allows refined sugarcane microwax with a high grade to be obtained.

7.5 Conclusion

The sugarcane processing industry generates a huge amount of bagasse, molasses and press mud, which have great economic value. Bioactive compounds with health benefits can be recovered from these byproducts and wastes. Technological development has facilitated the efficient extraction of these valuable compounds.

References

Asikin, Y., Takahashi, M., Mishima, T., Mizu, M., Takara, K. & Wada, K. 2013. Antioxidant activity of sugarcane molasses against 2,2′-azobis(2-amidinopropane) dihydrochloride-induced peroxyl radicals. *Food Chemistry*, **141**, 466–472.

Bravo, L. 1998. Polyphenols: chemistry, dietary sources, metabolism, and nutritional significance. *Nutrition Reviews*, **56**, 317–333.

Chonde, S. & Raut, P. D. 2012. Studies on extraction of sugarcane wax from press mud of sugar factories from Kolhapur District, MAHARASHTRA. *Journal of Environmental Research and Development*, **6**, 715–720.

Colombo R., Lanças F. M. & Yariwake, J. H. 2006. Determination of flavonoids in cultivated sugarcane leaves, bagasse, juice and in transgenic sugarcane by liquid chromatography-UV detection. *Journal of Chromatography A*, **1103**, 118–124.

Colombo, R., Yariwake, J. H., Queiroz, E. F., Ndjoko, K. & Hostettmann, K. 2009. On-line identification of minor flavones from sugarcane juice by LC/UV/MS and post-column derivatization. *Journal of the Brazilian Chemical Society*, **20**, 1574–1579.

De Lucas, A., García, A., Alvarez, A. & Gracia, I. 2007. Supercritical extraction of long chain n-alcohols from sugar cane crude wax. *Journal of Supercritical Fluids*, **41**, 267–271.

Del Río, J. C., Marques, G., Lino, A. G., Lima, C. F., Colodette, J. L. & Gutiérrez, A. 2015. Lipophilic phytochemicals from sugarcane bagasse and straw. *Industrial Crops and Products*, 77, 992–1000.

Duarte-Almeida, J. M., Novoa, A. V., Linares, A. F., Lajolo, F. M. & Genovese, M. I. 2006. Antioxidant activity of phenolics compounds from sugar cane (*Saccharum officinarum* L.) juice. *Plant Foods for Human Nutrition*, **61**, 187–192.

Duarte-Almeida, J. M., Negri, G., Salatino, A., De Carvalho, J. E. & LAJOLO, F. M. 2007. Antiproliferative and antioxidant activities of a tricin acylated glycoside from sugarcane (*Saccharum officinarum*) juice. *Phytochemistry*, **68**, 1165–1171.

Carter, G. B., Hunter, P. M. & Lamberton, J. A. 1954. Sugarcane wax. A survey of crude waxes and their bleaching with chromic acid. Studies in waxes. Part III. *Proceedings of the QSSCT Twentieth Conference*.

Chen, G. L., Guo, H. R., Wang, X. F. & Li, Y. R. 2007. Extraction and characterization of natural higher aliphatic alcohol from sugarcane wax *Sugar Tech*, **9**, 244–248.

Gnanaraj, R. A. 2012. Applications of sugarcane wax and its products: a review. *International Journal of ChemTech Research*, **4**, 705–712.

Guan, Y., Tang, Q., Fu, X., Yu, S., Wu, S. & Chen, M. 2014. Preparation of antioxidants from sugarcane molasses. *Food Chemistry*, **152**, 552–557.

Inarkar, M. B. & Lele, S. S. 2012. Extraction and characterization of sugarcane peel wax. *ISRN Agronomy*, **2012**.

Kerstiens, G. 1996. Cuticular water permeability and its physiological significance. *Journal of Experimental Botany*, **47**, 1813–1832.

Kong, F., Yu, S., Zeng, F. & Wu, X. 2016. Phenolics content and inhibitory effect of sugarcane molasses on α-glucosidase and α-amylase in vitro. *Sugar Tech*, 1–7.

Maryott, A. A. & Smith, E. R. 1951. Table of dielectric constants of pure liquids. National Bureau of Standards Circular 514. Washington, DC: US Government Printing Office.

Ou, S. Y., Luo, Y. L., Huang, C. H. & Jackson, M. 2009. Production of coumaric acid from sugarcane bagasse. *Innovative Food Science & Emerging Technologies*, **10**, 253–259.

Phukan, A. C. & Boruah, R. K. 1999. Extraction and evaluation of microcrystalline wax from press mud waste of the sugar industry. *Separation and Purification Technology*, **17**, 189–194.

Randhir, R., Lin, Y. T. & Shetty, K. 2004. Phenolics, their antioxidant and antimicrobial activity in dark germinated fenugreek sprouts in response to peptide and phytochemical elicitors. *Asia Pacific Journal Clinical Nutrition*, **13**, 295–307.

Riederer, M. & Schreiber, L. 2001. Protecting against water loss: analysis of the barrier properties of plant cuticles. *Journal of Experimental Botany*, **52**, 2023–2032.

Singh, A., Lal, U. R., Mukhtar, H. M., Singh, P. S., Shah, G. & Dhawan, R. K. 2015. Phytochemical profile of sugarcane and its potential health aspects. *Pharmacognosy Reviews*, **9**, 45–54.

Sittiwanichai, S., Boonyarattanakalin, S., Jarussophon, S., Sramala, I., Pinket, W. & Kasemwong, K. 2016. Purification of policosanol from crude sugarcane wax obtained from supercritical CO_2 extraction. In: Sripanidkulchai, B. (ed.). The 6th International Conference on Natural Products for Health and Beauty (NATPRO6), Khon Kaen, Thailand, 29–33.

Solomon, S. 2011. Sugarcane by-products based industries in India. *Sugar Tech*, **13**, 408–416.

Sun, J., He, X. M., Zhao, M. M., Li, L., Li, C. B. & Dong, Y. 2014. Antioxidant and nitrite-scavenging capacities of phenolic compounds from sugarcane (*Saccharum officinarum* L.) tops. *Molecules*, **19**, 13 147–13 160.

Takara, K., Ushijima, K., Wada, K., Iwasaki, H. & Yamashita, M. 2007. Phenolic compounds from sugarcane molasses possessing antibacterial activity against cariogenic bacteria. *Journal of Oleo Science*, **56**, 611–614.

Uppal, S. K. & Kaur, R. 2011. Hemicellulosic furfural production from sugarcane bagasse using different acids. *Sugar Tech*, **13**, 166–169.

Vila, F. C., Colombo, R., Lira, T. O. D. & Yariwake, J. H. 2008. HPLC microfractionation of flavones and antioxidant (radical scavenging) activity of *Saccharum officinarum* L. *Journal of the Brazilian Chemical Society*, **19**, 903–908.

Xu, F., Sun, R. C., Sun, J. X., Liu, C. F., He, B. H. & Fan, J. S. 2005. Determination of cell wall ferulic and p-coumaric acids in sugarcane bagasse. *Analytica Chimica Acta*, **552**, 207–217.

Yu, D., Chen, M. S. & Yu, S. J. 2016. Effect of sugarcane molasses extract on the formation of 2-amino-1-methyl-6-phenylimidazo[4,5-b]pyridine (PhIP) in a model system. *Food Chemistry*, **197**(Part A), 924–929.

Zhao, Y., Chen, M., Zhao, Z. & Yu, S. 2015. The antibiotic activity and mechanisms of sugarcane (*Saccharum officinarum* L.) bagasse extract against food-borne pathogens. *Food Chemistry*, **185**, 112–118.

8

Recovering Bioactive Compounds from Starch Wastes

Thang Trung Khong[1] *and Van Tang Nguyen*[2,3]

[1] *Nha Trang University, Nha Trang, Khanh Hoa, Vietnam*
[2] *School of Environmental and Life Sciences, Faculty of Science and Information Technology, University of Newcastle, Ourimbah, NSW, Australia*
[3] *Department of Food Technology, Faculty of Food Technology, Nha Trang University, Nha Trang, Khanh Hoa, Vietnam*

8.1 Introduction

Agricultural residues and byproducts/wastes are known as an abundant source of bioactive compounds, as well as of cellulose, hemicellulose and starch, which can all be converted into bioactive compounds. These bioactive compounds have applications in the nutraceutical, medical, pharmaceutical and cosmetic industries, as well as in the production of intermediate products and the development of biomaterials (Cardenas-Toro et al., 2014)

According to the Food and Agriculture Organization of the United Nations (FAO), the total harvested area (million ha), total production (million tonnes) and total gross production value (billion US$) of major starch crops in 2013 were 165, 741 and 429.27 for rice; 219, 716 and 242.25 for wheat; 185, 1020 and 382.34 for maize; 19.34, 376 and 149.51 for potatoes; 20.39, 277 and 47.31 for cassava; and 49.15, 144 and 36.30 for barley (FAOSTAT, 2015). Among natural-origin foods, starch products make up the largest part of human and animal diet. A large amount of residues/byproducts of the starch food industry are discarded to the environment, causing serious pollution if not treated properly. These wastes are considered one of the most abundant and cheapest raw materials on the planet.

Cardenas-Toro et al. (2014) indicated recovery of mono- and oligosaccharides from starch wastes or residues could provide a sustainable source of these substances. Recently, many valuable bioactive compounds have been derived from starch wastes, including polyphenols and alkaloids (Hossain et al., 2014, 2015; Huang & Lai, 2016; Maldonado et al., 2014; Singh & Saldaña, 2011; Vadivel & Brindha, 2015; Wanyo et al., 2016; Wijngaard et al., 2012), peptides (Dei Piu et al., 2014), γ-oryzanol and β-glucans (Baiano, 2014), xylose, lignin, solanesol, β-sitosterol, carotenoids and quinones (Santana-Méridas et al., 2012). The utilisation of starch wastes for the recovery of bioactive compounds as higher value-added products is of great interest to researchers and producers as a means of increasing the economic value of starch production crops and starch-origin processing and production, reducing pollution risks for the

Recovering Bioactive Compounds from Agricultural Wastes, First Edition. Edited by Van Tang Nguyen.

environmental and promoting the development of related industries, such as functional foods, medicines, pharmaceuticals and cosmetics.

8.2 Recovery of Bioactive Compounds from Potato Wastes

Figure 8.1 shows a flow chart of the potato chip and flour production process and its main wastes.

Hossain et al. (2014) reported that the potato processing industry generates about 70–140 thousand tons of peels worldwide annually, which contain a group of compounds that are toxic to normal dividing cells, known as steroidal alkaloids. These compounds may also possess beneficial properties, such as anticancer and anti-inflammatory effects. Thus, many studies have been performed to extract and isolate polyphenols from potato wastes. Singh & Saldaña (2011) applied subcritical

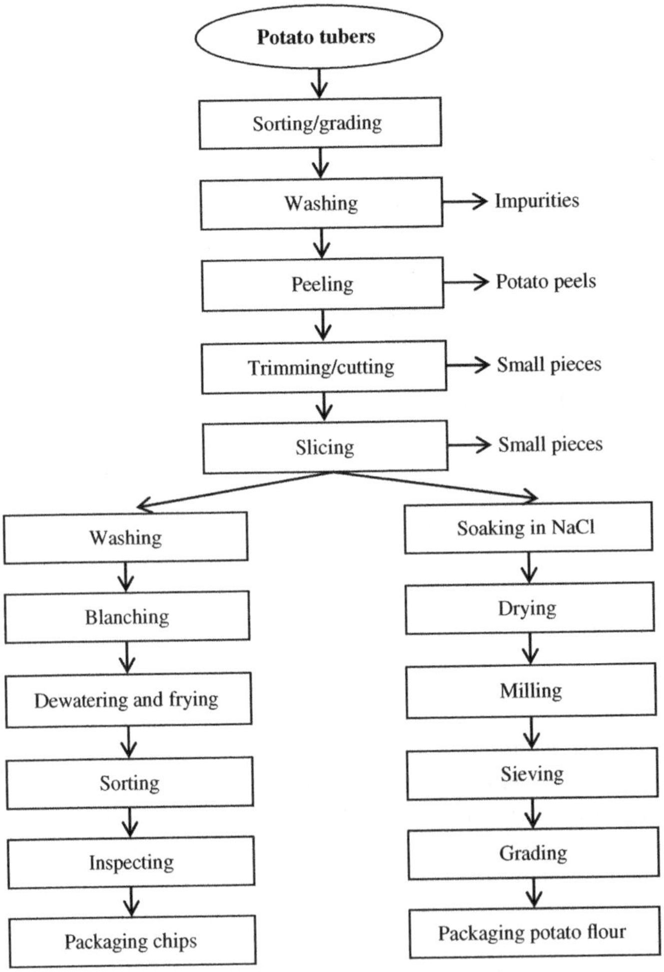

Figure 8.1 Flow chart of the potato chip and flour production process.

water extraction (SWE) to isolate the phenolic compounds from potato peel and found that an optimal extraction (81.83 mg/100 g wet basis) was obtained with a temperature of 180 °C, a pressure of 6 MPa, a flow rate of 2 ml/min and an extraction time of 30 minutes. By comparison, conventional extraction with pure methanol or 50% ethanol for 3 hours produced 46.59 and 29.5 mg/100 g wet basis, respectively (Tables 8.1 and 8.2). SWE also required about 50% less solvent (160 vs 300 ml). Gallic acid (29.56 mg/100 g wet basis) and chlorogenic acid (14.59 mg/100 g wet basis) were the main phenolic compounds obtained from potato peel by SWE at 180 °C (Table 8.3). The results show that SWE at 160–180 °C, 6 MPa and 60 minutes could be a good alternative to conventional extraction with organic solvents (methanol and ethanol) in order to obtain high levels of phenolic compounds from potato peel.

Wijngaard et al. (2012) optimised the conditions for the extraction of antioxidant compounds from potato peel using solid–liquid extraction (SLE) and pressurised liquid extraction (PLE). The findings showed that the optimal SLE conditions were 75% ethanol, 80 °C and 22 minutes, which resulted in the highest contents of phenolics and caffeic acid (394 mg gallic acid equivalents (GAE)/100 g dry weight and 651 μg/g dry weight, respectively) and an optimum antioxidant activity of 352 mg trolox equivalents (TE)/100 g dry weight, while the optimal PLE conditions were 70% ethanol and 125 °C, which resulted in the greatest levels of phenolics and caffeic acid (368 mg GAE/100 g dry weight and 563 μg/g dry weight, respectively) and an optimum antioxidant activity of 339 mg TE/100 g dry weight. Therefore, the use of PLE could not significantly improve the extraction efficiency of polyphenols from potato peel in comparison with SLE, but using aqueous ethanol as the extraction solvent could recover a higher level of polyphenols compared to the use of 100% methanol, because of the increased polarity of the solvent mixture (Table 8.4).

Maldonado et al. (2014) studied the extraction and fractionation of phenolic acids and glycoalkaloids from potato peels using acidified water/ethanol-based solvents and found that the methanolic extracts from potato peel contained mixtures of phenolic acids and glycoalkaloids. Potato peel extract (per 100 g fresh weight) consisted of 17.0 mg α-chaconine, 7.1 mg α-solanine, 0.1 mg solanidine, 4.8 mg caffeic acid, 13.3 mg

Table 8.1 Phenolic compounds (mg/100 g fresh weight) extracted from potato peel using different organic solvents. *Source:* Singh & Saldaña (2011). Reproduced with permission of Elsevier.

Phenolic compound	Methanol	Ethanol (50%)
CFA	12.22 ± 0.15	5.18 ± 0.10
CGA	28.56 ± 0.17	8.67 ± 0.12
CMA	0.83 ± 0.01	6.50 ± 0.12
FRA	1.50 ± 0.02	4.90 ± 0.10
GAC	0.46 ± 0.01	0.60 ± 0.01
PCA	1.94 ± 0.01	1.97 ± 0.02
SGA	0.85 ± 0.02	1.70 ± 0.02
Total	46.36	29.52

CFA, caffeic acid; CGA, chlorogenic acid; CMA, coumaric acid; FRA, ferulic acid; GAC, gallic acid; PCA, protocatechuic acid; PBA, *p*-hydroxyl benzoic acid; SGA, syringic acid.

Table 8.2 Total phenolic content (TPC) (mg/100 g) from various potato peel varieties. *Source:* Singh & Saldaña (2011). Reproduced with permission of Elsevier.

Potato varieties	Extraction solvent	Phenolic compound	Total phenolic compounds
Siecle, Vivaladi, Yukon, Purple, Majesty and Dakota Pearl	Methanol	CGA and CFA	151.0–333.0 (db)
Kennebec, Norchip, Russet Burbank, Red Norland, Red Pontiac and Viking	Methanol (95%)	GAC, PCA, VNA, CFA, CGA, PBA, CMA and FRA	48.0–821.0 (db)
Sieglinde Frances, Nicole, ISC1 4052 and ISCI 67	Methanol (80%)	CGA, CFA, CMA and FRA	221.01–427.0 (db)
Sumi	Ethanol (80%)	CGA and CFA	7.0–10.0 (wb)
Kufri chandermukhi	Ethanol	CGA	70.0 (wb)
Industry sample (n.i.)	Methanol	GAC, PAC, CGA and CFA	41.6 (wb)
Red	Methanol	CGA, CFA, PCA, GAC, SGA, CMA and FRA	46.4 (wb)
Red	Ethanol (50%)	CGA, CFA, PCA, GAC, SGA, CMA and FRA	29.5 (wb)

wb, wet base; dw, dry base; n.i., not indicated; CGA, chlorogenic acid; CFA, caffeic acid; GAC, gallic acid; PCA, protocatechuic acid; CMA, coumaric acid; FRA, ferulic acid; SGA, syringic acid; VNA, vanillic acid; PBA, *p*-hydroxy benzoic acid.

neochlorogenic acid and 77.6 mg chlorogenic acid, with more than 90% of these compounds recovered after two consecutive extractions. Hydrolysis followed by solid-phase fractionation of the crude potato peel extract allowed 139 µmol caffeic acid/100 g fresh weight to be recovered. Partial degradation of caffeic acid and glycoalkaloids was found during this process (chemical structures are shown in Figures 8.2 and 8.3), in which the degradation of caffeic acid was likely mitigated by the addition of antioxidants and metal chelators. The method developed in this study allows for the sustainable recovery of secondary plant metabolites from potato peels and their fractionations using food-grade water/ethanolic solvents, the potential application of phenolic-enriched extracts free of toxic glycoalkaloids for food preservation and the application of glycoalkaloid-enriched extracts for synthesis of pharmaceuticals.

Hossain et al. (2014) used both ultrasound-assisted extraction (UAE) and SLE with methanol as solvent to extract steroidal alkaloids from potato peel waste. They identified the optimal conditions for UAE of α-solanine, α-chaconine, solanidine and demissidine from potato peel as amplitude 61 µm and extraction time 17 minutes, which resulted in the recovery of 1102 µg steroidal alkaloids/g dried potato peel (DPP) (Table 8.5). SLE yielded 710.51 µg glycoalkaloid/g DPP under the same conditions. Individual glycoalkoid contents using UAE were found to be 273, 542.7, 231 and 55.3 µg/g DPP for α-solanine, α-chaconine, solanidine and demissidine, respectively – much higher than those achieved by SLE (180.3, 337.6, 160.2 and 32.4 µg/g DPP, respectively). Predicted models were highly significant for all parameters studied, and the values predicted by the models were not significantly different to the experimental values. This study indicates that UAE is a promising low-cost and environmentally friendly choice for the effective

Table 8.3 Phenolic compounds (mg/100 g wet basic) extracted with subcritical water at different temperatures. *Source:* Singh & Saldaña (2011). Reproduced with permission of Elsevier.

Compound	100 °C	120 °C	140 °C	160 °C	180 °C	200 °C	220 °C	240 °C
CFA	6.15 ± 0.49	8.98 ± 0.33	8.63 ± 0.63	8.73 ± 0.45	9.23 ± 0.73	5.83 ± 0.12	2.13 ± 0.06	2.10 ± 0.01
CGA	6.11 ± 0.26	12.11 ± 0.97	15.00 ± 0.75	15.75 ± 0.93	14.59 ± 0.98	11.21 ± 0.33	6.09 ± 0.10	6.15 ± 0.60
GAC	7.23 ± 0.97	11.17 ± 0.65	18.30 ± 0.97	20.80 ± 1.13	29.56 ± 0.79	25.11 ± 1.59	19.87 ± 1.15	17.27 ± 1.19
PCA	1.16 ± 0.07	3.31 ± 0.03	9.32 ± 0.10	9.41 ± 0.46	13.58 ± 0.13	–	–	–
PBA	–	0.69 ± 0.08	2.17 ± 0.06	4.12 ± 0.09	11.18 ± 0.16	–	–	–
SGA	1.91 ± 0.20	2.23 ± 0.09	3.63 ± 0.06	3.61 ± 0.63	3.69 ± 0.06	2.23 ± 0.21	3.63 ± 0.11	3.61 ± 0.01
Total	22.56	38.49	57.05	62.42	81.83	44.38	31.72	29.13

CFA, caffeic acid; CGA, chlorogenic acid; CMA, coumaric acid; FRA, ferulic acid; GAC, gallic acid; PCA, protocatechuic acid; PBA, *p*-hydroxyl benzoic acid; SGA, syringic acid.

Table 8.4 Optimal conditions for extracting antioxidant compounds using three different extraction methods: DPPH radical scavenging activity (DPPH, mg TE/100 g dry weight), phenolic content (FCR, mg GAE/100 g dry weight) and caffeic acid content (CA, μg caffeic acid/g dry weight). *Source:* Wijngaard et al. (2012). Reproduced with permission of Elsevier.

	Response 1		Response 2		Response 3	
	DPPH		FCR		CA	
	Pred.	Actual	Pred.	Actual	Pred.	Actual
Conventional extracts: methanolic extracts	NA	200 ± 4	NA	126 ± 7	NA	465 ± 13
Solid–liquid extraction (SLE): 75% ethanol, 80 °C, 22 minutes	355	352 ± 0	409	394 ± 1	606	651 ± 9
Pressurised liquid extraction (PLE): 70% ethanol, 125 °C	365	339 ± 5	431	368 ± 0	573	563 ± 10

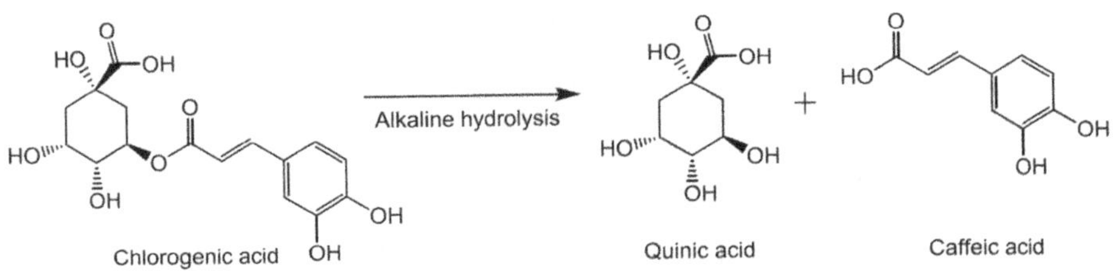

Figure 8.2 Products of the alkaline hydrolysis of chlorogenic acid. *Source:* Maldonado et al. (2014). Reproduced with permission of Elsevier.

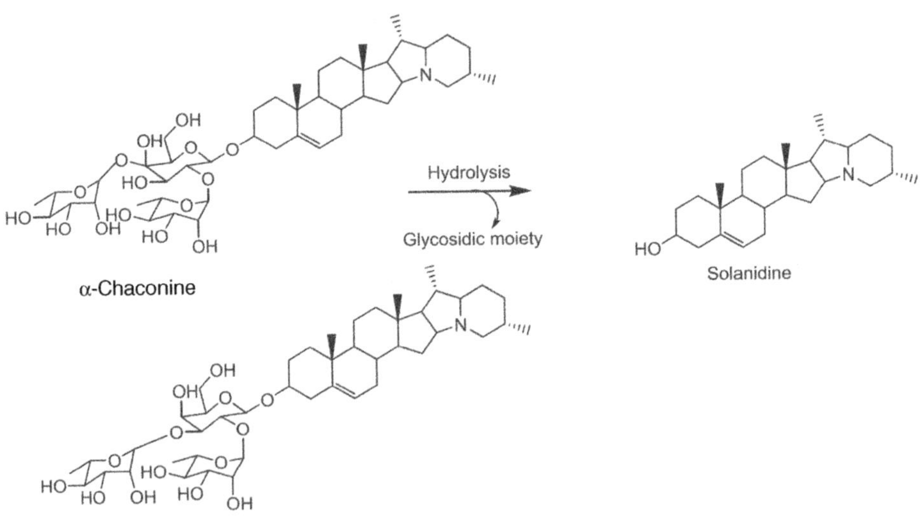

Figure 8.3 Product of the hydrolysis of α-chaconine and α-solanine. *Source:* Maldonado et al. (2014). Reproduced with permission of Elsevier.

Table 8.5 Steroidal alkaloid contents under optimal ultrasound-assisted extraction (UAE) conditions in comparison to conventional solid–liquid extraction (SLE). *Source:* Hossain et al. (2014). Reproduced with permission of Elsevier.

Parameter	Optimum UAE conditions for all parameters combined	Predicted values at optimal UAE	Desirability	Experimental values at optimal UAE	(*E*%)	Solid–liquid extraction values
α-solanine (μg/g DW)	61 μm and 17 minutes	263.0	0.934	273 ± 3.20	3.80	180.3 ± 2.11
α-chaconine (μg/g DW)		534.7		542.7 ± 5.12	1.49	337.6 ± 4.15
Solanidine (μg/g DW)		239.0		231 ± 4.32	3.34	160.2 ± 3.09
Demissidine (μg/g DW)		54.1		55.3 ± 1.56	2.21	32.41 ± 1.11
Total steroidal alkaloid (μg/g DW)		1090.8		1102 ± 1.02	2.32	710.5 ± 6.04
Yield (mg/g DW)		19.0		19.6 ± 0.56	3.15	14.56 ± 0.78

DW, dry weight.

extraction of steroidal alkaloids, which have applications in the medical and pharmaceutical industries.

Hossain et al. (2015) applied PLE to the recovery of steroidal alkaloids from potato peels and found that among eleven solvents tested, methanol obtained the highest amounts of α-solanine, solanidine, demissidine, total steroidal alkaloids and total glycoalkaloids (4.29, 2.40, 0.15 and 10.12 μg/mg dried extract and 761.02 μg/g DPP, respectively). A higher yield of glycoalkaloids could be achieved from potato peels using PLE (1.92 mg/g DPP) as compared to conventional SLE (0.981 mg/g DPP). The optimal PLE conditions for obtaining the highest levels of glycoalkaloids from potato peels in terms of temperature and solvent concentration (methanol) were 80 °C and 89% methanol. Under these conditions, the individual steroidal alkaloid contents achieved were 597, 873, 374 and 75 μg/g DPP for α-solanine, α-chaconine, solanidine and demissidine, respectively – 59, 46, 40 and 52% higher than the values obtained by SLE. It may therefore be a useful strategy to use the steroidal alkaloids for pharmaceutical applications.

8.3 Recovery of Bioactive Compounds from Rice and Wheat Wastes

Figure 8.4 shows a flow chart of the rice and rice flour production process and its main wastes.

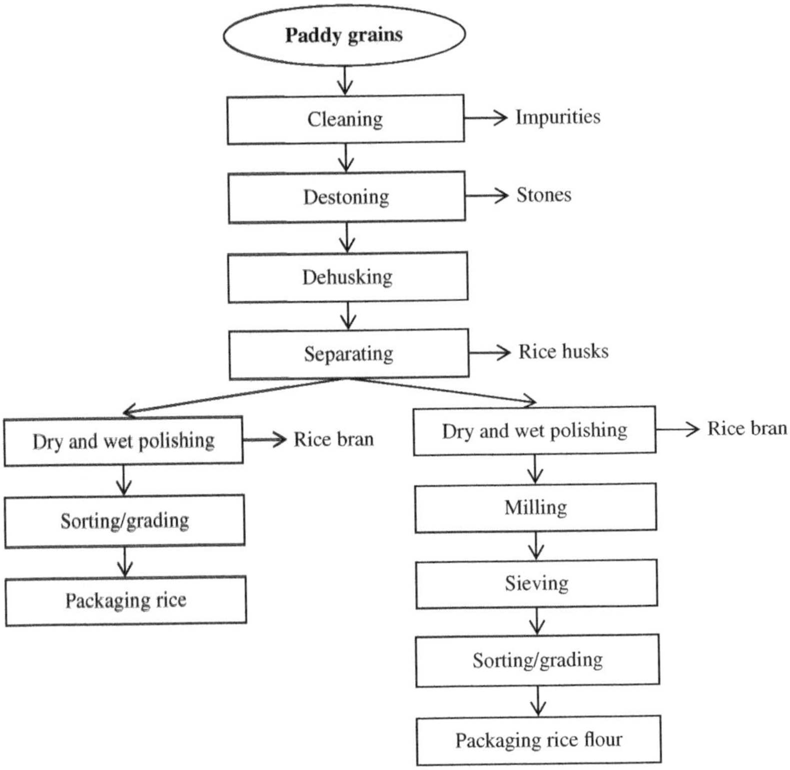

Figure 8.4 Flow chart of the rice and rice flour production process.

Vadivel & Brindha (2015) studied the recovery of polyphenols and antioxidant properties from rice husks using different solvent extraction conditions (Figure 8.5). They found that among the various methods used, grinding in a mortar and pestle with 75% ethanol achieved the greatest yield of total phenolics (240.14 mg ferulic acid equivalents (FAE)/l (Table 8.6). Alkali hydrolysis (932.14 mg FAE/l) recovered more phenolic compounds than acid hydrolysis (674.29 mg FAE/l) (Table 8.7). Alkali hydrolysate displayed higher antioxidant activity in terms of radical scavenging capacity against DPPH (78.96%), superoxide (85.22%), hydroxyl radicals (71.54%) and hydrogen peroxide (78.64%). Moreover, ferric-reducing power (0.739 absorption at 700 nm) and phosphomolybdate assays (1458 ascorbic acid equivalent antioxidant activity (AEAA)/mg sample) also proved the good antioxidant activity of alkali hydrolysate of rice hulls when compared to acid hydrolysate and solvent extracts. The rice hull extracts could reduce or prevent the oxidative spoilage of foodstuffs, and can therefore be applied as natural, safe and plant-based antioxidants for use by the food industry.

Huang & Lai (2016) analysed the bioactive compounds (including total phenolics (TPs), total flavonoids (TFs), anthocyanins (ACNs), proanthocyanidins (PAs), vitamin E and γ-oryzanol) and antioxidant activities of outer and inner rice bran from six different coloured rices and found that the free fraction of the extracts dominated the TPs (72–92%) and TFs (72–96%) of coloured rice bran as compared to the bound fraction. Overall, the TPs ranged from 17.89 to 36.70 or from 2.92 to 6.92 mg FAE/g

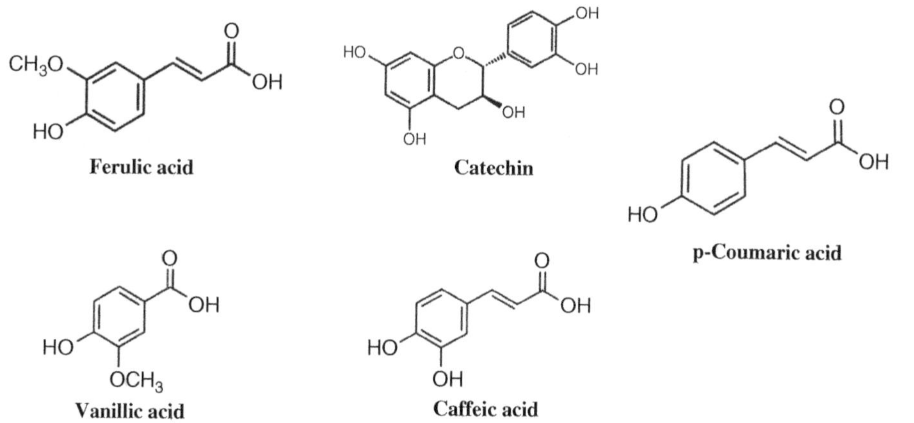

Figure 8.5 Chemical structures of some major phenolic compounds found in rice hull extracts. *Source:* Vadivel & Brindha (2015). Reproduced with permission of Elsevier.

dry weight in the free and bound fractions, respectively, and from 4.24 to 11.92 or from 0.90 to 3.82 mg FAE/g dry weight in the free and bound fractions, respectively. The TFs varied from 11.72 to 17.37 or from 0.79 to 2.85 mg catechin equivalents (CE)/g dry weight in the free and bound fractions, respectively, and from 1.20 to 3.29 or from 0.30 to 0.95 mg CE/g dry weight in the free and bound fractions, respectively. The TP of red rice bran (32.26–36.70 mg FAE/g dry weight) was 1.8 times higher than that of black rice bran (17.89–19.95 mg FAE/g dry weight). The PA content depended on the type of red rice bran, ranging from 12.16 to 19.13 or from 0.75 to 4.31 mg CE/g dry weight. The total ACNs in black rice bran were also 6.29–11.27 or 3.46–6.85 mg cyanidin-3-glucoside equivalents/g dry weight. In contrast, red rice bran contained only a small amount (0.31–0.38 or 0.20–0.28 mg cyanidin-3-glucoside equivalents/g dry weight). Most of the phenolic acids (83–97%) in coloured rice bran were present in the bound form. Higher amounts of phenolic acids were determined in the bound fractions than in the free factions of coloured rice bran. The major bound phenolic acids in black rice bran were ferulic acid, protocatechuic acid, vanillic acid and *p*-coumaric acid (1295–1601, 948–1043, 335–652 and 140–232 mg/g dry weight, respectively), while ferulic acid was also the predominant bound phenolic acid in red rice bran (1171–1998 mg/g dry weight), followed by *p*-coumaric acid (334–574 mg/g dry weight). Protocatechualdehyde was first identified in the bound fraction of red rice bran. The coloured rice bran contained the predominant vitamin Es, including α-tocopherol (13.12–67.30 mg/g dry weight), γ-tocotrienol (16.91–53.09 mg/g dry weight), γ-tocopherol (10.76–28.43 mg/g dry weight) and γ-oryzanol (3.59–7.72 mg/g dry weight). The antioxidant properties of the free fraction of the coloured rice bran were attributed to the proanthocyanidins in red-coloured rice and anthocyanins in black rice, while those of the bound fraction were mainly due to the phenolic acids.

Wanyo et al. (2016) determined the bioactive compounds and antioxidant properties of different solvent extracts derived from Thai rice byproducts and indicated that various solvents affected the contents and compositions of phenolic compounds and the antioxidant activities of the extracts from rice bran and husk. Among various solvents tested, heated water extracts provided the highest total phenolic content (TPC), total flavonoid

Table 8.6 Recovery of polyphenols from rice husk using different solvent extraction conditions. *Source:* Vadivel & Brindha (2015). Reproduced with permission of Elsevier.

Time	Total phenol concentration of extract (mg FAE/l)[a,b]	Total phenol yield on fresh weight basis (mg FAE/g sample)[b]	Total phenol yield on dry weight basis (mg FAE/g sample)[b]
Magnetic stirring with 75% ethanol at 150 rpm and 30 °C			
60 minutes	$81.57^{A} \pm 5.25$	5.71 ± 0.37	6.09 ± 0.39
120 minutes	$94.57^{A} \pm 5.45$	6.62 ± 0.38	7.06 ± 0.41
180 minutes	$179.79^{C} \pm 2.73$	12.59 ± 0.19	13.42 ± 0.20
240 minutes	$162.07^{B} \pm 10.00$	11.35 ± 0.70	12.10 ± 0.75
300 minutes	$154.36^{B} \pm 13.23$	10.81 ± 0.93	11.52 ± 0.99
Grinding in mortar and pestle with 75% ethanol			
15 minutes	$27.29^{D} \pm 5.25$	1.91 ± 0.37	2.04 ± 0.39
30 minutes	$144.29^{E} +5.05$	10.10 ± 0.35	10.77 ± 0.38
45 minutes	$196.57^{F} \pm 9.90$	13.76 ± 0.69	14.67 ± 0.74
60 minutes	$287.00^{H} \pm 9.29$	20.09 ± 0.65	21.42 ± 0.69
75 minutes	$240.14^{G} \pm 12.93$	16.81 ± 0.91	17.93 ± 0.97
Sohxlet extraction with 75% ethanol at 80 °C			
60 minutes	$4.29^{I} \pm 3.03$	0.30 ± 0.21	0.32 ± 0.23
120 minutes	$17.71^{I} \pm 3.84$	1.24 ± 0.27	1.32 ± 0.29
180 minutes	$31.29^{J} \pm 5.25$	2.19 ± 0.37	2.34 ± 0.39
240 minutes	$52.14^{K} \pm 4.04$	3.65 ± 0.28	3.89 ± 0.30
300 minutes	$53.21^{K} \pm 3.54$	3.73 ± 0.25	3.97 ± 0.26
Ultrasonic extraction with 75% ethanol			
05 minutes	$4.29^{L} \pm 3.03$	0.30 ± 0.21	0.32 ± 0.23
10 minutes	$6.43^{L} \pm 2.02$	0.45 ± 0.14	0.48 ± 0.15
15 minutes	$29.29^{M} \pm 6.06$	2.05 ± 0.42	2.19 ± 0.45
20 minutes	$48.57^{N} \pm 3.03$	3.40 ± 0.21	3.63 ± 0.23
25 minutes	$77.14^{O} \pm 1.01$	5.40 ± 0.07	5.76 ± 0.08

a) Values with different letters in this column are statistically different according to analysis of variance (ANOVA) and Turkey's post hoc test ($p < 0.01$).
b) Values are reported as mean ± SD of three separate determinations.

content, DPPH radical scavenging capacity and ferric-reducing antioxidant power in all samples studied. The antioxidant activity of the extracts increased with increasing water temperature from 50 to 70 °C, but decreased again at 80 °C. The contents of bound phenolics in all the samples studied were greater than those of free phenolics in the forms of phenolic acids and flavonoids. Acetone extract achieved the highest amounts of γ-oryzanol and tocopherols in all samples. A significant increase in the amounts of extracted phenolic acids, flavonoids and antioxidant properties was obtained by reducing the particle size of the rice husk. Regarding the wheat wastes/residues, Figure 8.6

Table 8.7 Recovery of polyphenols from rice husk using acid and alkali hydrolysis. *Source:* Vadivel & Brindha (2015). Reproduced with permission of Elsevier.

Time	Total phenol concentration of extract (mg FAE/l)[a,b]	Total phenol yield on fresh weight basis (mg FAE/g sample)[b]	Total phenol yield on dry weight basis (mg FAE/g sample)[b]
Acid hydrolysis with 2% sulphuric acid at 90 °C			
30 minutes	$217.14^{A} \pm 3.03$	15.20 ± 0.21	16.21 ± 0.23
60 minutes	$471.43^{B} \pm 7.07$	33.00 ± 0.49	35.19 ± 0.53
90 minutes	$575.71^{C} \pm 1.01$	40.30 ± 0.07	42.98 ± 0.08
120 minutes	$702.14^{D} \pm 2.02$	49.15 ± 0.14	52.42 ± 0.15
150 minutes	$674.29^{D} \pm 1.01$	47.20 ± 0.07	50.35 ± 0.08
Alkali hydrolysis with 2 N NaOH at 150 rpm			
30 minutes	$845.71^{E} \pm 7.07$	59.20 ± 0.49	63.13 ± 0.53
60 minutes	$953.57^{F} \pm 4.04$	66.75 ± 0.28	71.18 ± 0.30
90 minutes	$1228.57^{G} \pm 21.21$	86.00 ± 1.48	91.71 ± 1.58
120 minutes	$1067.86^{G} \pm 18.18$	74.75 ± 1.27	79.72 ± 1.36
150 minutes	$932.14^{F} \pm 8.08$	65.25 ± 0.57	69.59 ± 0.60

a) Values with different letters in this column are statistically different according to analysis of variance (ANOVA) and Turkey's post hoc test ($p < 0.01$).
b) Values are reported as mean ± SD of three separate determinations.

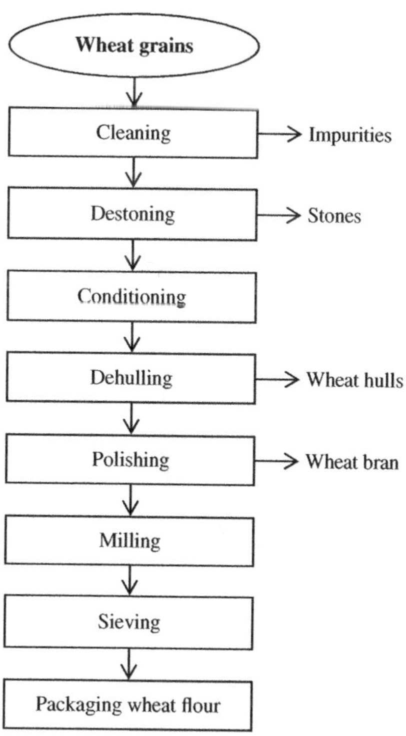

Figure 8.6 Flow chart of the wheat flour production process.

provides a flow chart of the wheat flour production process and the main wastes discarded from it. Of these, wheat bran consists of pericarp, an aleurone layer and germ, and is discarded during the early stages of durum wheat milling (Durante et al., 2012). Wang et al. (2008) optimised UAE conditions for phenolic compounds found in wheat bran and found that ethanol was the best solvent for the extraction of phenolics. Among three extraction parameters studied, extraction time had the greatest effect on the UAE process. The optimum extraction conditions were 64% ethanol, 60 °C extraction temperature and 25 minutes' extraction time. Under these conditions, the experimental TPC was 3.12 mg GAE/g wheat bran, which is similar to the predicted value from the model (3.08 mg GAE/g wheat bran).

Jonnala et al. (2010) determined the TPC, total phenolic acid content (TPAC) and phenolic acid composition of brans from two waxy wheat sets (Svevo: durum and N11: bread wheat) and four triticale (nonwaxy wheat) sets (GDS7, Trim, Rhino and Rigel) (Figures 8.7 and 8.8). The samples were defatted twice by hexane at a 1 : 4 ratio (w/v)

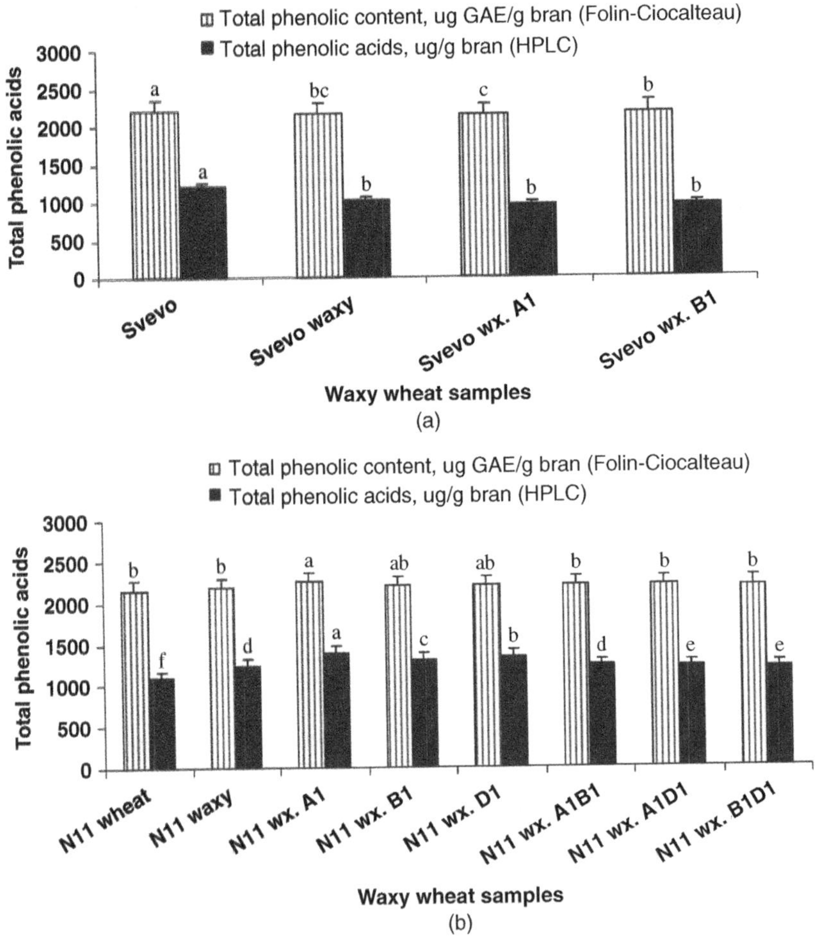

Figure 8.7 Total phenolic content (TPC) of bran from two waxy wheat genotypes: (a) Svevo set and (b) N11 set. *Source:* Jonnala et al. (2010). Reproduced with permission of Elsevier.

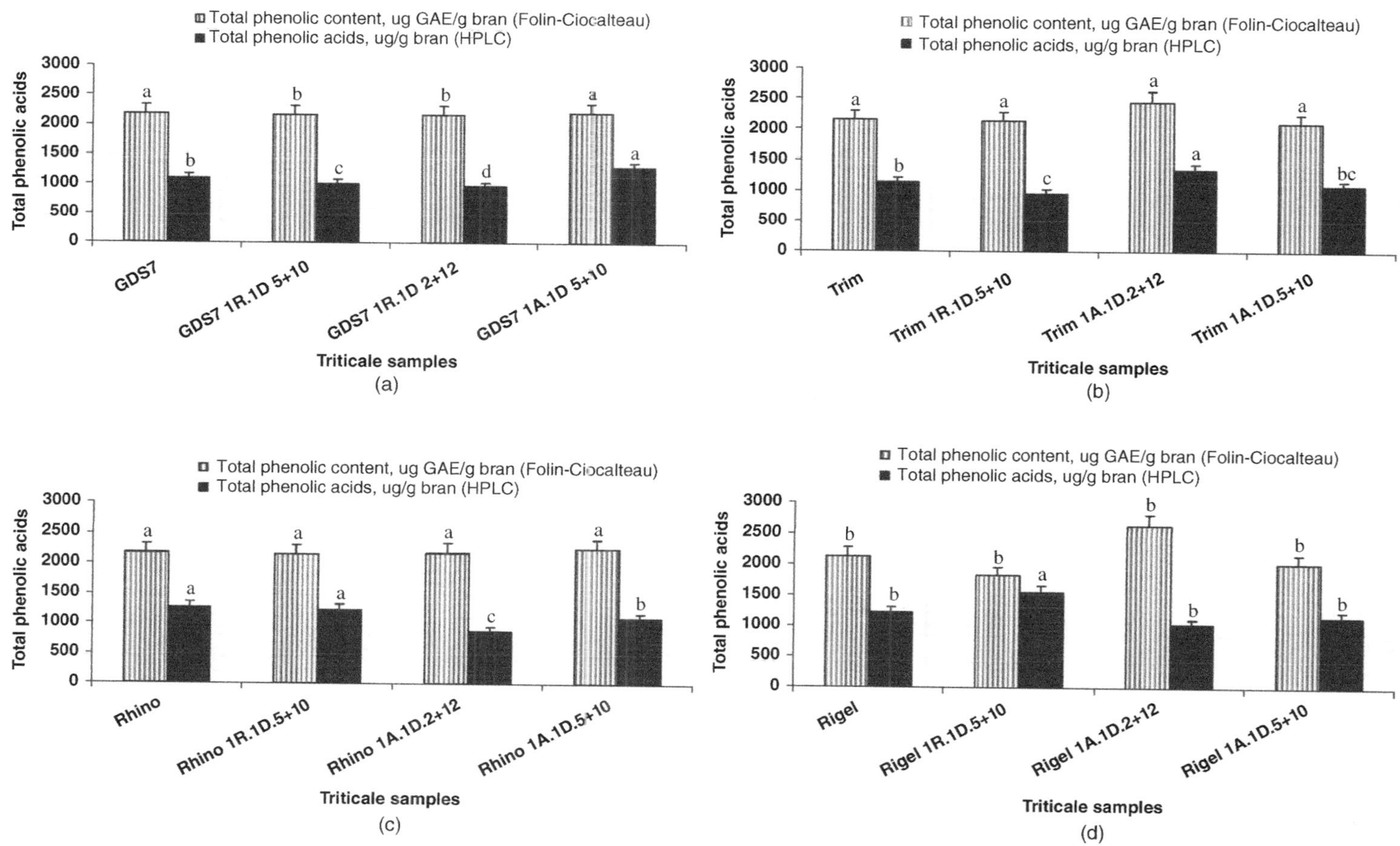

Figure 8.8 Total phenolic content (TPC) of bran from four triticale genotypes: (a) GDS7 set (b) Trim set (c) Rhino set and (d) Rigel set. *Source:* Jonnala et al. (2010). Reproduced with permission of Elsevier.

Table 8.8 Phenolic acid composition (μg/g sample) of waxy wheat bran. *Source:* Jonnala et al. (2010). Reproduced with permission of Elsevier.

Sample name	FA	*p*-CM	VA	SA	Others	Total
Svevo set						
Svevo wheat	1116^{A}	44.3^{A}	16.4^{A}	4.8^{A}	18.1^{A}	1200^{A}
Svevo waxy	948^{B}	39.5^{B}	14.2^{AB}	4.3^{A}	14.0^{C}	1020^{B}
Svevo waxy A1⁻	899^{C}	28.1^{C}	13.3^{B}	4.1^{A}	15.2^{BC}	960^{B}
Svevo waxy B1⁻	891^{C}	26.9^{C}	13.2^{B}	4.2^{A}	17.1^{AB}	952^{B}
N11 set						
N11 wheat	1006^{H}	28.7^{E}	28.1^{AB}	5.4^{E}	20.9^{AB}	1089^{F}
N11 waxy	1150^{D}	31.2^{E}	29.4^{A}	5.7^{E}	19.3^{B}	1236^{D}
N11 waxy A1⁻	1253^{A}	75.1^{A}	26.0^{B}	9.5^{AB}	22.1^{AB}	1386^{A}
N11 waxy B1⁻	1199^{C}	45.1^{B}	26.3^{B}	9.9^{A}	23.4^{A}	1304^{C}
N11 waxy D1⁻	1239^{B}	46.8^{B}	20.2^{CD}	9.5^{AB}	23.6^{A}	1339^{B}
N11 waxy A1⁻/B1⁻	1142^{E}	44.4^{B}	19.3^{D}	6.6^{DE}	22.2^{AB}	1235^{D}
N11 waxy A1⁻/D1⁻	1126^{F}	38.0^{D}	22.1^{C}	8.0^{BC}	19.6^{B}	1214^{E}
N11 waxy B1⁻/D1⁻	1113^{G}	41.5^{C}	18.4^{D}	7.4^{CD}	21.7^{AB}	1202^{E}

Means with the same letter in the same column of the same group are not significantly different at $p < 0.05$. FA, Ferulic acid; *p*-CM, p-coumaric acid; VA, vanillic acid; SA, syringic acid. 'Others' indicates the total of *t*-cinnamic acid, caffeic acid and *p*-hydroxybenzoic acid.

for 1 hour at room temperature and hydrolysed by 2 M NaOH in 80% methanol at a 1 : 5 ratio (w/v) for 3 hours. The findings showed that among the waxy and nonwaxy wheats analysed, ferulic acid was the major phenolic acid (891–1253 and 845–1501 μg/g bran in waxy and nonwaxy wheats, respectively), followed by *p*-coumaric (26.9–75.1 and 17.0–42.5 μg/g bran), vanillic acids and syringic acid (13.2–29.4 and 4.1–9.9 μg/g bran) (Tables 8.8 and 8.9). The TPC and TPAC in the wheat bran ranged from 2000 to 2500 μg GAE/g bran and from 900 to 1600 μg/g bran, respectively (Tables 8.7 and 8.8), with the highest levels being found in the waxy lines, rather than in the N11 set. The Svevo set presented an opposite trend. TPAC of waxy bread wheats had a positive correlation with its amylose fractions in the order of complete waxy, double waxy nulls and single waxy nulls. In each group of triticales, the lines with HMW-GS 2 + 12 had lower TPC and TPAC than other lines (except the Rigel and Trim sets for TPC and Trim set for TPAC). Moreover, TPC and TPAC in triticale lines of wheat bran had a negative correlation with bran yields, indicating that the phenolics in the wheat bran were varied not only in terms of bran yields but also in terms of genotypic differences.

Durante et al. (2012) reported that durum wheat byproducts/wastes are considered a rich sources of valuable nutrients (vitamin E, carotenoids and unsaturated fatty acids). Among these wastes, the durum wheat germ was the most abundant source of vitamin E, in the forms of tocopherols and tocotrienols (Figure 8.9) – lipophilic molecules that differ in the degree of saturation in their side chains. Table 8.10 shows that β/γ-tocotrienols were the most abundant vitamin E forms in durum wheat bran oil (4.4–4.6 mg/g oil), followed by α-tocopherol, β/γ-tocopherols and α-tocotrienol (2.3–3.5, 1.6–2.0 and 0.6–0.8 mg/g oil, respectively). No differences were observed

Table 8.9 Phenolic acid composition (μg/g sample) of triticales. *Source:* Jonnala et al. (2010). Reproduced with permission of Elsevier.

Sample name	Ferulic	*p*-coumaric	Others	Total
GDS7 set				
GDS7	1053^{B}	28.8^{B}	26.9^{C}	1109^{B}
GDS7 1R.1D. 5+10	956^{C}	21.7^{C}	28.6^{BC}	1006^{C}
GDS7 1R.1D. 2+12	911^{D}	23.7^{C}	31.4^{B}	966^{D}
GDS7 1A.1D. 5+10	1210^{A}	37.3^{A}	35.7^{A}	1283^{A}
Trim set				
Trim	1087^{AB}	29.7^{C}	17.3^{AB}	1134^{B}
Trim 1R.1D. 5+10	920^{B}	21.1^{D}	13.5^{C}	955^{C}
Trim 1A.1D. 2+12	1286^{A}	42.5^{A}	19.4^{A}	1348^{A}
Trim 1A.1D. 5+10	1039^{B}	33.1^{B}	14.6^{BC}	1087^{BC}
Rhino set				
Rhino	1226^{A}	28.2^{A}	14.9^{A}	1269^{A}
Rhino 1R.1D. 5+10	1186^{AB}	31.8^{A}	14.2^{A}	1232^{A}
Rhino 1A.1D. 2+12	845^{C}	17.0^{B}	13.3^{A}	875^{C}
Rhino 1A.1D. 5+10	1063^{B}	17.0^{B}	14.4^{A}	1094^{B}
Rigel set				
Rigel	1180^{B}	30.2^{B}	14.9^{A}	1225^{B}
Rigel 1R.1D. 5+10	1501^{A}	41.7^{A}	13.0^{A}	1556^{A}
Rigel 1A.1D. 2+12	984^{B}	33.7^{AB}	13.5^{A}	1031^{B}
Rigel 1A.1D. 5+10	1120^{B}	18.4^{C}	14.9^{A}	1153^{B}

Means with the same letter in the same column of the same group are not significantly different at $p < 0.05$.
'Others' indicates the total of *t*-cinnamic acid, caffeic acid, syringic acid and vanillic acid.

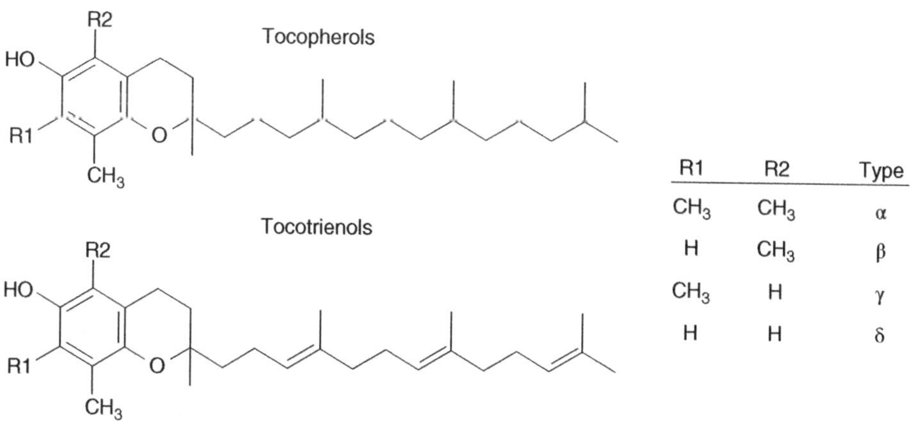

Figure 8.9 Chemical structures of naturally occurring forms of vitamin E: tocopherols and tocotrienols. *Source:* Durante et al. (2012). Reproduced with permission of Springer.

Table 8.10 Chemical composition of oil extracted from durum wheat bran by supercritical carbon dioxide (SC-CO_2) and Soxhlet. *Source:* Durante et al. (2012). Reproduced with permission of Springer.

	SC-CO_2	Soxhlet
Vitamin E (mg/g oil)		
α-tocopherol	2.3 ± 0.5^{A}	3.5 ± 0.3^{A}
$\beta\gamma$-tocopherol	2.0 ± 0.2^{A}	1.6 ± 0.1^{A}
α-tocotrienol	0.8 ± 0.7^{A}	0.6 ± 0.1^{A}
$\beta\gamma$-tocotrienol	4.4 ± 0.8^{A}	4.6 ± 0.9^{A}
Carotenoid (μg/g oil)		
Lutein	4.1 ± 1.5^{A}	9.6 ± 0.2^{B}
Zeaxanthin	1.6 ± 0.2^{A}	2.1 ± 0.1^{A}
β carotene	1.9 ± 0.4^{A}	2.6 ± 0.1^{A}
Quinone isoprenoid (mg/g oil)		
Coenzyme Q8	0.2 ± 0.1^{A}	0.2 ± 0.1^{A}
Coenzyme Q9	0.4 ± 0.1^{A}	0.8 ± 0.1^{A}
Coenzyme Q10	0.1 ± 0.1^{A}	0.2 ± 0.1^{A}
Lipid classes (mg/g oil)		
Triglycerides	683.0 ± 52.0^{A}	576.0 ± 28.0^{A}
Diglycerides	127.0 ± 18.0^{A}	198.0 ± 16.0^{A}
Free fatty acids	88.8 ± 3.3^{A}	110.6 ± 6.6^{A}
Fatty acid (mg/g oil)		
C16:1 ω7	1.1 ± 0.1^{A}	1.0 ± 0.3^{A}
C16:0	139.0 ± 2.5^{A}	141.0 ± 2.0^{A}
C18:2 ω6	366.0 ± 8.2^{A}	368.2 ± 4.0^{A}
cC18:1 ω9	192.0 ± 4.6^{A}	199.0 ± 3.2^{A}
*t*C18:1 ω9	9.3 ± 0.1^{A}	8.8 ± 0.3^{A}
C18:0	7.6 ± 0.7^{A}	7.1 ± 1.0^{A}
C18:3 ω3	10.4 ± 0.7^{A}	9.1 ± 0.4^{A}

Values represent the mean of three independent experiments ± standard error (SE). Data were submitted to one-way analysis of variance (ANOVA), with differences among groups detected using multiple comparison procedures (Tukey Test). Different letters indicate significant differences ($p < 0.05$).

between tocopherols and tocotrienols in oils obtained by SC-CO_2 and Soxhlet extraction processes. Lutein was the most abundant carotenoid, followed by zeaxanthin and β-carotene. The content of lutein in durum wheat bran oil extracted by Soxhlet was 2.3-fold higher than that extracted by SC-CO_2, due to a greater solubility of lutein in hexane as a nonpolar solvent.

Moreover, durum wheat bran oil contained mainly triglycerides, with approximately 80% unsaturated fatty acids and 20% saturated fatty acids. The main fatty acid was linoleic acid (nearly 50% of the total), followed by oleic, palmitic and linolenic acids.

The fatty acid composition of durum wheat bran oil extracted by SC-CO_2 was not significantly different from that extracted by Soxhlet with hexane, and the contents of other compounds were similar in both extraction processes.

Wang et al. (2014) applied ultrasound-assisted enzymatic extraction (UAEE) for optimisation of arabinoxylan (the major dietary fibre component) from wheat bran and indicated that the UAEE technique was more effective in improving the extraction yield of arabinoxylan from wheat bran than either enzymatic extraction or UAE. The optimal UAEE conditions were found to be raw material concentration 50 g/l, enzyme (Sunzymes) dose 4.5 g/l, extraction temperature 50 °C, extraction time 70 minutes and ultrasonic power 180 W. The yield of arabinoxylan from wheat bran obtained under these conditions was 14.26%.

8.4 Recovery of Bioactive Compounds from Other Starch Wastes

Figures 8.10–8.12 show the barley, maize and cassava flour production processes and the main wastes discarded from them.

Panfili et al. (2008) extracted the bioactive compounds (tocols and β-glucans) from barley pearling byproducts (hulled and pearled kernels) using n-hexane/ethyl acetate

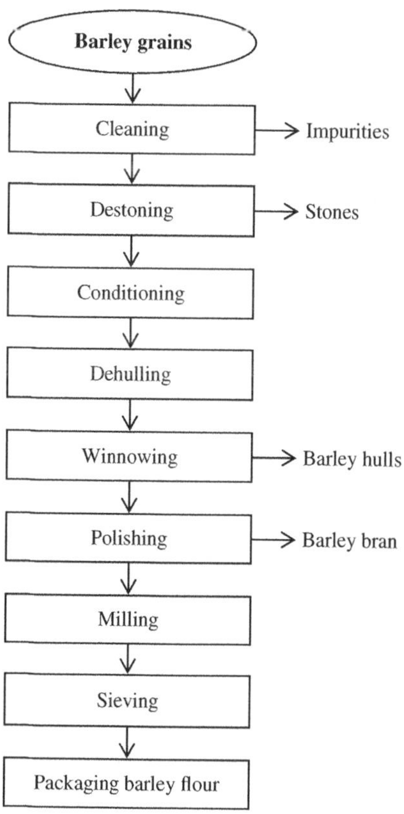

Figure 8.10 Flow chart of the barley flour production process.

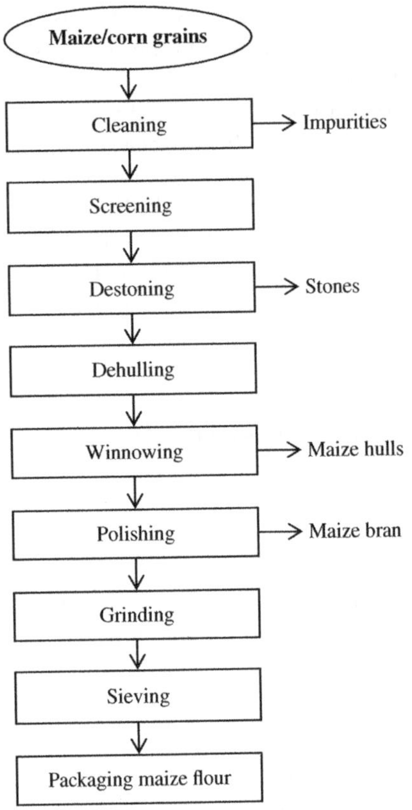

Figure 8.11 Flow chart of the maize flour production process.

(9 : 1, v/v) and found that the β-glucan content in hulled kernels (I–VI) varied with the type of fraction (0.35–4.50 g/100 g wet basis) but was always lower than that in pearled kernels (4.82 g/100 g wet basis) (Table 8.11), while the amounts of tocopherols and tocotrieols in hulled kernels (10.1–54.5 and 38.0–205.4 mg/kg wet basis, respectively) were significantly higher than those in pearled kernels (1.8 and 6.2 mg/kg wet basis, respectively). Total tocols in hulled kernels (48.1–247.4 mg/kg wet basis) were significantly higher than in pearled kernels (8.0 mg/kg wet basis). Table 8.12 shows the contents of individual tocols in terms of α, β, γ and δ-tocopherols and α, β, γ and δ-tocotrienols in hulled kernels. This also depends on the type of fraction but is always higher than in pearled kernels. The ratio of tocotrienol to tocopherol (T3/T) varies widely between 1.7 and 4.9 in hulled kernels, versus 3.5 in pearled kernels (Table 8.12). Izydorczyk & Dexter (2008) indicated that barley β-glucans have positive correlation with lowering plasma cholesterol, reducing glycaemic index and reducing the risk of colon cancer. Based on their rheological characteristics, the potential application of β-glucans as food hydrocolloids has been recommended. In addition, β-glucans have been shown to gel under certain conditions because of their solution viscosity enhancement.

Gupta et al. (2010) found that barley spent grains account for about 31% of the original malt weight, and that a number of value-added products can be produced from barley byproducts, such as organic acids, amino acids and vitamins, with potential applications

Figure 8.12 Flow chart of the cassava flour production process.

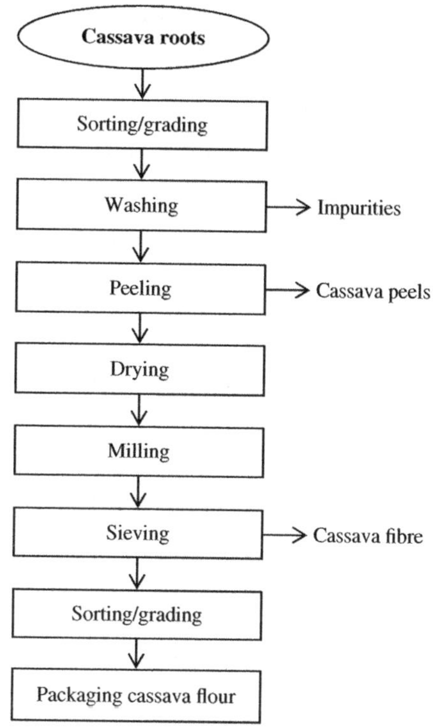

Table 8.11 Protein, lipid, ash (% wet basis), β-glucans (g/100 g wet basis) and tocols (mg/kg wet basis) in barley byproducts. *Source:* Panfili et al. (2008). Reproduced with permission of Elsevier.

Stock and fraction	% kernel weight removed	% total cumulative weight removed	Protein in fraction	Lipid in fraction	Ash in fraction	β-glucans	Sum of toco-pherols	Sum of toco-trienols	Total tocols	Tocopherol equivalents
HK			8.6	2.1	1.6	4.34^{A}	10.1^{A}	38.0^{A}	48.1^{A}	17.2^{A}
I	6.8	6.8	5.1	3.0	5.7	0.35^{B}	29.3^{B}	51.0^{B}	80.4^{B}	33.4^{B}
II	3.2	10.0	15.1	4.0	5.6	0.62^{C}	41.0^{C}	177.4^{C}	218.4^{C}	68.7^{C}
III	4.3	14.3	16.7	7.4	5.4	2.78^{D}	42.0^{D}	205.4^{D}	247.4^{D}	83.9^{D}
IV	4.3	18.6	16.8	6.3	5.1	3.46^{E}	54.4^{E}	155.7^{E}	210.1^{C}	$84.9^{C,D}$
V	4.2	22.8	15.4	5.1	3.9	3.79^{F}	35.1^{B}	113.4^{F}	148.5^{E}	55.9^{C}
VI	7.2	30.0	14.0	3.9	3.1	4.50^{G}	25.4^{B}	65.1^{G}	90.6^{B}	37.0^{B}
PK			6.4	1.0	0.6	4.82^{H}	1.8^{F}	6.2^{H}	8.0^{F}	2.7^{F}

Values with the same letter in the same column of the same group are not significantly different at $p < 0.05$.
HK, hulled kernel; PK, pearled kernel.

in the food industry and other fields. Verardo et al. (2011) reported that phenolic compounds are concentrated mainly in the bran. In the grain refining process, the bran is removed, resulting in the loss of a significant amount of dietary fibre, vitamins, minerals, lignans, phyto-estrogens and phenolic compounds. Thus, the outermost fractions of barley grains contain the highest phenolic content, and they are considered a rich source of bioactive compounds.

Table 8.12 Contents of individual tocols (mg/kg wet basis) and tocotrienol/tocopherol (T3/T) ratios in barley byproducts. *Source:* Panfili et al. (2008). Reproduced with permission of Elsevier.

Stock and fraction	Tocols								
	α-T	α-T3	β-T	γ-T	β-T3	γ-T3	δ-T	δ-T3	T3/T
HK	6.9	28.6	0.2	2.9	2.9	6.3	0.1	0.2	3.7
I	14.4	38.6	0.3	14.1	3.1	8.8	0.4	0.5	1.7
II	28.2	136.7	0.8	11.6	9.6	29.6	0.4	1.4	4.3
III	30.9	157.2	0.8	9.7	11.7	35.2	0.5	1.2	4.9
IV	44.1	117.7	0.9	8.9	9.7	27.1	0.5	1.1	2.9
V	26.9	83.0	0.6	6.9	7.3	22.2	0.6	0.9	3.2
VI	18.7	49.6	0.6	5.8	4.1	11.0	0.2	0.4	2.6
PK	1.3	3.9	0.1	0.4	1.2	0.9	0.0	0.1	3.5

HK, hulled kernel; I–VI, pearling byproducts; PK, pearled kernel; T, tocopherol; T3, tocotrienol.

Regarding maize/corn residues/wastes, Ashour et al. (2013) isolated phytochemical compounds from the extracts of corncobs (about 15% of total corn production) using methanol as solvent at room temperature, and identified a new lignan named 7,7′-dihydroxy-3′-*O*-demethyl-4-methoxymatairesinol, along with seven other compounds: β-sitosterol, β-sitosteryl-β-D-glucoside, 6β-hydroxycampest-4-en-3-one, 5α, 8α-epidioxyergosta-6,22-dien-3β-ol, tricin, kaempferol and *p*-coumaric acid.

Sanni et al. (2005) examined the opportunities and challenges offered by value-added cassava products, such as organic acids, flavour and aroma compounds. Ubalua (2007) reported that cassava peels are usually divided into a thin brown outer covering and a thicker, leathery, parenchymatous inner covering. These peels account for about 20–35% of the total weight of the tuber and are regarded as wastes, normally being discarded to the environment. Aro (2008) found that the tannin, saponin and total alkaloid contents of cassava starch residues range from 0.04 to 0.09, 0.01 to 0.04 and 0.03 to 5.44% by weight, respectively.

In order to exploit the valuable bioactive compounds obtained from barley wastes, such as tocols, β-glucans, phenolic compounds, dietary fibre, vitamins, lignans and phyto-estrogens; those from maize wastes, such as lignin, kaempferol and phenolic compounds; and those from cassava wastes, like tannins, saponins and alkaloids, it is necessary to study and apply advanced extraction techniques such as UAE, microwave-assisted extraction (MAE), supercritical fluid extraction (SFE) and electric pulsed field (EPF) (Nguyen et al., 2015, 2016a, 2016b, 2016c; Vuong et al., 2014a, 2014b). These techniques not only improve the extraction efficiency of target compounds but also reduce the amount of solvent required, the extraction time and the energy consumption, which has a great influence on the cost of manufacture and the selling price of the final products. An economic analysis of extraction techniques is presented in Chapter 10.

Purification of the crude extracts obtained from starch wastes should be considered in order to achieve pure compounds. A wide range of purification technologies can be used, including liquid–liquid extraction, crystallisation, membrane technique, capillary electrophoresis and chromatography (size-exclusion, ion-exchange, gel filtration, high-performance liquid chromatography (HPLC) coupled with an auto fraction collector; Vuong et al., 2014a).

8.5 Conclusion

Rice, wheat, maize, barley, potato and cassava are the main starch sources in human and animal consumption. The processing and production of starch-based products from these sources discard a huge amount of residues and byproducts as wastes. Interestingly, starch wastes are rich in bioactive compounds possessing a wide range of potential biological activities. Therefore, the utilisation of starch wastes as abundant, inexpensive and sustainable sources for the recovery of bioactive compounds is highly attractive to researchers and producers. Exploitation of these wastes would help increase the economic value of agricultural crops, reduce pollution risks and provide support for related industries, such as nutraceuticals, medicines, pharmaceuticals and cosmetics, particularly with regard to sustainable development.

References

Aro, S. O. 2008. Improvement in the nutritive quality of cassava and its by-products through microbial fermentation. *African Journal of Biotechnology*, **7**.

Ashour, A., Amer, M., Marzouk, A., Shimizu, K., Kondo, R. & El-Sharkawy, S. 2013. Corncobs as a potential source of functional chemicals. *Molecules*, **18**, 13 823–13 830.

Baiano, A. 2014. Recovery of biomolecules from food wastes-a review. *Molecules*, **19**, 14 821–14 842.

Cardenas-Toro, F. P., Alcazar Alay, S. C., Forster-Carneiro, T. & Meireles, M. A. A. 2014. Obtaining oligo-and monosaccharides from agroindustrial and agricultural residues using hydrothermal treatments. *Food and Public Health*, **4**, 123–139.

Dei Piu, L., Tassoni, A., Serrazanetti, D. I., Ferri, M., Babini, E., Tagliazucchi, D. & Gianotti, A. 2014. Exploitation of starch industry liquid by-product to produce bioactive peptides from rice hydrolyzed proteins. *Food Chemistry*, **155**, 199–206.

Durante, M., Lenucci, M. S., Rescio, L., MITA, G. & CARETTO, S. 2012. Durum wheat by-products as natural sources of valuable nutrients. *Phytochemistry Reviews*, **11**, 255–262.

FAOSTAT 2015. FAO Statistic Division. http://faostat3.fao.org/home/E (last accessed March 20, 2017).

Gupta, M., Abu-Ghannam, N. & Gallaghar, E. 2010. Barley for brewing: characteristic changes during malting, brewing and applications of its by-products. *Comprehensive Reviews in Food Science and Food Safety*, **9**, 318–328.

Hossain, M. B., Tiwari, B. K., Gangopadhyay, N., O'Donnell, C. P., Brunton, N. P. & Rai, D. K. 2014. Ultrasonic extraction of steroidal alkaloids from potato peel waste. *Ultrasonics Sonochemistry*, **21**, 1470–1476.

Hossain, M. B., Rawson, A., Aguiló-Aguayo, I., Brunton, N. P. & Rai, D. K. 2015. Recovery of steroidal alkaloids from potato peels using pressurized liquid extraction. *Molecules*, **20**, 8560–8573.

Huang, Y. P. & Lai, H. M. 2016. Bioactive compounds and antioxidative activity of colored rice bran. *Journal of Food and Drug Analysis*, **24**, 564–574.

Izydorczyk, M. S. & Dexter, J. E. 2008. Barley β-glucans and arabinoxylans: molecular structure, physicochemical properties, and uses in food products – a review. *Food Research International*, **41**, 850–868.

Jonnala, R. S., Irmak, S., Macritchie, F. & Bean, S. R. 2010. Phenolics in the bran of waxy wheat and triticale lines. *Journal of Cereal Science*, **52**, 509–515.

Maldonado, A. F. S., Mudge, E., Gänzle, M. G. & Schieber, A. 2014. Extraction and fractionation of phenolic acids and glycoalkaloids from potato peels using acidified water/ethanol-based solvents. *Food Research International*, **65**, 27–34.

Nguyen, V. T., Bowyer, M. C., Vuong, Q. V., Van Altena, I. A. & Scarlett, C. J. 2015. Phytochemicals and antioxidant capacity of Xao tam phan (*Paramignya trimera*) root as affected by various solvents and extraction methods. *Industrial Crops and Products*, **67**, 192–200.

Nguyen, V. T., Pham, H. N. T., Bowyer, M. C., Van Altena, I. A. & Scarlett, C. J. 2016a. Influence of solvents and novel extraction methods on bioactive compounds and antioxidant capacity of *Phyllanthus amarus*. *Chemical Papers*, **70**, 556–566.

Nguyen, V. T., Bowyer, M. C., Van Altena, I. A. & Scarlett, C. J. 2016b. Optimisation of microwave-assisted extraction from *Phyllanthus amarus* for phenolic compounds-enriched extracts and antioxidant capacity. *Chemical Papers*, **70**, 713–725.

Nguyen, V. T., Vuong, Q. V., Bowyer, M. C., Van Altena, I. A. & Scarlett, C. J. 2016c. Microwave-assisted extraction for saponins and antioxidant capacity from Xao tam phan (*Paramignya trimera*) root. *Journal of Food Processing and Preservation*, doi: 10.1111/jfpp.12851.

Panfili, G., Fratianni, A., Di Criscio, T. & Marconi, E. 2008. Tocol and β-glucan levels in barley varieties and in pearling by-products. *Food Chemistry*, **107**, 84–91.

Sanni, L., Maziya-Dixon, B., Patino, M., Akoroda, M., Ezedinma, C., Okechukwu, R., Lemchi, J., Ogbe, F., Mukumbira, J. & Ilona, P. 2005. Value addition to cassava in Africa: challenges and opportunities. *African Crop Science Conference Proceedings*, 583–590.

Santana-Méridas, O., González-Coloma, A. & Sánchez-Vioque, R. 2012. Agricultural residues as a source of bioactive natural products. *Phytochemistry Reviews*, **11**, 447–466.

Singh, P. P. & Saldaña, M. D. A. 2011. Subcritical water extraction of phenolic compounds from potato peel. *Food Research International*, **44**, 2452–2458.

Ubalua, A. O. 2007. Cassava wastes: treatment options and value addition alternatives. *African Journal of Biotechnology*, **6**, 2065–2073.

Vadivel, V. & Brindha, P. 2015. Antioxidant property of solvent extract and acid/alkali hydrolysates from rice hulls. *Food Bioscience*, **11**, 85–91.

Verardo, V., Gómez-Caravaca, A. M., Messia, M. C., Marconi, E. & Caboni, M. F. 2011. Development of functional spaghetti enriched in bioactive compounds using barley coarse fraction obtained by air classification. *Journal of Agricultural and Food Chemistry*, **59**, 9127–9134.

Vuong, Q. V., Nguyen, V. T., Goldsmith, C. D., Dang, T. T., Bhuyan, D. J., Sadeqzadeh, E., Scarlett, C. J. & Bowyer, M. C. 2014a. Optimization of ultrasound-assisted extraction conditions for euphol from the medicinal plant, *Euphorbia tirucalli*, using response surface methodology. *Industrial Crops and Products*, **63**, 197–202.

Vuong, Q. V., Goldsmith, C. D., Dang, T. T., Nguyen, V. T., Bhuyan, D. J., Sadeqzadeh, E., Scarlett, C. J. & Bowyer, M. C. 2014b. Optimisation of ultrasound-assisted extraction conditions for phenolic content and antioxidant capacity from *Euphorbia tirucalli* using response surface methodology. *Antioxidants*, **3**, 604–617.

Wang, J., Sun, B., Cao, Y., Tian, Y. & Li, X. 2008. Optimisation of ultrasound-assisted extraction of phenolic compounds from wheat bran. *Food Chemistry*, **106**, 804–810.
Wang, J., Sun, B., Liu, Y. & Zhang, H. 2014. Optimisation of ultrasound-assisted enzymatic extraction of arabinoxylan from wheat bran. *Food Chemistry*, **150**, 482–488.
Wanyo, P., Kaewseejan, N., Meeso, N. & Siriamornpun, S. 2016. Bioactive compounds and antioxidant properties of different solvent extracts derived from Thai rice by-products. *Applied Biological Chemistry*, **59**, 373–384.
Wijngaard, H. H., Ballay, M. & Brunton, N. 2012. The optimisation of extraction of antioxidants from potato peel by pressurised liquids. *Food Chemistry*, **133**, 1123–1130.

9

Recovering Bioactive Compounds from Other Agricultural Wastes

Giovani L. Zabot[1] *and Fiorella P. Cárdenas-Toro*[2]

[1] *Federal University of Santa Maria (UFSM), Cachoeira do Sul, RS, Brazil*
[2] *Department of Engineering, Section of Industrial Engineering, Pontifical Catholic University of Peru, San Miguel, Lima, Peru*

9.1 Introduction

The recovery of bioactive compounds is commonly accomplished by a unit operation known as extraction. Such unit operations typically comprise a physicochemical separation process based on the contact between a fixed bed of comminuted material and a solvent, which removes a solute or mixture of solutes from the solid phase. Several parameters influence the yield and concentration of compounds using different extraction techniques, including the solvent type, temperature, pressure, ratio of solvent to feed, particle size of the solids, processing time, solvent flow rate (when a continuous or semicontinuous mode is fulfilled), solid moisture content and energy density. The selection of suitable conditions requires knowledge of the vegetable matrix properties. Edaphoclimatic aspects, morphology, the location of the solute in the matrix, shape, size, porosity and part of the plant used (stalks, stems, seeds, leaves, rinds, etc.) are some of the raw material characteristics that affect the extraction process, because they have direct effects on the mass transfer rate.

Plants parts are often discarded as waste or are only partially used in the processing of value-added products in agrifood industries. However, some researchers are beginning to prove that the entire part of several botanical matrices can be used in obtaining bioactive compounds (complete use) (Mokhtar et al., 2016; Moraes et al., 2015; Silva et al., 2015; Zabot et al., 2014b, 2015). Wastes such as peels and seeds can contain substances that inhibit oxidation reactions (Lee et al., 2014) and/or act as potential sources of natural dyes and antimicrobial agents (Ksibi et al., 2015). Polyphenol-rich extracts, edible oils and volatile oils can also be recovered (Katsampa et al., 2015; Silva et al., 2016; Taghvaei et al., 2014; Zabot et al., 2014a). Therefore, some wastes have attracted the interest of researchers into food-related or pharmaceutical-related products, especially those that are discarded in large amounts, with ongoing projects dealing with the optimisation of the recovery of bioactive compounds from them. In this chapter, we discuss the characteristics of technologies used for the recovery of compounds from pepper, onion and cotton wastes, attempting to demonstrate the technical, economical and environmental aspects of each.

Recovering Bioactive Compounds from Agricultural Wastes, First Edition. Edited by Van Tang Nguyen.

9.2 Recovering Bioactive Compounds from Pepper Waste

9.2.1 Bioactive Compounds found in Pepper Waste

Recent trends in the global market indicate an increasing demand for natural products as a consequence of increasing concern over the use of synthetic products. Production of natural extracts containing bioactive compounds is becoming ever more notorious because of their nutraceutical properties. Thus, researchers around the world are engaged in conducting studies into the replacement of synthetic products by natural ones. Furthermore, they are discovering new natural sources for the extraction of a broad range of bioactive compounds, especially from wastes generated through the several processes developed in the agrifood industries.

Among the wastes generated, pepper waste is one of the most prominent sources of health-promoting compounds. According to Romo-Hualde et al. (2012), after processing of the piquillo red pepper variety (*Capsicum annuum* L.) in the Spanish region of Navarre, approximately 65% of the weight of raw material is disposed of as municipal solid waste. The same is likely true in other regions, and it constitutes a tremendous loss of easily available functional compounds, because peppers are rich source of phytochemicals. Pepper fruits are most used in culinary preparations, with the seeds generally considered an industrial byproduct. The waste material often constitutes a promising source of bioactive compounds, such as flavonoids (quercetin, luteolin, kaempferol, apigenin and myricetin) (Bae et al., 2012), capsaicinoids (capsaicin) (Santos et al., 2015), carotenoids (β-carotene, β-cryptoxanthin and zeaxanthin) (Hervert-Hernandez et al., 2010), sterols (campesterol, stigmasterol and β-sitosterol), triterpenes (betulin), monoterpenes (menthol, myrtenal and (Z)-β-ocimene), sesquiterpenes (carotol) and ketones (benzophenone) (Silva et al., 2013). These are used as biological agents against diseases. Extracts containing capsaicinoids present antimicrobial activity, being active against Gram-positive bacteria, except the two beneficial bacteria *Lactobacillus rhamnosus* and *Bifidobacterium longum*. Pepper extracts are reported to have cytotoxic effects on tumour cell lines and peripheral blood mononuclear cells, reducing cell proliferation and inducing apoptotic death (Mokhtar et al., 2016). Therefore, using pepper waste to obtain such bioactive compounds is a promising alternative revenue stream for food- and pharmaceutical-related companies.

9.2.2 Major Techniques for the Recovery of Bioactive Compounds from Pepper Waste

9.2.2.1 Conventional Extraction

One agricultural waste product is the red pepper stalk, which has high phenolic, flavonoid and capsaicin contents. Mostly, red pepper stalks are discarded at landfills and into rivers, where they oxidise and harm the local ecology. Thus, any attempt to reutilise pepper stalk waste would be useful. One conventional extraction technique involves the use of organic solvents under low-pressure conditions (room pressure close to 0.1 MPa) to transfer the solute from the solid to the liquid phase. The operation steps employed by Chen & Kang (2013) using such a technique are summarised in Figure 9.1. After performing solid–liquid extraction (SLE), approximately 11 mg of capsaicin/g dry waste was recovered. Total phenolic and flavonoid contents were 71 mg gallic acid equivalents (GAE)/g dry waste and 70 mg of naringin equivalents (NE)/g dry

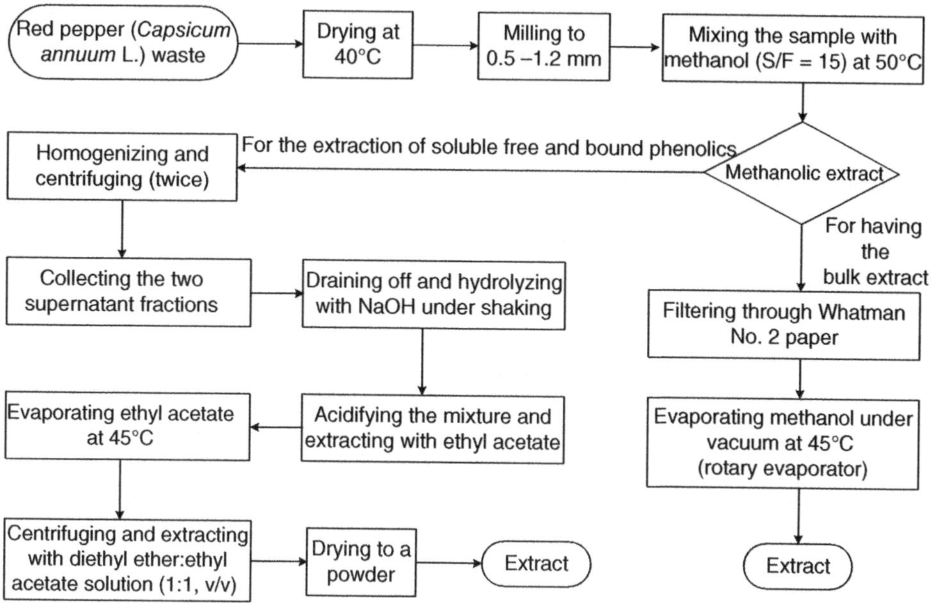

Figure 9.1 Simplified flow sheet for obtaining extract from pepper waste through solid–liquid extraction (SLE). S/F, solvent to feed mass ratio.

waste, respectively (Chen & Kang, 2013). The authors observed that the stalk extract contained the highest amounts of phenolics, capsaicin and flavonoids when compared with pericarp and placenta parts. Furthermore, as agricultural waste, red pepper stalk yielded extracts presenting compounds that acted as effective inhibitors of nitric oxide production in microphages.

The characteristics of the extraction process include mass transfer and practical equilibrium. A liquid solvent (ethanol, methanol, ethyl acetate, hexane, acetone, etc.) or a mixture of solvents is used to dissolve one or more solutes from the solid matrix. Literature reports refer to the unit operation of SLE as 'leaching', 'lixiviation', 'elution' or 'decoction', each of which matches some particular characteristics of the process. For instance, elution is used when the soluble solids are at the surface of the solid and decoction is used when the solvent is at the boiling temperature. When using pepper waste with a considerable moisture content, preparation is required, especially drying and milling. Drying eliminates the excess water, which could compete with the solvent in dissolving target compounds, negatively influencing the transport of the solute. Milling increases the contact area between the solid and the particles. Selection of the solvent depends on its physicochemical properties, cost, availability and toxicity. Regarding the toxicity, traces of organic solvents in the extracts could restrict their application in food and pharmaceutical derivatives. The selection of the solvent also includes the selectivity, the ability to dissolve the solute, the viscosity and the interfacial tension.

Attempting to determine the best solvent for the extraction of phenolic compounds, Ksibi et al. (2015) recovered phenolic compounds from pepper waste using ethanol, acetone and water. Pepper waste was constituted by seeds, skin leftovers and stems. The extraction procedure for the dried and grounded material involved loading the samples in closed flasks in a laboratory autoclave machine at 50 °C. Binary or ternary mixtures were prepared and added to each flask in order to keep the plant materials fully immersed in the solvent for 1 hour. A solvent to feed mass ratio (S/F) of 50 was used for the extraction of the phenolics. The authors found that the phenolics extracted from agrifood waste presented antimicrobial activity against *Pseudomonas aeruginosa* and *Staphilococcus aureus*. Furthermore, acetone and water (68 : 32, v/v) mixture extracted more phenolics than did other mixtures (>200 mg/l). Addition of water to the solvent mixture improved the extraction rate, but too high water content (pure water as solvent) brought an increased concomitant extraction of other compounds and lower phenol concentration in the extracts (32.9 mg/l). Comparison of all solvent systems revealed acetone (50–70%, v/v) to be the optimum extraction medium, regarding both total phenolic and total flavonoid contents from pepper waste (Ksibi et al., 2015). An equal mixture of acetone and water (50 : 50%, v/v) yielded a total phenolic content of 233 mg/l.

Extracts with antioxidant capacity were also taken from pepper waste. The extraction technique involved stirring the dried samples with H_2SO_4 for 30 minutes and then evaporating to dryness under reduced pressure (Silva et al., 2013). Therefore, pepper waste is a potential source of bioactive compounds, increasing the possibility of profitting from the great wastes produced by the pepper processing industry. The bioactive compounds present potential health benefits, being able to produce added-value products as functional food ingredients or food supplements.

Although SLE as Soxhlet extraction is a conventional technique widely used for the extraction of alkaloids, terpenes, terpenoids and phenolic compounds from several raw materials, including pepper waste, the long processing times and high temperatures required provide drawbacks. SLE can generate undesirable residues and extracts may undergo oxidative transformations during the solvent removal step. To overcome this, other operations are being studied and developed.

9.2.2.2 Microwave-Assisted and Ultrasound-Assisted Extraction

Microwave-assisted extraction (MAE) consists in the direct application of non-ionising electromagnetic energy with a frequency of 0.3–300.0 GHz to the raw material. The energy penetrates into the solid matrix and interacts with polar molecules, mostly water, to generate heat; the heat expands and disrupts the cells, favouring the extraction of intracellular phytochemical compounds. According to Gogus et al. (2015), discarded red pepper fruits were used in the recovery of volatile components as terpenes (cumene, limonene, α-copaene, o-cymene, etc.). MAE was used as an alternative to conventional extraction methods, mainly due to its considerable savings in processing time, solvent consumption and energy. MAE conditions commonly reported in the literature are presented in Table 9.1. The operation steps are shown in Figure 9.2, where the process is usually performed in batch mode.

Ultrasound-assisted extraction (UAE) consists in the use of mechanical vibrations caused by ultrasonication with frequencies higher than 20 kHz to cause expansion and compression cycles in the extraction medium, creating bubbles which collapse and cause cavitation, instantly creating a high local pressure and intense local heating. The fast changes induce disruption and thinning of the cellular structure, favouring the transference of the solute from the solid matrix to the solvent. The advantages of using UAE include shortening the extraction time, reducing solvent consumption, avoiding degradation of phytochemicals and being able to use simple equipment. The processing conditions are given in Table 9.2 and the operation steps are shown in Figure 9.3.

Table 9.1 Microwave-assisted extraction (MAE) conditions applied in the recovery of compounds from pepper waste.

Compounds	Pepper waste mass (g)	Solvent	P (W)	F (MHz)	T (°C)	Time (min)	S/F (-)	Y (g/100 g pepper)	Reference
Terpenes, fatty acids, aldehydes and ketones	20	Water	250	2450	95	10	2	-	Gogus et al. (2015)
Terpenes, fatty acids, aldehydes and ketones	20	N-hexane	250	2450	55	10	2	-	Gogus et al. (2015)
Capsaicinoids	-	Olive oil	500	-	70	1	-	-	Paduano et al. (2014)
Capsaicin	2	Ethanol	90–600	-	-	5–20	200	0.65	Chuichulcherm et al. (2013)
Tocopherols	5	Ethanol	1000	2450	40-80	30	8	0.3–0.4	Xiao et al. (2009)
Ascorbic acid	2	MPA	175–595	-	-	2–15	6–12	0.12–0.15	Yu et al. (2009)
Capsaicin	0.5	Acetone	150	2450	120	15	30	0.67	Nazari et al. (2007)
Capsaicinoids	1	Ethanol-water	500	-	50	5	15	-	Barbero et al. (2006)
Piperine	2	Ethanol	750	2450	-	2	5	0.9	Raman & Gaikar (2002)

T, temperature; F, frequency; P, power; Y, yield of compounds; MPA, metaphosphoric acid solution (20 wt.%)

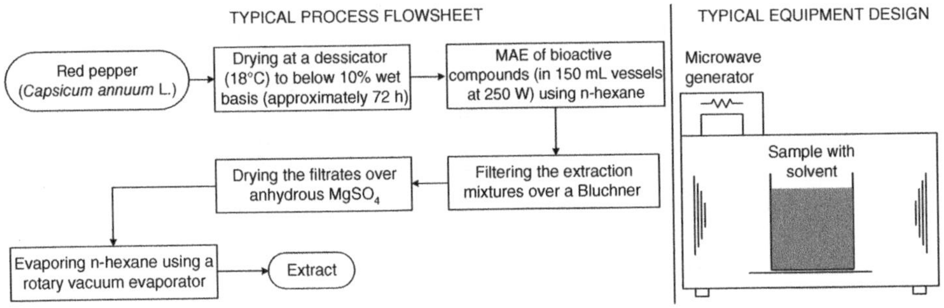

Figure 9.2 Flow sheet of microwave-assisted extraction (MAE) in the recovery of bioactive compounds from discarded pepper fruits and a diagram of the equipment commonly used in MAE processes.

9.2.2.3 Supercritical Fluid Extraction

There is a point in the phase diagram where a pure substance exists as a fluid comprising simultaneously the properties of both liquids and gasses, in particular high solubility and low viscosity. This is the critical point, which is set by the pair of critical temperature (T_C) and critical pressure (P_C) in the pressure-versus-temperature ($P \times T$) diagram. A pure substance is in the supercritical condition when it is heated beyond its T_C and pressurised beyond its P_C. Supercritical fluids present singular characteristics that enable them to be used as solvents due to their high solvation power, and they are useful in the recovery of volatile oils. Recent trends in the global market also indicate their use in microbial decontamination to avoid fast degradation of foods, for analytical purposes in supercritical fluid chromatography, in biotechnological processes and in processing engineering in the nanoscience and materials industries. Supercritical carbon dioxide (SC-CO_2) is used extensively to perform supercritical fluid extraction (SFE) because it allows the operation to be conducted under mild conditions of temperature. Overall, SC-CO_2 is used to extract nonpolar to moderately polar bioactive compounds.

In the case of pepper waste, SFE has been applied to improve the quality of extracts and the content of bioactive compounds recovered. In the traditional process, the dried and ground *Capsicum* fruit is extracted by organic solvents, even though a great amount of capsaicinoids remains in the residue. Perva-Uzunali et al. (2004) proposed instead using SFE with SC-CO_2, recovering approximately 95% of the capsaicinoids at 60 °C and 10 MPa. Likewise, Romo-Hualde et al. (2012) carried out SFE using pepper (*Capsicum annum* L. variety Piquillo) byproducts to recover a red-coloured oil. Approximately 97% of the vitamin E and 68% of the provitamin A were recovered at 60 °C and 24 MPa using SC-CO_2 without modifier in a total process time of only 90 minutes. In addition, the bioactive compounds extracted from pepper waste could be entrapped in microcapsules to extend their storage stability.

Berries of black pepper (*Piper nigrum*) that are discarded from culinary preparations are a source of important phytochemicals. In order to recover such phytochemicals, Perakis et al. (2005) performed SFE in a bench-scale apparatus. They studied the influence of pressure, temperature and solvent flow rate on the yields of pepper extracts, finding they were significantly enhanced as pressure increased or as temperature decreased. Furthermore, the extraction rate increased with increasing CO_2 flow rate. They analysed the composition of the pepper oil and compared it to that obtained using a hydrodistillation technique (Table 9.3). The essential oil obtained by

Table 9.2 Ultrasound-assisted extraction (UAE) conditions applied in the recovery of compounds from pepper waste.

Compounds	Mass of pepper waste (g)	Solvent	P (W)	F (kHz)	T (°C)	Time (min)	S/F (-)	Yield	Reference
Capsaicinoids	0.2	Methanol	360	-	50	15	125	1.8 μmol/g pepper	Barbero et al. (2014)
Capsaicinoids	5	Ethanol	360	20	-	10	20	15.6 g/100 g pepper	Aguiar et al. (2014)
Capsaicinoids	5	Methanol	360	20	-	10	20	33.4 g/100 g pepper	Aguiar et al. (2014)
Capsinoids	5	Acetone	360	20	-	10	20	9.6 mg/g extract	Aguiar et al. (2014)
Capsinoids	5	Hexane	360	20	-	10	20	8.9 mg/g extract	Aguiar et al. (2014)
Capsaicinoids	1	Acetone–methanol	-	40	24	2.9	25	1.9 mg/g pepper	Sganzerla et al. (2014)
Capsaicinoids	1	Methanol	-	40	24	10	25	2.9 mg/g pepper	Sganzerla et al. (2014)
Capsaicin	-	Olive oil	-	28.5	-	10	-	0.2 mg/g pepper	Paduano et al. (2014)
Capsaicinoids	2	Ethanol	320	40	-	20	100	4.0 mg/g pepper	Chuichulcherm et al. (2013)
Capsaicinoids	25	Ethanol	600	35	45	180	8	87.4%[a]	Boonkird et al. (2008)
Capsaicinoids	25	Ethanol	600	35	45	180	5	84.9%[a]	Boonkird et al. (2008)
Capsaicinoids	25	Ethanol	600	26	45	5	5	68%[a]	Boonkird et al. (2008)
Capsaicinoids	2	Methanol	360	20	50	10	8	98%[a]	Barbero et al. (2008)
Capsaicinoids	2	Ethanol	360	20	50	10	8	99%[a]	Barbero et al. (2008)
Capsaicinoids	2	Acetonitrile	360	20	50	10	8	78%[a]	Barbero et al. (2008)
Capsaicinoids	2	Water	360	20	50	10	8	5%[a]	Barbero et al. (2008)

a) Recovery of compounds.
T, temperature; F, frequency; P, power; S/F, solvent to feed mass ratio.

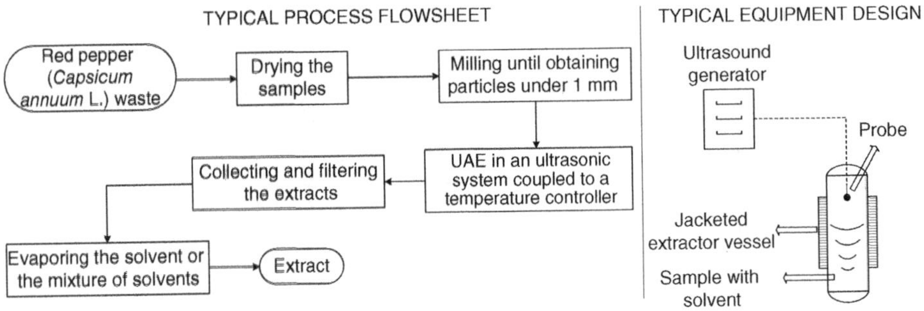

Figure 9.3 Flow sheet of ultrasound-assisted extraction (UAE) in the recovery of bioactive compounds from pepper waste and a diagram of the equipment commonly used in UAE processes.

Table 9.3 Composition (% of total peak area) of black pepper essential oil obtained by supercritical fluid extraction (SFE) and hydrodistillation.
Source: Perakis et al. (2005). Reproduced with permission of Elsevier.

Bioactive compound	Area (%)	
	SFE	Hydrodistillation
Monoterpenes		
α-thujene	0.43	0.45
α-pinene	1.77	2.15
Sabinene	5.46	4.81
β-pinene	3.47	4.21
δ-3-carene	4.91	4.34
Limonene	6.81	6.32
Oxygenated monoterpenes		
Linalool	1.28	0.59
Myrtenol	-	0.27
Carvone	-	0.29
Sesquiterpenes		
δ-elemene	0.23	0.30
α-cubebene	0.52	0.38
α-copaene	6.83	5.42
Calarene	2.64	2.18
α-gurjunene	0.49	0.30
β-caryophyllene	12.43	8.91
α-humulene	3.31	2.36
γ-gurjunene	0.59	0.48
α-selinene	3.68	2.64
δ-cadinene	3.73	2.14
Oxygenated sesquiterpenes		
Elemol	6.39	7.95
Caryophillene oxide	7.94	12.48
β-eudesmol	1.80	3.79

SFE presented higher levels of sesquiterpene hydrocarbons, yielding higher sesquiterpene/monoterpene ratios. The extract was enriched in sesquiterpene hydrocarbons as the extraction time increased, mainly in the second separator vessel of the operational apparatus.

The SFE process can be more effective than conventional techniques in selectively recovering bioactive compounds from wastes and reducing extraction times. The extraction of target compounds involves at least two phases, where a solvent (commonly CO_2) continuously flows through the extraction bed, denoting a process developed in semicontinuous mode (Figure 9.4). The solvent fills the bed void fraction and penetrates through the pores of the solid particles of the waste material to solubilise the solutes. At the outlet of the extraction bed, the solvent is more concentrated in bioactive compounds (Zabot et al., 2014c). According to some data presented for the solubility of pepper compounds in SC-CO_2 (Gupta & Shim, 2007), interactions of CO_2 or a mixture of solvents with solutes depend on the characteristics of the compounds found in the pepper waste, yielding different solubilities. For instance, the solubility of

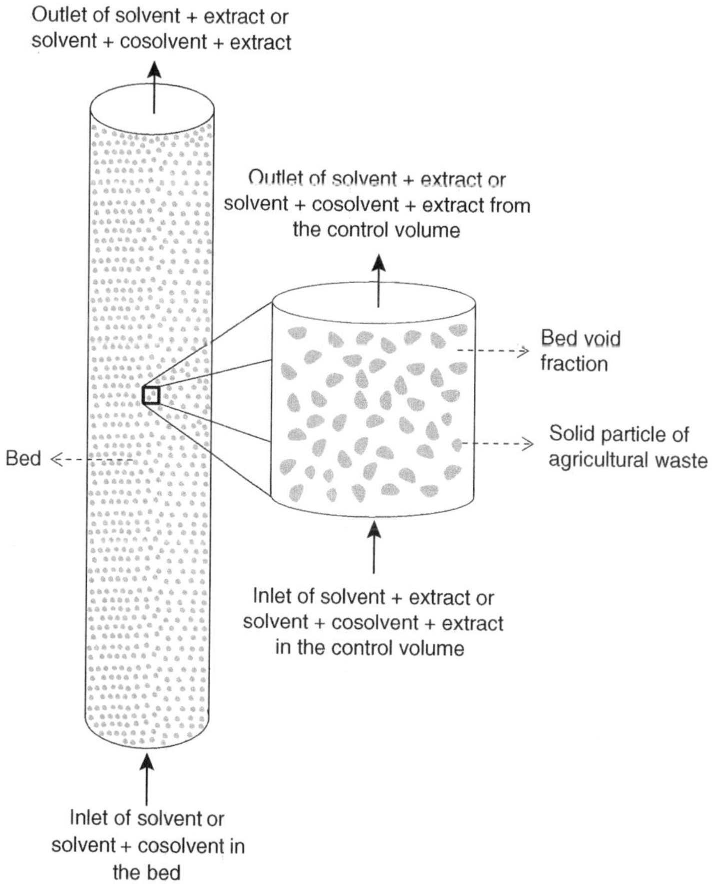

Figure 9.4 Simplified scheme of an extraction bed with continuous flow of a supercritical solvent (or mixture).

limonene in CO_2 at 40 °C and 15 MPa is 24 500 mmol/kmol CO_2, while the solubility of capsaicin in CO_2 at 40 °C and 14.7 MPa is 101 mmol/kmol CO_2.

Extracts from pepper waste made up of several parts (stalks, seeds, entire discarded fruits, etc.) are multicomponent systems. The large number of components improves the complexity of the system and influences understanding of the mass transfer mechanisms. Usually, when describing the SFE process, the system is considered to be biphasic: fluid phase + solid phase. The fluid phase comprises the solvent (or the mixture of solvent and cosolvent) and the bulk extract. The solid phase comprises the target extract and the inert material (non-extractable structure). The fluid and the solid phase are both pseudobinary systems, while the whole extraction system is considered a pseudoternary system (extract + solvent + non-extractable structure).

One way of increasing the transference of the solute from the solid to the fluid phase of pepper waste is to associate the ultrasound technique with SFE. This was done by Santos et al. (2015), who recovered capsaicinoids and phenolics from malagueta pepper (*Capsicum frutescens* L.) after using ultrasonic waves in the extraction bed during SFE (40 °C and 15 MPa). SFE without ultrasound yielded a recovery of 58% of the part of

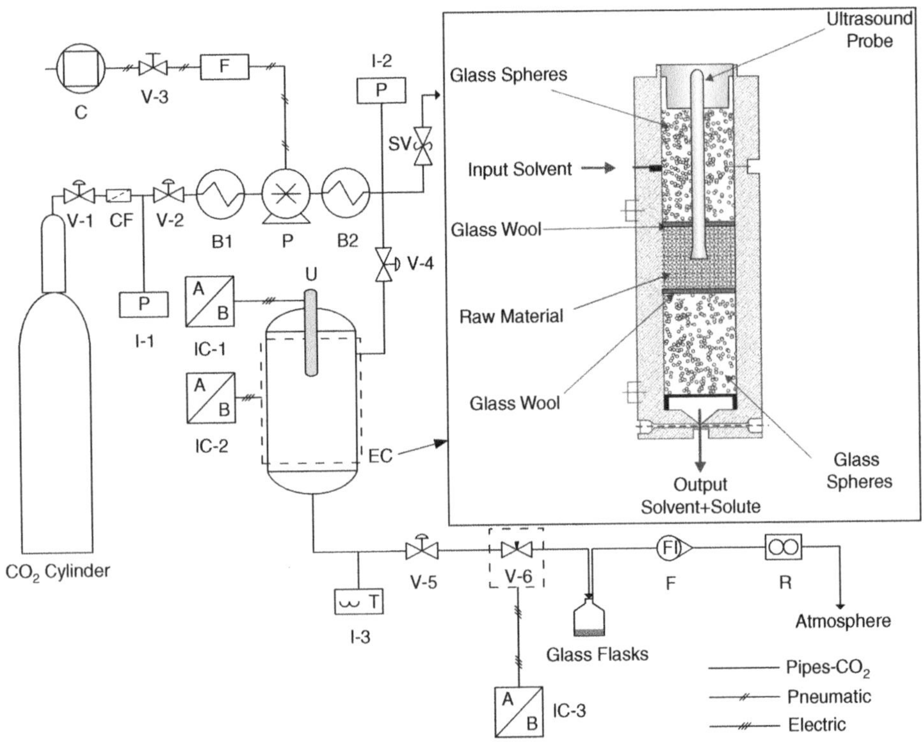

Figure 9.5 Diagrams of a supercritical fluid extraction (SFE) + ultrasound (US) unit and of the internal configuration of the extraction bed used in kinetic experiments. V-1, V-2, V-3, V-4 and V-5, control valves; V-6, micrometer valve; SV, safety valve; C, compressor; F, compressed air filter; CF, CO[2] filter; B1, cooling bath; P, pump; B2, heating bath; I-1 and I-2, pressure indicators; I-3, temperature indicator; IC-1, IC-2 and IC-3, indicators and controllers of ultrasound power, temperature of the extraction column and temperature of the micrometer valve, respectively; U, ultrasound probe; R, flow totaliser; F, flow meter; EC, extraction column. *Source:* Santos et al. (2015). Reproduced with permission of Elsevier.

the extract containing bioactive compounds. However, when ultrasonic waves of high power (360 W and 55.7 kJ/cm^2) were applied, the recovery increased to 75%. Regarding the operational apparatus, the ultrasound probe was installed inside the extractor vessel. Figure 9.5 shows a process diagram of an important model used to develop practical runs for the recovery of bioactive compounds from pepper waste.

9.3 Recovering Bioactive Compounds from Onion Waste

9.3.1 Bioactive Compounds found in Onion Waste

The amount of onion waste produced annually within the European Union is almost half a million tons (Andersson et al., 2012). Onion (*Allium cepa* L.) waste is a renewable raw material that contains several types of phytonutrients. The flavonoid content in onion peels is expressively higher than that identified in onion bulbs (Sharma et al., 2015). A great amount of antioxidant polyphenol content was recovered from onion solid wastes (Katsampa et al., 2015). Quercetin, a prominent naturally occurring antioxidant, is a flavonoid mostly found in onion peels (Wach et al., 2007), which are discarded as waste material all around the world. Onion peels and tunic (the papery outer part that protects the bulb from damage and drying out) are only a portion of the mass discarded after using the flower bud. Apical bud and basal plate also constitute waste materials (Figure 9.6). Some reports dealing with onion waste as a rich source of bioactive compound sare summarised in Table 9.4.

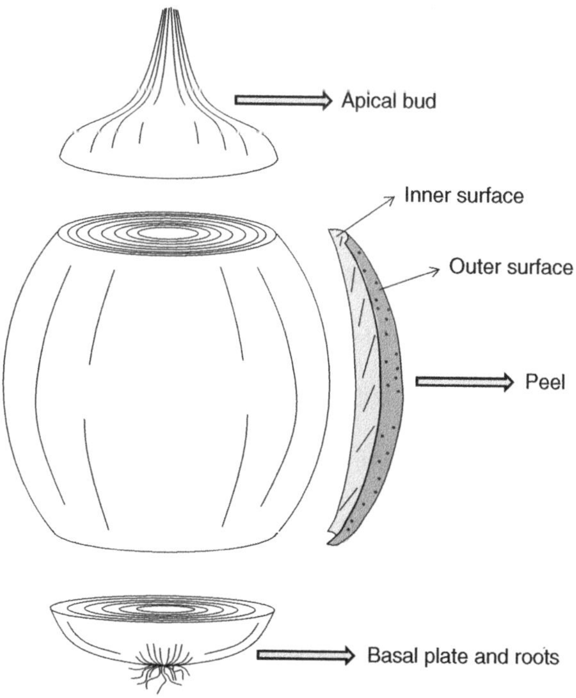

Figure 9.6 Parts of the onion bulb considered as solid wastes.

Table 9.4 Bioactive compounds recovered from onion waste.

Bioactive compound	Content	Extraction technique	Reference
Quercetin	6.8 mg/g extract	Solid–liquid with water	Wach et al. (2007)
	5.6 mg/g extract	Solid–liquid with methanol	
	5.3 mg/g extract	Solid–liquid with ethanol	
	4.0 mg/g extract	Solid–liquid with ethyl acetate	
	1.5 μmol/g dw	Solid–liquid with methanol	Sharma et al. (2015)
	26 g/100 g mp	PLE followed by SAS precipitation	Zabot & Meireles (2016)
	8.1 mg/g dw	UAE with ethanol	Jang et al. (2013)
	62.4 mg/g dw	Solid–liquid with ethanol	Lee et al. (2014)
	25.8 mg/g dw	Solid–liquid with hot water	
	44.4 mg/g dw	Subcritical water extraction	
Total polyphenols	64.9 mg GAE/g dw	UAE with aqueous glycerol	Katsampa et al. (2015)
	21.0 mg GAE/g dw	Solid–liquid with acidified aqueous ethanol	Makris & Kefalas (2015)
	19.3 mg GAE/g dw	Solid–liquid with acidified ethanol	Khiari et al. (2009)
	93.0 mg GAE/g dw	Solid–liquid with acidified aqueous ethanol	Kiassos et al. (2009)
	372.5 mg GAE/g extract	Solid–liquid with ethanol	Lee et al. (2014)
	120.6 mg GAE/g extract	Solid–liquid with hot water	
	218.7 mg GAE/g extract	Subcritical water extraction	
Carotenoid	35 μg/g dw	Solid–liquid with ether	Gupta & Ahmad (2013)
Tocopherol	8 μg/g dw	Solid–liquid with petroleum ether:ethanol	Gupta & Ahmad (2013)
Anthocyanin	198.9 mg CyE/100 g dw	Maceration followed by solid–liquid extraction	Makris (2010)
Flavonols	34.7 mg/g dw	Solid–liquid with acidified aqueous ethanol	Khiari et al. (2008)
Total flavonoids	184.0 mg QE/g extract	Solid–liquid with ethanol	Lee et al. (2014)
	54.5 mg QE/g extract	Solid–liquid with hot water	
	119.5 mg QE/g extract	Subcritical water extraction	

GAE: gallic acid equivalents; dw: dry weight of raw material; UAE: ultrasound-assisted extraction; mp: microparticles; PLE: pressurised liquid extraction; SAS: supercritical antisolvent; CyE: cyaniding 3-O-glucoside equivalents; QE: quercetin equivalents.

9.3.2 Major Techniques for the Recovery of Bioactive Compounds from Onion Waste

9.3.2.1 Conventional Extraction

As for pepper waste, conventional extraction of bioactive compounds from onion waste commonly uses a solvent at room pressure. This method was used by Lee et al. (2014) to obtain phenolics from orange-coloured onion peels. The waste was mixed with ethanol and heated for 3 hours at 60 °C. The post-extraction steps for the recovery of the extracts involved filtering the sample through filter paper and evaporating the solvent using a rotary evaporator. Great concentrations of total phenolics and flavonoids were obtained, and the extracts presented potent antioxidant activity in different *in vitro* assays, including a DPPH (2,2-diphenyl-1-picrylhydrzil) free-radical scavenging test and a thiobarbituric acid (TBA) test (Lee et al., 2014). A simplified protocol of operation steps for the recovery of bioactive compounds from onion waste is given in Figure 9.7, comprising ethanol extraction and hot water extraction, based on procedures developed by Kim et al. (2011). One disadvantage of using water as a solvent for the recovery of phytonutrients from wastes, as compared to organic solvents, is the difficulties in concentrating the extracts. Drying of aqueous extracts involves further steps around concentrating the bioactive compounds. Water can be removed by freeze or hot-air drying, which is more time and energy consuming than evaporating organic solvents (Andersson et al., 2012).

9.3.2.2 Supercritical Fluid Extraction and Subcritical Water Extraction

Several studies have undertaken the isolation and identification of quercetin and its glycosides from onion waste (Lee et al., 2014; Sharma et al., 2015; Wach et al., 2007). Most use simple conventional extraction techniques that require the disposal of organic solvents, although nonconventional extraction techniques have been proposed as well. A study on the recovery of quercetin from red and yellow onion peels (broken into small flakes) used a commercial SFE apparatus (Autoclave Engineers, Erie, PA) with CO_2 and 5% ethanol as cosolvent (molar concentration) (Martino & Guyer, 2004). The modifier had an important influence on free quercetin recovery. The greater the amount of ethanol collected in the trap, the greater the amount of

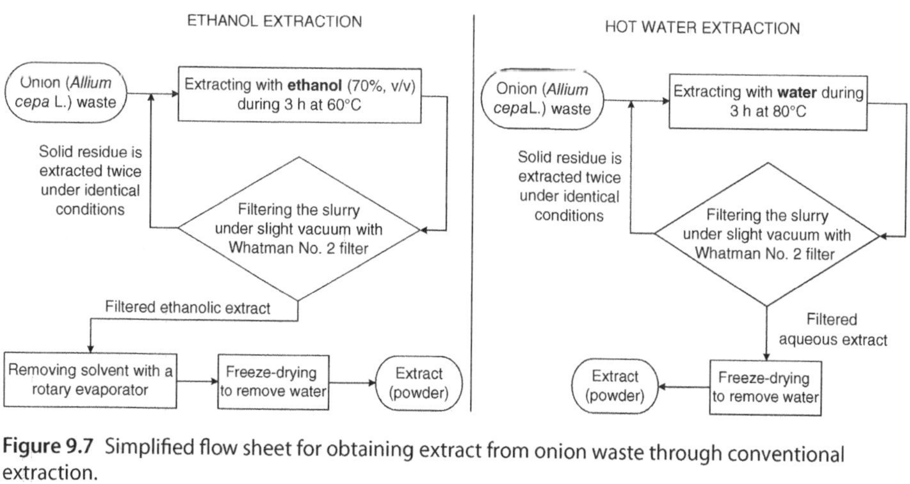

Figure 9.7 Simplified flow sheet for obtaining extract from onion waste through conventional extraction.

free quercetin recovered. Fresh onion stalk extract was also obtained by SFE with CO_2. Four compounds were identified: (+)-catechin, (−)-epicatechin, astragalin and 3-O-β-D(2-O-β-D-glucopyranosyl)-glucopyranosides of kaempferol (Fu et al., 2010). Few reports on this subject are available in the literature, because the solubility of quercetin in CO_2 is extremely low. For instance, at 45 °C and 10.1 MPa, the solubility of quercetin is 5.4 mmol/kmol CO_2. If 5% (molar basis) ethanol is added to the mixture, the solubility of quercetin increases approximately 100 times (520 mmol/kmol CO_2 + ethanol) (Gupta & Shim, 2007).

With several researchers attempting to use procedures that are 'green', we also see many applications for the recovery of bioactive compounds from onion agricultural waste using other solvents than CO_2. One of these environmentally sustainable green processes uses only water as solvent through subcritical water extraction (SWE). Turner et al. (2006) attempted to obtain quercetin and other flavonoid molecular species from onion waste using SWE at 5 MPa and three temperatures of 40, 80 and 120 °C. They reached their goal in that the quercetin yields were higher than those using the conventional method, and the total processing time per sample was reduced. Lee et al. (2014) applied SWE to recover flavonoids and phenolics from onion waste. When the process was developed at lower temperature (110 °C), the quercetin concentration was improved, demonstrating that onion peels extracts produced by SWE have great potential application as rich sources of antioxidants. A simplified operation protocol for the recovery of bioactive compounds from onion waste by SWE is presented in Figure 9.8.

9.3.2.3 On-Line Process: Pressurised Liquid Extraction and Particle Formation in Sequence

A novel process for particle production named 'extraction and particle formation on-line' (EPFO) has been reported (Zabot & Meireles, 2016). This involves the hyphenated on-line association of pressurised liquid extraction (PLE) with supercritical antisolvent (SAS) precipitation, performed using the same multipurpose equipment. Using this process, dried microparticles rich in quercetin and other bioactive compounds were obtained as a fine powder from onion peels (waste material). Optimal conditions (40 °C and 12 MPa for the extraction step, 40 °C and 10 MPa for the precipitation step) yielded 4.1 ± 0.6 g microparticles/100 g dried onion peels and the highest quercetin content in the microparticles (26 ± 2% by weight). The highlight of the process is reducing the degradation of sensible bioactive compounds, because the manipulation of extracts from onion waste is reduced as a consequence of the direct flow of the solvent–extract solution to the precipitation vessel. Consequently, light and

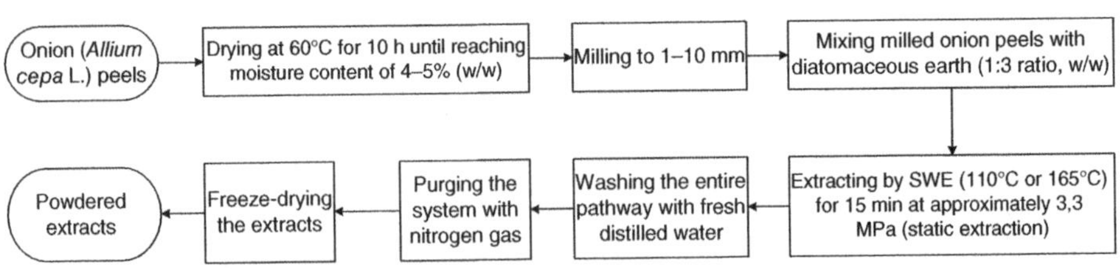

Figure 9.8 Simplified flow sheet for obtaining extract from onion peels through subcritical water extraction (SWE).

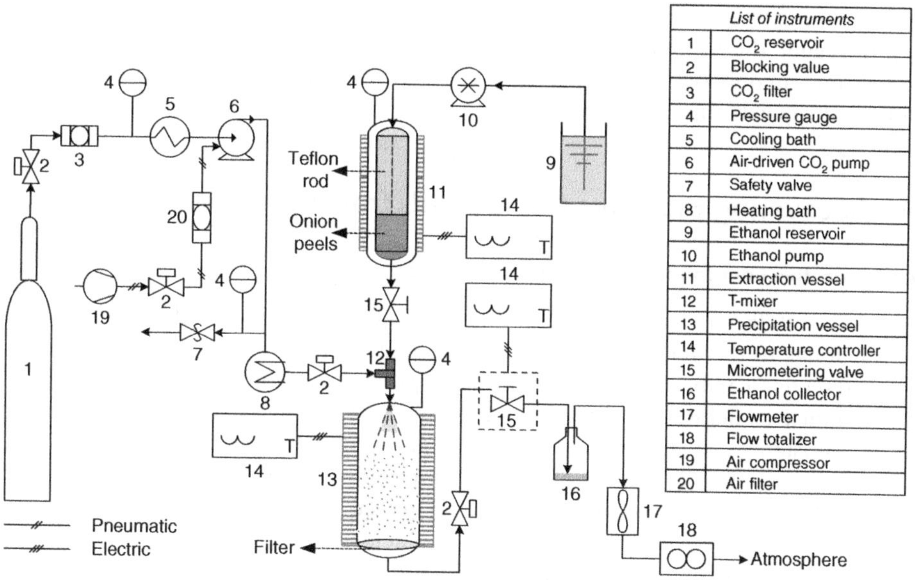

Figure 9.9 Schematic drawing of extraction and particle formation on-line (EPFO) equipment. *Source:* Zabot & Meireles (2016). Reproduced with permission of Elsevier.

oxygen exposure of bioactive compounds are also reduced. Figure 9.9 illustrates the use of the EPFO process for onion waste.

A similar on-line process that uses the concept of process intensification was developed to add value to onion waste (Andersson et al., 2012). Quercetin, quercetin-3, 4′-diglucoside and quercetin-4′-glucoside (Figure 9.10) were the major compounds

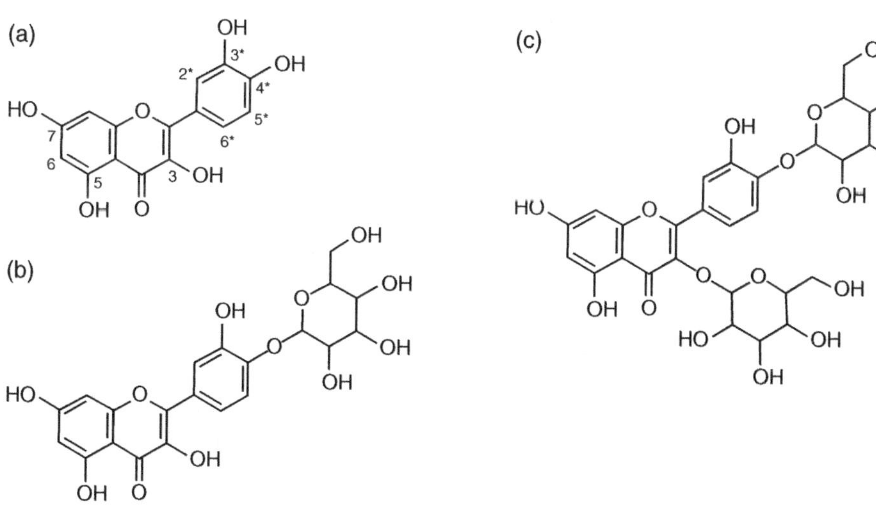

Figure 9.10 Chemical structures of (a) quercetin, (b) quercetin-49-glycoside and (c) quercetin-3,49-diglycoside. *Source:* Turner et al. (2006). Reproduced with permission of the Royal Society Of Chemistry.

present in the extract and in the microparticles formed in an expansion-drying chamber using nitrogen. The process could be considered a suitable and promising alternative for obtaining, in only one step, fine and dried particles (smaller than 4 μm) with intact antioxidant capacity directly from onion waste.

9.4 Recovering Bioactive Compounds from Cotton Waste

9.4.1 Bioactive Compounds found in Cotton Waste

Cotton is a soft fibre that grows around the seed and the main part of the cotton plant, which belongs to the genus *Gossypium*. The remaining parts – the stalks, hulls, root barks and seeds – can be used to produce refined products, having practical applications in some industrial fields. Cottonseeds, as byproducts of cotton harvesting, contain a lot of useful substances, such as phenolic compounds in the extracted edible oil (Taghvaei et al., 2014), tocopherols (Zhu et al., 2012), flavonoids (Xiao et al., 2007), sterols and gossypol (Zhu et al., 2014). Cottonseed oil recovered from cottonseed hull contains approximately 30% by weight of saturated fatty acids, but also phenolic compounds (Taghvaei et al., 2014). Oskoueian et al. (2011) found the total phenolic and flavonoid contents of cottonseed meal were 1.5 mg GAE/g dry weight and 1.2 mg rutin equivalents (RE)/g dry weight, respectively. Tocopherols and flavonoids are substances with wide application in the pharmaceutical and food fields as powerful antioxidants, alleviating oxidative stress and defending the body by quenching free radicals.

Gossypol is a polyphenolic compound present in cotton plant tissues, including the seed and flower petals, in the form of tiny glands that protect the plant from insects, pests and diseases. This polyphenol is toxic if consumed in high quantities in cottonseed edible oil. However, it has been reported to have a wide spectrum of pharmacological activities when used at appropriate levels. According to Lin et al. (2016), gossypol can induce pyroptosis in mouse macrophages via a noncanonical inflammasome pathway. Other therapeutic properties have also been reported, such as antitumour, anti-inflammatory, anticancer, antifertility, antioxidant and antifungal activities (Keshmiri-Neghab & Goliaei, 2014; Moon et al., 2011; Yang et al., 2015).

9.4.2 Main Techniques for the Recovery of Bioactive Compounds from Cotton Waste

A recent study evaluated the influence of MAE on the oxidative stability of cottonseed oil, focusing on the phenolic compounds (Taghvaei et al., 2014). Three levels of moisture content (1, 7 and 14% by fresh weight) were applied, prior to extraction with n-hexane. The MAE system consisted of a volumetric flask (500 ml) coupled with a condenser at the top and a magnetic stirrer beneath. The microwave output was 900 W, with 2450 MHz frequency. The total phenolic content of the oil increased with increasing irradiation time and moisture content. Phenolic compounds that were exposed to microwave irradiation for long periods had a pronounced recovery. As n-hexane is a nonpolar solvent, which is not influenced by microwaves, increasing the amount of n-hexane in the extraction flask led to poor penetration of microwaves into the plant cells and lower efficiency. The highest phenolic content measured in the recovered oil was 70 ppm GAE. Taghvaei et al. (2014) also assessed the influence of phenolic compounds contained in cottonseed hull on the oxidative stability of the oil, finding that increasing the total phenolic content increased the thermal stability. Furthermore, the thermal stability of blank cottonseed oil was greater than that of refined soybean oil, with or without 100 or 200 ppm butylated hydroxytoluene (BHT). Therefore, the authors concluded that MAE is a suitable process for the rapid recovery of compounds (approximately 3.5 minutes) from cotton waste, yielding an extract with favourable physical properties and oxidative stability.

Oskoueian et al. (2011) also recovered phenolics and flavonoids from cottonseed using MAE (280 W for 4 minutes) and found that the contents of kaempferol, naringenin and rutin were 114, 179 and 209 μg/g dry weight, respectively. The extracts displayed antioxidant activity, with an IC_{50} (compound concentration causing 50% inhibition of the desired activity) of 191 μg/ml for DPPH radical scavenging activity.

Commonly, the steps taken to recover compounds from cotton waste using MAE technique involve preparing the samples (setting the ideal moisture content and milling to fine particles), selecting the solvent, defining the best values for the operational parameters, effectively extracting the compounds, removing the miscella by filtration followed by centrifugation, removing the remaining solvent under vacuum using a rotary evaporator and collecting the extracts. With UAE, on the other hand, the steps are preparing the samples (drying and milling to fine particles), defining the solvent or the mixture of solvents, loading the sample in the extractor vessel, effectively extracting the compounds under stirring and controlled temperature, collecting and filtering the extracts, washing the filter cake with the solvents and separating (liquid–liquid) the compounds and solvents using a separatory funnel. Afterwards, the upper phase is evaporated under vacuum at constant temperature, submitted to transesterification and then submitted to crystallisation for recovery of sterols. Otherwise, gossypol is recovered from the lower phase using the unit operations of crystallisation and evaporation.

An integrated approach was developed at laboratory scale for the recovery of several products from cottonseed, using a biorefinery concept. The recovery of gossypol using UAE was greater than 80% at 35 °C and 20 minutes' processing time. The separation and concentration of gossypol involved the unit operations of liquid–liquid extraction, crystallisation and evaporation. When the extraction temperature was over 35 °C, the gossypol yield decreased with increasing temperature, because the effects of permeability and solubility did not exceed the effect of oxidation. Moreover, the yields of gossypol in later stages declined markedly, and the solution became too dark (rather brown than red) because of the degradation and photo-oxidation of gossypol over a too-long process time. Crude gossypol was purified using recrystallisations. Regarding sterols, their extraction by UAE and transesterification with methanol yielded a final recovery of 75.4% by weight, giving a sterol purity of 89.8% by weight (Zhu et al., 2014).

As has been seen, gossypol and sterols are recovered in different phases during the processing of cottonseed. Sterols are concentrated in the lighter phase, together with the oil, while gossypol is concentrated in the heavier phase. Several attempts using different solvents have been made to extract gossypol from the seeds. After extracting the cottonseed edible oil, the leftover cake contains high amounts of gossypol. Therefore, a study was undertaken to extract gossypol from the defatted cottonseed meal using ethanol (considered Generally Recognised As Safe, GRAS) acidified by mild organic oxalic acid at 70 °C. The results indicated the objective of the study – to improve the extraction efficiency – had been met. Approximately 92% by weight of gossypol could be recovered from the defatted cottonseed meal using acidic ethanol, versus approximately 62% by weight with pure ethanol (Saxena et al., 2012). Purification of the gossypol was also reported, with packed microtip columns in combination with high-performance liquid chromatography (HPLC) providing 98% recovery. Gossypol content varied with the cotton waste tissues used: 11.0 mg/g dry weight for seeds, 0.8 mg/g dry weight for leaves and 0.3 mg/g dry weight for stems (Meyer et al., 2004).

Another bioactive compound found in cotton waste is α-tocopherol, the extraction of which involves SLE with n-hexane as solvent. Acting as chain-breaking antioxidants, tocopherols react with lipid radicals to convert them into more stable products. Mariod et al. (2011) found cottonseed oil containing tocopherols (α-, β-, γ- and δ-isomers) was extracted from the ground material with n-hexane at 50–60 °C in a Soxhlet apparatus for 6 hours. The total content of tocopherols in the cottonseeds was approximately 75 mg/100 g, divided between γ-tocopherol (61.0%), α-tocopherol (38.0%), β-tocopherol (0.5%) and δ-tocopherol (0.5%). Sterols were also recovered (Table 9.5). However, a possible need for a more polar extraction solvent mixture, in order to improve the extractability of tocopherols, has observed. Petroleum ether and methanol can be used for the recovery of α-tocopherol from cotton waste. Cottonseed oil and tocopherols can be dissolved in petroleum ether, while gossypol, water, raffinose, colloids and so on are dissolved in methanol. The total content of α-tocopherol in cottonseeds is approximately 110 mg/100 g fresh weight, achieving approximately 97% recovery. When increasing the extraction temperature from 25 to 45 °C, the extraction rate of α-tocopherol increases from 88.6 to 95.5%. Furthermore, increasing

Table 9.5 Tocopherol composition (mg/100 g cottonseed) and phytosterol composition (mg/100 g oil) after recovery of bioactive compounds from cotton waste. *Source:* Adapted from Mariod et al. (2011).

Compound	Content
Tocopherols	
α-tocopherol	28.6 ± 0.2
β-tocopherol	0.4 ± 0.1
γ-tocopherol	45.9 ± 0.3
δ-tocopherol	0.3 ± 0.1
Phytosterols	
Campesterol	43.9 ± 0.3
Stigmasterol	5.3 ± 0.1
β-sitosterol	403.3 ± 0.5
δ5-avenasterol	15.5 ± 0.3
δ7-avenasterol	2.1 ± 0.1
δ7-estigmasterol	3.4 ± 0.1
Others[a]	15.7 ± 0.3

a) Others include 24-methylcholesterol, campestanol, clerosterol, sitostanol and 5,24-stigmastadienol.

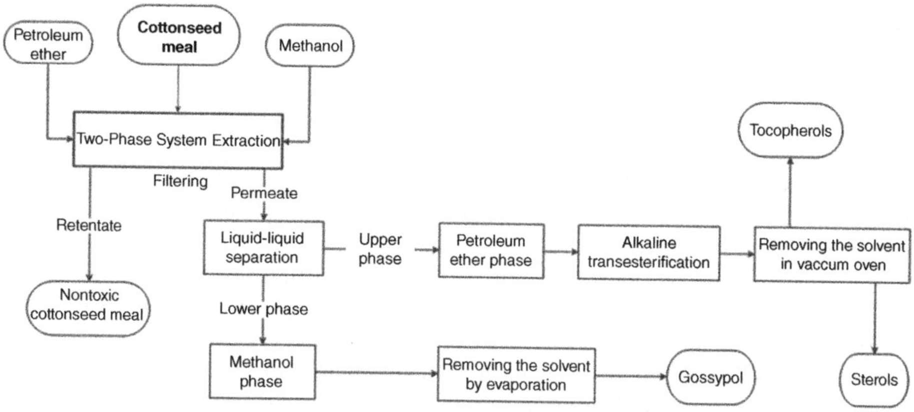

Figure 9.11 Process for dealing with cottonseed using a two-phase system (TPS).

the petroleum ether/methanol volume ratio from 1 : 6 to 1 : 3 increased the amount of α-tocopherol dissolved in petroleum ether (Zhu et al., 2012). The system previously mentioned corresponds to a two-phase system, where two solvents separate into two immiscible phases. Some value-added compounds can be obtained by following a suggested schematic diagram for cottonseed meal processing (Figure 9.11).

9.5 Conclusion

Pepper, onion and cotton wastes can be used for the recovery of flavonoids, capsaicinoids, terpenes, terpenoids, tocopherols, carotenoids, polyphenols and phytosterols, among other compounds. The SFE technique is preferable in most cases, providing greater yields and selective extractions. This trend is seen because the use of huge amounts of toxic organic solvents in the recovery of bioactive compounds from agricultural wastes is no longer feasible due to global health and environmental issues. SFE has proved its suitability when it comes to the extraction of nonpolar compounds, representing a business opportunity for producers of extracts from nonconventional raw materials as wastes. The economic appeal comes from the low or null cost of acquiring the wastes. In several cases, the only costs before processing are related to transport and pre-processing (drying, milling and storing). As with SFE, SWE, PLE, MAE and UAE have demonstrated applicability, mainly due to the considerable savings they provide in processing time, solvent consumption and energy. Therefore, combining the use of easily available wastes with a suitable extraction technique can represent an ideal association for the industrial manufacture of value-added substances.

Acknowledgements

The authors are grateful to FAPESP (2015/13299-0) for financial support. Giovani L. Zabot thanks FAPESP (2014/15685-1) for the postdoctoral scholarship. Giovani L. Zabot thanks FIPE-ARD-UFSM (2016/042956) for the assistantship.

References

Aguiar, A. C., Santos, P., Coutinho, J. P., Barbero, G. F., Godoy, H. T. & Martínez, J. 2014. Supercritical fluid extraction and low pressure extraction of Biquinho pepper (*Capsicum chinense*). *LWT – Food Science and Technology*, **59**, 1239–1246.

Andersson, J. M., Lindahl, S., Turner, C. & Rodriguez-Meizoso, I. 2012. Pressurised hot water extraction with on-line particle formation by supercritical fluid technology. *Food Chemistry*, **134**, 1724–1731.

Bae, H., Jayaprakasha, G. K., Jifon, J. & Patil, B. S. 2012. Extraction efficiency and validation of an HPLC method for flavonoid analysis in peppers. *Food Chemistry*, **130**, 751–758.

Barbero, G. F., Palma, M. & Barroso, C. G. 2006. Determination of capsaicinoids in peppers by microwave-assisted extraction–high-performance liquid chromatography with fluorescence detection. *Analytica Chimica Acta*, **578**, 227–233.

Barbero, G. F., Liazid, A., Palma, M. & BARROSO, C. G. 2008. Ultrasound-assisted extraction of capsaicinoids from peppers. *Talanta*, **75**, 1332–1337.

Barbero, G. F., Ruiz, A. G., Liazid, A., Palma, M., Vera, J. C. & Barroso, C. G. 2014. Evolution of total and individual capsaicinoids in peppers during ripening of the Cayenne pepper plant (*Capsicum annuum* L.). *Food Chemistry*, **153**, 200–206.

Boonkird, S., Phisalaphong, C. & Phisalaphong, M. 2008. Ultrasound-assisted extraction of capsaicinoids from *Capsicum frutescens* on a lab- and pilot-plant scale. *Ultrasonics Sonochemistry*, **15**, 1075–1079.

Chen, L. & Kang, Y. H. 2013. Anti-inflammatory and antioxidant activities of red pepper (*Capsicum annuum* L.) stalk extracts: comparison of pericarp and placenta extracts. *Journal of Functional Foods*, **5**, 1724–1731.

Chuichulcherm, S., Prommakort, S., Srinophakun, P. & Thanapimmetha, A. 2013. Optimization of capsaicin purification from *Capsicum frutescens* Linn. with column chromatography using Taguchi design. *Industrial Crops and Products*, **44**, 473–479.

Fu, Q., Liu, J., Zhang, C., Zhou, Y., Zhang, G., MA, D. & Yang, X. 2010. Separation and identification of flavonoids from fistular onion stalk (*Allium fisturosum* L. var. Caespitosum Makio). *Journal of Huazhong University of Science and Technology – Medical Science*, **30**, 255–257.

Gogus, F., Ozel, M. Z., Keskin, H., Yanik, D. K. & Lewis, A. C. 2015. Volatiles of fresh and commercial sweet red pepper pastes: processing methods and microwave assisted extraction. *International Journal of Food Properties*, **18**, 1625–1634.

Gupta, A. K. & Ahmad, M. 2013. Effect of refinery waste effluent on tocopherol, carotenoid, phenolics and other antioxidants content in *Allium cepa*. *Toxicology and Industrial Health*, **29**, 652–661.

Gupta, R. B. & Shim, J. J. 2007. *Solubility in Supercritical Carbon Dioxide*. Boca Raton, FL: CRC Press.

Hervert-Hernandez, D., Sayago-Ayerdi, S. G. & Goñi, I. 2010. Bioactive compounds of four hot pepper varieties (*Capsicum annuum* L.), antioxidant capacity, and intestinal bioaccessibility. *Journal of Agricultural Food Chemistry*, **58**, 3399–3406.

Jang, M., Asnin, L., Nile, S. H., Keum, Y. S., Kim, H. Y. & Park, S. W. 2013. Ultrasound-assisted extraction of quercetin from onion solid wastes. *International Journal of Food Science and Technology*, **48**, 246–252.

Katsampa, P., Valsamedou, E., Grigorakis, S. & Makris, D. P. 2015. A green ultrasound-assisted extraction process for the recovery of antioxidant polyphenols and pigments from onion solid wastes using Box–Behnken experimental design and kinetics. *Industrial Crops and Products*, **77**, 535–543.

Keshmiri-Neghab, H. & Goliaei, B. 2014. Therapeutic potential of gossypol: an overview. *Pharmaceutical Biology*, **52**, 124–128.

KHIARI, Z., MAKRIS, D. P. & KEFALAS, P. 2008. Recovery of bioactive flavonols from onion solid wastes employing water/ethanol-based solvent systems. *Food Science and Technology International*, **14**, 497–502.

Khiari, Z., Makris, D. P. & Kefalas, P. 2009. An investigation on the recovery of antioxidant phenolics from onion solid wastes employing water/ethanol-based solvent systems. *Food and Bioprocess Technology*, **2**, 337–343.

Kiassos, E., Mylonaki, S., Makris, D. P. & Kefalas, P. 2009. Implementation of response surface methodology to optimise extraction of onion (*Allium cepa*) solid waste phenolics. *Innovative Food Science and Emerging Technologies*, **10**, 246–252.

Kim, W. J., Lee, K. A., Kim, K. T., Chung, M. S., Cho, S. & Paik, H. D. 2011. Antimicrobial effects of onion (*Allium cepa* L.) peel extracts produced via subcritical water extraction against *Bacillus cereus* strains as compared with ethanolic and hot water extraction. *Food Science and Biotechnology*, **20**, 1101–1106.

Ksibi, I. E., Slama, R. B., Faidi, K., Ticha, M. B. & M'Henni, M. F. 2015. Mixture approach for optimizing the recovery of colored phenolics from red pepper (*Capsicum annum* L.) by-products as potential source of natural dye and assessment of its antimicrobial activity. *Industrial Crops and Products*, **70**, 34–40.

Lee, K. A., Kim, K. T., Kim, H. J., Chung, M. S., Chang, P. S., Park, H. & Paik, H. D. 2014. Antioxidant activities of onion (*Allium cepa* L.) peel extracts produced by ethanol, hot water, and subcritical water extraction. *Food Science and Biotechnology*, **23**, 615–621.

Lin, Q. R., Li, C. G., Zha, Q. B., Xu, L. H., Pan, H., Zhao, G. X., Ouyang, D. Y. & He, X. H. 2016. Gossypol induces pyroptosis in mouse macrophages via a non-canonical inflammasome pathway. *Toxicology and Applied Pharmacology*, **292**, 56–64.

Makris, D. P. 2010. Optimisation of anthocyanin recovery from onion (*Allium cepa*) solid wastes using response surface methodology. *Electronic Journal of Environmental, Agricultural and Food Chemistry*, **9**.

Makris, D. P. & Kefalas, P. 2015. Kinetic modelling for polyphenol extraction from onion (*Allium cepa*) solid wastes using acidified water/ethanol mixture. *Acta Alimentaria*, **44**, 482–492.

Mariod, A., Mattha, B. & Hussein, I. H. 2011. Fatty acids, tocopherols and sterols of *Cephalocroton cordofanus* in comparison with sesame, cotton, and groundnut oils. *Journal of the American Oil Chemists' Society*, **88**, 1297–1303.

Martino, K. G. & Guyer, D. 2004. Supercritical fluid extraction of quercetin from onion skins. *Journal of Food Process Engineering*, **27**, 17–28.

Meyer, R., Vorster, S. & Dubery, I. A. 2004. Identification and quantification of gossypol in cotton by using packed micro-tips columns in combination with HPLC. *Analytical and Bioanalytical Chemistry*, **380**, 719–724.

Mokhtar, M., Russo, M., Cacciola, F., Donato, P., Giuffrida, D., Riazi, A., Farnetti, S., Dugo, P. & Mondello, L. 2016. Capsaicinoids and carotenoids in *Capsicum annuum* L.: optimization of the extraction method, analytical characterization, and evaluation of its biological properties. *Food Analytical Methods*, **9**, 1381–1390.

Moon, D. O., Choi, Y. H., Moon, S. K., Kim, W. J. & Kim, G. Y. 2011. Gossypol decreases tumor necrosis factor-a-induced intercellular adhesion molecule-1 expression via suppression of NF-κB activity. *Food and Chemical Toxicology*, **49**, 999–1005.

Moraes, M. N., Zabot, G. L. & Meireles, M. A. A. 2015. Extraction of tocotrienols from annatto seeds by a pseudocontinuously operated SFE process integrated with low-pressure solvent extraction for bixin production. *Journal of Supercritical Fluids*, **96**, 262–271.

Nazari, F., Ebrahimi, S. N., Talebi, M., Rassouli, A. & Bijanzadeh, H. R. 2007. Multivariate optimisation of microwave-assisted extraction of capsaicin from *Capsicum frutescens* L. and quantitative analysis by ^{1}H-NMR. *Phytochemical Analysis*, **18**, 333–340.

Oskoueian, E., Abdullah, N., Hendra, R. & Karimi, E. 2011. Bioactive compounds, antioxidant, xanthine oxidase inhibitory, tyrosinase inhibitory and anti-inflammatory activities of selected agro-industrial by-products. *International Journal of Molecular Sciences*, **12**, 8610–8625.

Paduano, A., Caporaso, N., Santini, A. & Sacchi, R. 2014. Microwave and ultrasound-assisted extraction of capsaicinoids from chili peppers (*Capsicum annuum* L.) in flavored olive oil. *Journal of Food Research*, **3**, 51–59.

Perakis, C., Louli, V. & Magoulas, K. 2005. Supercritical fluid extraction of black pepper oil. *Journal of Food Engineering*, **71**, 386–393.

Perva-Uzunali, A., Skerget, M., Weinreich, B. & Knez, Z. 2004. Extraction of chilli pepper (var. Byedige) with supercritical CO_2: effect of pressure and temperature on capsaicinoid and colour extraction efficiency. *Food Chemistry*, **87**, 51–58.

Raman, G. & Gaikar, V. G. 2002. Microwave-assisted extraction of piperine from *Piper nigrum*. *Industrial and Engineering Chemistry Research*, **41**, 2521–2528.

Romo-Hualde, A., Yetano-Cunchillos, A. I., González-Ferrero, C., Sáiz-Abajo, M. J. & González-Navarro, C. J. 2012. Supercritical fluid extraction and microencapsulation of bioactive compounds from red pepper (*Capsicum annum* L.) by-products. *Food Chemistry*, **133**, 1045–149.

Santos, P., Aguiar, A. C., Barbero, G. F., Rezende, C. A. & Martínez, J. 2015. Supercritical carbon dioxide extraction of capsaicinoids from malagueta pepper (*Capsicum frutescens* L.) assisted by ultrasound. *Ultrasonics Sonochemistry*, **22**, 78–88.

Saxena, D. K., Sharma, S. K. & Sambi, S. S. 2012. Kinetics and thermodynamics of gossypol extraction from defatted cottonseed meal by ethanol acidified by oxalic acid. *International Journal of Science and Research*, **4**, 1967–1971.

Sganzerla, M., Coutinho, J. P., Melo, A. M. T. & Godoy, H. T. 2014. Fast method for capsaicinoids analysis from *Capsicum chinense* fruits. *Food Research International*, **64**, 718–725.

Sharma, K., Assefa, A. D., Ko, E. Y., Lee, E. T. & Park, S. W. 2015. Quantitative analysis of flavonoids, sugars, phenylalanine and tryptophan in onion scales during storage under ambient conditions. *Journal of Food Science and Technology*, **52**, 2157–2165.

Silva, L. R., Azevedo, J., Pereira, M. J., Valentão, P. & Andrade, P. B. 2013. Chemical assessment and antioxidant capacity of pepper (*Capsicum annuum* L.) seeds. *Food and Chemical Toxicology*, **53**, 240–248.

Silva, E. K., Zabot, G. L. & Meireles, M. A. A. 2015. Ultrasound-assisted encapsulation of annatto seed oil: retention and release of a bioactive compound with functional activities. *Food Research International*, **78**, 159–168.

Silva, E. K., Zabot, G. L., Cazarin, C. B. B., Maróstica, M. R. & Meireles, M. A. A. 2016. Biopolymer-prebiotic carbohydrate blends and their effects on theretention of bioactive compounds and maintenance of antioxidantactivity. *Carbohydrate Polymers*, **144**, 149–158.

Taghvaei, M., Jafari, S. M., Assadpoor, E., Nowrouzieh, S. & Alishah, O. 2014. Optimization of microwave-assisted extraction of cottonseed oil and evaluation of its oxidative stability and physicochemical properties. *Food Chemistry*, **160**, 90–97.

Turner, C., Turner, P., Jacobson, G., Almgren, K., Sjöberg, M. W. P., Karlssonb, E. N. & Markides, K. E. 2006. Subcritical water extraction and b-glucosidase-catalyzed hydrolysis of quercetin glycosides in onion waste. *Green Chemistry*, **8**, 949–959.

Wach, A., Pyrzyńska, K. & Biesaga, M. 2007. Quercetin content in some food and herbal samples. *Food Chemistry*, **100**, 699–704.

Xiao, Y. H., Zhang, Z. S., Yin, M. H., Luo, M., Li, X. B., Hou, L. & Pei, Y. 2007. Cotton flavonoid structural genes related to the pigmentation in brown fibers. *Biochemical and Biophysical Research Communications*, **358**, 73–78.

Xiao, X. H., Wang, J. X., Wang, G., Wang, J. Y. & Li, G. K. 2009. Evaluation of vacuum microwave-assisted extraction technique for the extraction of antioxidants from plant samples. *Journal of Chromatography A*, **1216**, 8867–8873.

Yang, D., Qu, J., Qu, X., Cao, Y., Xu, L., Hou, K., Feng, W. & Liu, Y. 2015. Gossypol sensitizes the antitumor activity of 5- FU through down- regulation of thymidylate synthase in human colon carcinoma cells. *Cancer Chemotherapy and Pharmacology*, **76**, 575–586.

Yu, Y., Chen, B., Chen, Y., Xie, M., Duan, H., Li, Y. & Duan, G. 2009. Nitrogen-protected microwave-assisted extraction of ascorbic acid from fruit and vegetables. *Journal of Separation Science*, **32**, 4227–4233.

Zabot, G. L. & Meireles, M. A. A. 2016. On-line process for pressurized ethanol extraction of onion peelsextract and particle formation using supercritical antisolvent. *Journal of Supercritical Fluids*, **110**, 230–239.

Zabot, G. L., Moraes, M. N. & Meireles, M. A. A. 2014a. Influence of the bed geometry on the kinetics of rosemary compounds extraction with supercritical CO_2. *Journal of Supercritical Fluids*, **94**, 244–254.

Zabot, G. L., Moraes, M. N., Petenate, A. J. & Meireles, M. A. A. 2014b. Influence of the bed geometry on the kinetics of the extraction of clove bud oil with supercritical CO_2. *Journal of Supercritical Fluids*, **93**, 56–66.

Zabot, G. L., Moraes, M. N., Rostagno, M. A. & Meireles, M. A. A. 2014c. Fast analysis of phenolic terpenes by high-performance liquid chromatography using a fused-core column. *Analytical Methods*, **6**, 7457–7468.

Zabot, G. L., Moraes, M. N., Carvalho, P. I. N. & Meireles, M. A. A. 2015. New proposal for extracting rosemary compounds: process intensification and economic evaluation. *Industrial Crops and Products*, **77**, 758–771.

Zhu, Q. L., Zang, L. Y., Zhang, L. & Yun, Z. 2012. Two approaches in preparation for cogeneration α-tocopherol and biodiesel from cottonseed. *Canadian Journal of Chemical Engineering*, **90**, 171–179.

Zhu, Q. L., Shao, R., Dong, R. & Yun, Z. 2014. An integrated approach for obtaining biodiesel, sterols, gossypol, and raffinose from cottonseed on a biorefinery concept. *Energy*, **70**, 149–158.

10

Economics and Market for Recovered Bioactive Compounds from Agricultural Wastes

Van Tang Nguyen[1,2] *and Thang Trung Khong*[3]

[1] *School of Environmental and Life Sciences, Faculty of Science and Information Technology, University of Newcastle, Ourimbah, NSW, Australia*
[2] *Department of Food Technology, Faculty of Food Technology, Nha Trang University, Nha Trang, Khanh Hoa, Vietnam*
[3] *Nha Trang University, Nha Trang, Khanh Hoa, Vietnam*

10.1 Introduction

During development, researchers are always looking for new products with higher quality and lower prices, as well as more effective technologies with lower costs of manufacture (COMs) and greater safety in terms of human health and the environment, while producers want novel technologies capable of making higher-value products with better quality, lower production costs and higher safety level in terms of human health and the environment. To meet these requirements, it is necessary to perform an analysis of the economics and the market for recovered bioactive compounds from agricultural wastes.

A wide range of recovery processes are presented in this chapter, including determination of the scale-up process, chemical composition and economic feasibility of grape seed oil (*Vitis vinifera* L.) using supercritical fluid extraction (SFE) (Prado et al., 2012); determination of the economic feasibility of large-scale operations for the recovery of phenolic compounds from grape bagasse using SFE (Farías-Campomanes et al., 2013); evaluation of the extraction yield, composition and economic feasibility of antioxidant compounds from jabuticaba (*Myrciaria cauliflora*) skins (Santos et al., 2010); optimisation of extraction conditions and economic evaluation of pressurised liquid extraction (PLE) of phenolic compounds from jabuticaba skins (Santos et al., 2012); comparison and economic evaluation of the PLE and low-pressure solvent extraction (LPSE) (Soxhlet extraction and percolation) of carotenoids from pressed palm fibre (PPF) (Cardenas-Toro et al., 2015); determination of extraction yields, extraction kinetics, mathematical modelling and economic evaluation of *Anacardium occidentale* L. leaves using SFE (Leitão et al., 2013); investigation of the chemical compositions and economic evaluation of natural antioxidant extracts obtained by ultrasound-assisted extraction (UAE) and agitated-bed extraction (ABE) from jussara pulp (*Euterpe edulis*) (Vieira et al., 2013); process integration and economic evaluation for the extraction of curcuminoids from deflavoured turmeric (*Curcuma longa* L.) using pressurised liquids (Osorio-Tobón et al., 2014); and economic evaluation of the production of phenolic-rich extracts from Brazilian plants using SFE and subcritical

Recovering Bioactive Compounds from Agricultural Wastes, First Edition. Edited by Van Tang Nguyen.

fluid extraction (Veggi et al., 2014). In addition, we also discuss the determination of the market for recovered bioactive compounds from agricultural wastes.

10.2 Economic Analysis of Recovered Bioactive Compounds

An economic analysis of recovered bioactive compounds from agricultural wastes is undertaken in order to determine the economic feasibility of recovery processes, including COM (production cost), selling price and affordability (redeem, payback time or return time). With regard to developing countries, adding value to indigenous raw materials using environmentally friendly technologies gives the possibility of increasing their competitiveness in the global market.

The main cost is the COM, including total capital investment cost and operating cost. The total capital investment cost represents the fixed capital investment (FCI), working capital and start-up cost. The FCI involves expenses for equipment, installation, territorial taxes, engineering and so on, the working capital cost represents the operating liquidity available to the business and the start-up cost is associated with the beginning of operation and validation of the process. The operating cost represents direct costs that are directly dependent on the production rate; it is composed of the cost of raw materials (CRM), the cost of the solvent lost during the process, the cost of utilities (CUT) relating to the demand for steam and cooling water in the evaporator and condenser, electricity and the cost of operational labour (COL). Therefore, the COM may vary according to many factors, as described in Equation 10.1 (Cavalcanti et al., 2012; Deenu et al., 2013; Osorio-Tobón et al., 2014; Vieira et al., 2013).

$$\text{COM} = 0.304 \times \text{FCI} + 2.73 \times \text{COL} + 1.23 \times (\text{CUT} + \text{CWT} + \text{CRM}) \quad (10.1)$$

The relations between processes conducted at bench, pilot and industrial scales cannot always be simply approached or predicted. The differences observed in procedures conducted in equipment of significantly different sizes must be carefully studied and evaluated in order to avoid rough mistakes when scaling up a process. When validating scale-up criteria, it is necessary to assess their applicability to different types of raw materials, as the mass transfer mechanisms may differ among species and among the parts of the plant used for extraction; for instance, clove is a flower bud rich in volatile oil, sugarcane residue contains fatty alcohols and grape seeds contain vegetable oil, while lemon verbena leaves contain both volatile oil and oleoresin (Cardenas-Toro et al., 2015).

Prado et al. (2012) determined the economics of the SFE process for extracting oil from grape seed (*Vitis vinifera* L.), a waste discarded from wine production. The economic parameters used for the estimation of COM and the economic evaluation, with raw material costs US\$0 (donated by wineries) and US\$2.70/ton, are shown in Tables 10.1–10.3. Pre-processing costs included drying and comminution of raw material. Their findings indicated that it is viable to establish an SFE plant to process grape seed oil based on a technical–economic evaluation. Optimal parameters of 313 K/35 MPa, 240-minute extraction time and solvent to feed/sample ratio of 6.6 achieved the best yield and cost. Under these parameters, COM of grape seed oil using SFE at different scales was greatly decreased with increasing extraction time from 60 to 240 minutes and increasing extractor volume from 5 to 500 l (from approximately US\$290 to US\$10/kg of extract), when compared to cold pressing (US\$40 to US\$80/kg

Table 10.1 Economic parameters used for the estimation of cost of manufacture (COM) of grape seed oil. *Source:* Prado et al. (2012). Reproduced with permission of Elsevier.

Industrial units		**Raw materials**	
2 extractors of 5 l	US$100 000.00	Wet grape seed	0 and US$2.70/ton
2 extractors of 50 l	US$300 000.00	Pre-processing	US$40.00/ton
2 extractors of 500 l	US$1 150 000.00	CO_2 (2% loss)	US$0.15/kg
Depreciation rate	10%/year		
Labour	US$4.00/h	**Utilities**	
2 extractors of 5 l	1 operator	Electricity	US$0.092/kWh
2 extractors of 50 l	2 operators	Cooling water	US$0.19/ton
2 extractors of 500 l	3 operators	Steam	US$4.20/ton

Table 10.2 Economic evaluation of grape seed oil using supercritical fluid extraction (SFE) with raw material cost at zero. *Source:* Prado et al. (2012). Reproduced with permission of Elsevier.

Time (min)	Productivity (kg/year)			Operation cost (US$/year)	COM (US$/kg extract)	CRM (%)	COL (%)	FCI (%)	CQC (%)	CUT (%)	Return (years)
	S_1	S_2	Total								
5 l											
60	382	35	417	121 000	290.17	0.98	60.32	29.06	9.05	0.59	NV[g]
120	503	49	552	120 000	217.39	0.67	60.47	29.12	9.07	0.67	NV
180	562	78	640	120 000	187.50	0.53	60.53	29.15	9.08	0.70	NV
240	577	90	667	120 000	179.91	0.42	60.57	29.20	9.09	0.73	NV
300	569	95	664	120 000	180.72	0.40	60.57	29.20	9.09	0.74	NV
50 l											
60	3815	349	4164	292 000	70.12	4.06	49.93	36.09	7.49	2.43	NV
120	5032	490	5522	289 000	52.34	2.80	50.43	36.43	7.56	2.78	NV
180	5616	779	6395	288 000	45.04	2.23	50.64	36.59	7.60	2.94	46.93
240	5773	899	6672	287 000	43.02	1.77	50.82	36.75	7.62	3.04	21.88
300	5690	947	6637	286 000	43.09	1.68	50.83	36.76	7.62	3.10	23.18
500 l											
60	38 153	3485	41,638	844 000	20.27	14.02	25.88	47.80	3.88	8.41	2.92
120	50 319	4904	55,223	816 000	14.78	9.90	26.78	49.45	4.02	9.85	1.92
180	56 160	7789	63,949	804 000	12.57	7.98	27.19	50.20	4.08	10.54	1.58
240	57 731	8995	66,726	793 000	11.88	2.57	27.56	50.92	4.13	11.00	1.49
300	56 903	9472	66,375	792 000	11.93	6.09	27.58	50.97	4.14	11.22	1.50

COM, cost of manufacture; CRM, cost of raw materials; COL, cost of labour; FCI, fixed capital investment; CQC, cost of quality control; CUT, cost of utilities; NV, nonviable.

Table 10.3 Economic evaluation of grape seed oil using supercritical fluid extraction (SFE) with raw material cost at US$2.70/ton. *Source:* Prado et al. (2012). Reproduced with permission of Elsevier.

Time (min)	Productivity (kg/year)			Operation cost (US$/year)	COM (US$/kg extract)	CRM (%)	COL (%)	FCI (%)	CQC (%)	CUT (%)	Return (years)
	S1	S2	Total								
5 l											
60	382	35	417	121 000	290.17	1.28	60.14	28.98	9.02	0.59	NV[g]
120	503	49	552	121 000	219.20	0.86	60.35	29.07	9.05	0.67	NV
180	562	78	640	121 000	189.06	0.67	60.45	29.11	9.07	0.70	NV
240	577	90	667	120 000	179.91	0.53	60.50	29.16	9.08	0.72	NV
300	569	95	664	120 000	180.72	0.49	60.52	29.18	9.08	0.74	NV
50 l											
60	3815	349	4164	295 000	70.85	5.23	49.32	35.65	7.40	2.40	NV
120	5032	490	5522	291 000	52.70	3.57	50.03	36.14	7.50	2.76	NV
180	5616	779	6395	289 000	45.19	2.80	50.35	36.37	7.55	2.93	55.62
240	5773	899	6672	288 000	43.17	2.22	50.59	36.58	7.59	3.03	23.22
300	5690	947	6637	287 000	43.24	2.06	50.64	36.62	7.60	3.09	24.41
500 l											
60	38 153	3485	41,638	880 000	21.13	17.54	24.82	45.85	3.72	8.06	3.03
120	50 319	4904	55,223	839 000	15.19	12.36	26.05	48.10	3.91	9.58	1.95
180	56 160	7789	63,949	821 000	12.84	9.87	26.64	49.17	4.00	10.33	1.59
240	57 731	8995	66,726	806 000	12.08	7.92	27.10	50.08	4.07	10.82	1.50
300	56 903	9472	66,375	803 000	12.10	7.37	27.20	50.28	4.08	11.06	1.51

COM, cost of manufacture; CRM, cost of raw materials; COL, cost of labour; FCI, fixed capital investment; CQC, cost of quality control; CUT, cost of utilities; NV, nonviable.

of extract) (Figure 10.1). The SFE process starts becoming economically viable in the 50 l plant, depending on the selling price of the product. For a selling price of US$40.00/kg, only a 500 l plant is viable, with a minimum COM of US$12.00/kg extract. The return time, for an extractor volume of 500 l and an extraction time of 240 minutes, was determined to be just 1.49 and 1.50 years for raw material cost at US$0 and US$2.7/ton, respectively. Linoleic acid was found to be the major component of the extract, while palmitic, stearic and oleic acids were also detected.

Farías-Campomanes et al. (2013) found that the extraction of grape bagasse (one of the main residues produced after pressing in Pisco production (about 20% of total byproducts), which is composed mainly of stems (25%), skins (25%) and seeds (50%)) using supercritical carbon dioxide (SC-CO_2) containing 10% ethanol (w/w) as co-solvent at 313 K/20 MPa, an extraction time of 180 minutes and a ratio of solvent to feed/sample of 47 produced extracts with higher phenolic concentrations than are produced using conventional techniques. Under these conditions, the extracts contained major compounds such as syringic, vanillic, gallic, p-hydroxybenzoic, protocatechuic, p-coumaric acids and quercetin. An evaluation of the economics of the process indicates the feasibility

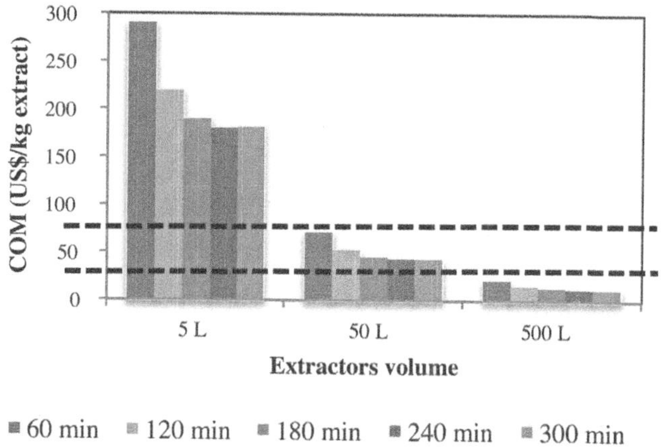

Figure 10.1 Estimation of cost of manufacture (COM) for grape seed oil obtained by supercritical fluid extraction (SFE) at different extraction times and extractor volumes, with grape seed cost at US$2.70/ton. Dotted lines are the selling prices of oil obtained by cold pressing in Brazilian (US$40) and international (US$80) markets. *Source:* Prado et al. (2012). Reproduced with permission of Elsevier.

of an industrial SFE plant with a capacity of 0.5 m^3 for producing an extract with an expected phenolic concentration of approximately 23 g/kg extract at an estimated COM of US$133.16/kg (Figure 10.2).

Jabuticaba (*Myrciaria cauliflora*) is grape-like in appearance and texture, but its skin is thicker and tougher. Santos et al. (2010) found that the COM for the extraction of antioxidant compoiunds from jabuticaba was inversely proportional to the extractor capacity, such that increasing the extractor capacity led to a decrease in COM, indicating an advantage for investment on a large industrial scale. In this instance, the COM was decreased by between 67 and 50% and by between 50 and 25% when the extractor capacity was increased from 0.05 to 0.1 m^3 and from 0.05 to 0.3 m^3, respectively. The results also showed that the variation of the percentage of phenolic compound and anthocyanin fraction in the extracts directly affected the specific cost, in which the extracts with higher amounts of each fraction had lower specific COMs. The combined UAE and ABE process significantly decreased the cost of the crude extracts, from US$422.18 to US$387.20/kg, when compared to the ABE process alone, despite the increase in processing cost. The combined process also gave a COM of US$1.15/g for the phenolic compounds fraction and US$7.3/g for anthocyanins (Table 10.4 and Figure 10.3). This study revealed that ultrasonic treatment in conventional extraction improved commercial extract production without considerably increasing costs, so ultrasound may be a viable option for scale-up purposes, particularly with shorter processing times before a conventional solvent extraction.

Santos et al. (2012) found the optimal PLE conditions for phenolic compounds from jabuticaba skins to be 5 MPa, 553 K and 9-minute extraction time using ethanol as solvent. Under these conditions, the contents of anthocyanins and total phenolic compounds extracted by PLE were 2.15- and 1.66-fold higher, respectively, than those obtained by conventional LPSE, while the COM of the extract using PLE was 40-fold lower than that using LPSE (Tables 10.5 and 10.6 and Figure 10.4).

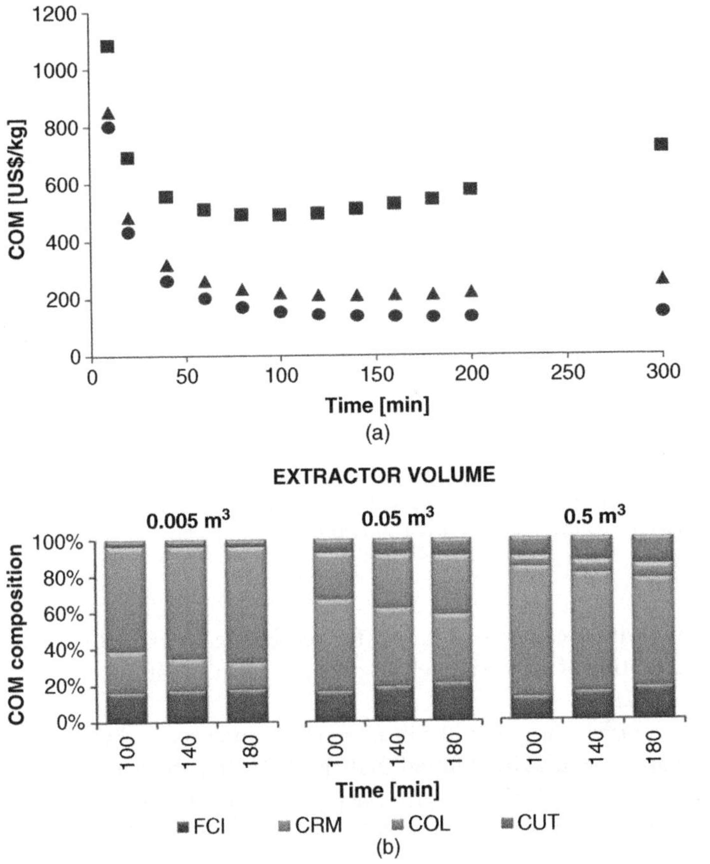

Figure 10.2 (a) Estimated cost of manufacture (COM) for bagasse extracts with different capacities of extractor at 0.005 (■), 0.05 (▲) and 0.5 (•) m^3. (b) Contribution of each component in the COM for different extractor capacities and extraction times. FCI, fixed capital investment; CRM, cost of raw materials; COL, cost of labour; CUT, cost of utilities. *Source:* Farías-Campomanes et al. (2013). Reproduced with permission of Elsevier.

The findings in this study also indicated that an increase in the extractor capacity led to an increase in the productivity and a decrease in the COM. Therefore, the use of higher-capacity extractors is more profitable. In this instance, the price of jabuticaba skins was considered to be the price of the whole fruit, including pulp, skin and seed (US$1.8/kg of fruit), and was compared to the price of glycolic extracts from jabuticaba (US$9.90/kg). However, the price might be lower if the skins were collected from the production of jabuticaba wine and jabuticaba pulp as residues.

Cardenas-Toro et al. (2015) compared the extraction yield and carotenoid composition, and evaluated the economic feasibility of PLE and LPSE, of carotenoids from PPF, one of the most abundant residues of palm oil extraction and a rich source of α- and β-carotene. The economic parameters used for COM estimation are shown in Table 10.7.

The findings indicated that the extraction yield obtained by Soxhlet extraction (LPSE-SOX) for 6 hours (96 mg extract/g PPF) was higher than that obtained by percolation (LPSE-PE) (74 mg extract/g PPF at 35 °C, 2.4 g/min) or PLE over a dynamic

Table 10.4 Estimated cost of manufacture (COM) for jabuticaba (*Myrciaria cauliflora*) skin extract using UAE, ABE, combined UAE and ABE, Soxhlet with ethanol and Soxhlet with acidified ethanol_pH 3. *Source:* Santos et al. (2010). Reproduced with permission of Elsevier.

Extraction method	Extractor capacity (m^3)	Global yield (%)	COM for crude extract (US$/kg)	Content of phenolic compounds (mg/g)	COM for phenolic compounds fraction (US$/g)	Content of anthocyanins (mg/g)	COM for anthocyanin fraction (US$/g)
UAE	0.05		794.46		3.70		26.50
	0.10	11.93	530.22	26.335	2.46	3.673	17.67
	0.30		401.21		1.86		13.37
ABE	0.05		1016.88		3.80		14.86
	0.10	9.01	666.50	24.103	2.50	6.168	9.74
	0.30		422.18		1.58		.17
UAE + ABE	0.05		932.95		2.77		17.60
	0.10	10.08	614.28	33.977	1.80	5.339	11.60
	0.30		387.20		1.15		7.30
Soxhlet	0.05		3020.00		9.23		60.40
	0.10	9.92	1800.00	32.433	5.50	4.993	36.00
	0.30		778.42		2.38		15.57
Soxhlet_pH 3	0.05		3776.00		10.00		33.20
	0.10	9.50	1862.00	35.845	4.94	5.416	16.36
	0.30		1001.00		2.65		8.80

UAE, ultrasound-assisted extraction; ABE, agitated-bed extraction.

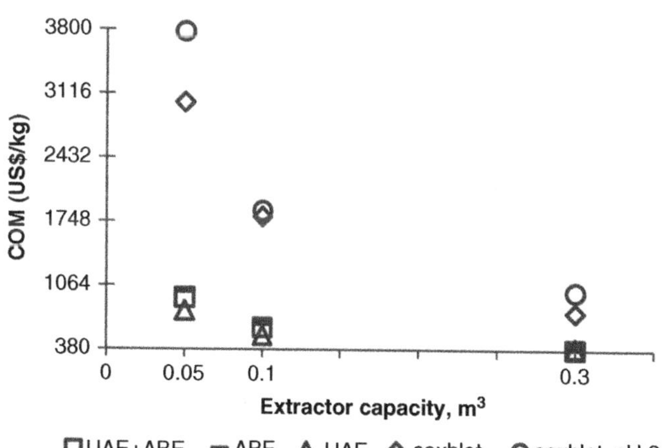

(a)

Figure 10.3 (a) Effect of extractor capacity on cost of manufacture (COM) for jabuticaba (*Myrciaria cauliflora*) skin extract using various methods. (b) Contribution of each component in the COM using combined ultrasound-assisted extraction (UAE) and agitated-bed extraction (ABE) at different extractor capacities. CUT, cost of utilities; CRM, cost of raw materials; FCI, fixed capital investment; COL, cost of operational labour. *Source:* Santos et al. (2010). Reproduced with permission of Elsevier.

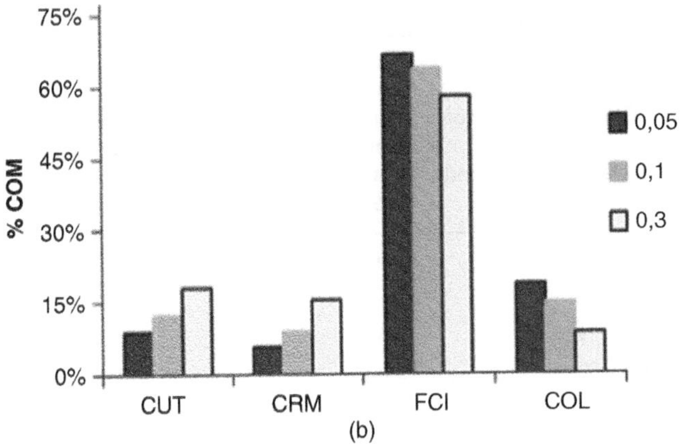

Figure 10.3 (*Continued*)

Table 10.5 Cost of manufacture (COM) for jabuticaba (*Myrciaria cauliflora*) skin crude extract, phenolic compounds and anthocyanins using pressurised liquid extraction (PLE) with different extractor capacities over 9 minutes. *Source:* Santos et al. (2012). Reproduced with permission of Elsevier.

Extractor capacity (m^3)	Global yield (%)	COM for crude extract (US$/kg)	Content of phenolic compounds (mg/g)	COM for phenolic compounds fraction (US$/g)	Content of anthocyanins (mg/g)	COM for anthocyanin fraction (US$/g)
0.05	13.0	19.26	7.8	0.32	2.4	1.06
0.10		17.24		0.29		0.95
0.30		15.53		0.26		0.85

Table 10.6 Contribution of each component to the cost of manufacture (COM) of jabuticaba (*Myrciaria cauliflora*) skin extracts using pressurised liquid extraction (PLE) with an extractor capacity of 0.3 m^3. *Source:* Santos et al. (2012). Reproduced with permission of Elsevier.

Extraction time (minutes)	Global yield (%)	FCI (%)	COL (%)	CUT (%)	CRM (%)
4	8.7	12.0	1.5	1.5	85.0
8	11.7	14.5	3.0	1.5	81.0
9	13.0	15.3	3.3	1.4	80.0
12	13.7	17.5	4.2	1.3	77.0
16	14.7	18.9	5.3	1.3	74.5
20	15.4	20.8	6.4	1.2	71.6

FCI, fixed capital investment; COL, cost of labour; CUT, cost of utilities; CRM, cost of raw material.

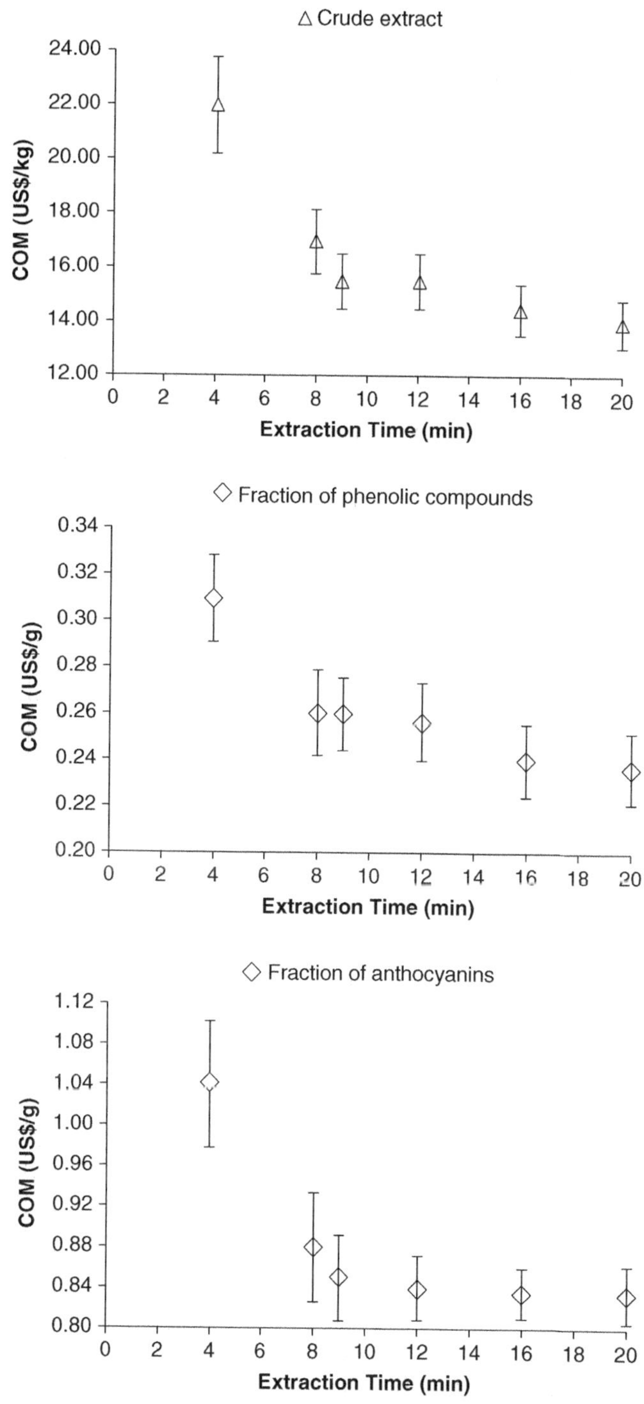

Figure 10.4 Cost of manufacture (COM) for jabuticaba (*Myrciaria cauliflora*) skin crude extract, anthocyanins and phenolic compounds using pressurised liquid extraction (PLE) with an extractor capacity of 0.3 m^3 across 9 minutes. *Source:* Santos et al. (2012). Reproduced with permission of Elsevier.

Table 10.7 Economic parameters used to estimate the cost of manufacture (COM) of extracts from pressed palm fibre (PPF). *Source:* Cardenas-Toro et al. (2015). Reproduced with permission of Elsevier.

Industrial units	
2 extractors, 0.005 m^3	US$95 000.00
2 extractors, 0.05 m^3	US$380 000.00
2 extractors, 0.5 m^3	US$1 500 000.00
Distiller column, 0.005 m^3	US$20 000.00
Distiller column, 0.05 m^3	US$53 000.00
Distiller column, 0.5 m^3	US$76 000.00
Depreciation rate	10%/year
Labour	US$4.00/h
2 extractors, 0.005 m^3	1 operator
2 extractors, 0.05 m^3	2 operators
2 extractors, 0.5 m^3	3 operators
Raw materials	
Pre-processing of PPF	US$40.00/ton
Apparent density of PPF[a)]	304.83 kg/m^3
Ethanol	US$0.716/l
Utilities	
Electricity	US$0.092/kWh
Cooling water	US$0.190/ton
Steam	US$4.20/ton

a) Determined experimentally.

extraction time of 17 minutes (44 mg extract/g PPF at 55 °C, 4 MPa, 2.4 g/min); however, the carotenoid yield achieved by PLE (305 and 713 μg α- and β-carotene/g extract, respectively) was higher than that obtained by LPSE-SOX (142 and 317 μg α- and β-carotene/g extract, respectively).

The COM decreased with increasing extraction capacity, due to increased productivity. The use of extractors with a capacity of 0.5 m^3 resulted in the lowest COM, revealing an advantage for investments at high capacities. In this instance, the COM decreased from 42 to 87% and from 88 to 98% when the extractor capacity increased from 0.005 to 0.05 m^3 and from 0.05 to 0.5 m^3, respectively. LPSE-SOX resulted in the highest COM (US$45.1 to US$2346.6/kg extract) because of the high extraction time and low productivity, while the lowest COM was obtained by PLE (US$29.2 to US$248.9/kg extract) and LPSE-PE (US$13.4 to US$285.7/kg extract) with a 0.5 m^3 vessel capacity (Table 10.8), indicating that PLE was the best choice for extraction of carotenoids when compared to LPSE techniques. Regarding the selling price, a bottle containing 60 soft-gel capsules costs between US$10 and US$30. Each capsule contains 250 mg of oil extract with a carotenoid content of 13 mg (0.052%, w/w). The estimated selling price of carotenoid-rich extract therefore varies between US$667 and US$2000/kg extract,

Table 10.8 Cost of manufacture (COM) for carotenoid-rich extract obtained from pressed palm fibre (PPF) by Soxhlet extraction (LPSE-SOX), percolation (LPSE-PE) and pressurised liquid extraction (PLE) with extraction units of 2 × 0.005 m^3, 2 × 0.05 m^3 and 2 × 0.5 m^3. *Source:* Cardenas-Toro et al. (2015). Reproduced with permission of Elsevier.

Economic parameter	LPSE-SOX			LPSE-PE			PLE			Commercial extract (SHP, 2014)
	0.005 m^3	0.05 m^3	0.5 m^3	0.005 m^3	0.05 m^3	0.5 m^3	0.005 m^3	0.05 m^3	0.5 m^3	
FCI (%)	49.2	45.4	40.3	44.2	34.9	15.9	51.3	77.1	50.5	
CRM (%)	0.2	1.5	9.9	2.0	11.8	40.2	2.9	5.1	25.3	
COL (%)	45.8	47.5	39.8	47.7	41.5	17.0	38.5	13.2	9.8	
CUT (%)	0.2	0.9	5.9	1.3	7.7	25.2	1.5	2.6	12.9	
CQC (%)	4.6	4.7	4.0	4.8	4.1	1.7	5.8	2.0	1.5	
COP × 10^3 (US$/year)	415	533	796	398	611	1865	275	1600	3231	
Production (kg/year)	180	1804	18 036	1428	14 286	142 885	1143	11 428	114 273	
COM (US$/kg extract)	2346.6	301.8	45.1	285.7	43.9	13.4	248.9	144.7	29.2	667–2000
COM (US$/g carotenoid)	5112.3	657.5	98.1	1042.7	160.2	48.9	333.7	194.0	39.1	12.8–38.5

FCI, fixed capital investment; CRM, cost of raw material; COL, cost of labour; CUT, cost of utilities; CQC, cost of quality control; COP, cost of operation.

corresponding to US$12.8–38.4/g carotenoid. Table 10.8 shows that the calculated COM of extracts (US$/kg extract) was lower than the selling price at the highest capacity of 0.5 m^3. In addition, the COM (US$/g carotenoid) of these extracts for PLE extraction was similar to that calculated for the commercial extracts, suggesting that competitive prices can be obtained using an extractor capacity of 0.5 m^3. However, taxes and extraction unit profits also need to be considered in order to make a better cost comparison.

Leitão et al. (2013) determined the extraction yields, extraction kinetics, mathematical modelling and economic evaluation of *Anacardium occidentale* L. leaves using SFE. The economic parameters used for the COM estimation are shown in Table 10.9. Overall, the COM decreased with increasing scale or extractor capacity (from 5 to 500 l), representing an advantage for investment on a large industrial scale (Figure 10.5). The major cost share was moved from the COL to the FCI when the extractor or plant capacity increased, while the CRM was the lowest. However, the CRM was increased with increasing extractor capacity (rising from 0.113 to 0.168% when extractor capacity increased from 5 to 500 l), while the FCI was the major component of the COM, which increased with increasing scale (Figure 10.6). The COM was reduced when the extraction time increased, indicating a positive effect of extraction time on COM (Figure 10.7). This study showed that *A. occidentale* leaves are considered a byproduct that can be used as an important source of nutraceutical compounds, including essential

Table 10.9 Economic parameters used to estimate the cost of manufacture (COM) of extracts from *Anacardium occidentale* L. leaves using supercritical fluid extraction (SFE). *Source:* Leitão et al. (2013). Reproduced with permission of Elsevier.

Industrial units[a)]	European market
2 extractors of 5 l	US$125 000
2 extractors of 50 l	US$490 000
2 extractors of 500 l	US$2 625 000
Depreciation rate	10%/year
Labour	US$4.00/h
2 extractors of 5 l	1 operator
2 extractors of 50 l	2 operators
2 extractors of 500 l	3 operators
Raw materials	
A. occidentable leaves	
Pre-processing	US$40.00/ton
CO_2 (2% loss)	US$2.30/kg
Utilities	
Electricity	US$0.091/kWh
Cooling water	US$5.15/ton
Steam	US$20.00/ton

a) Direct quotation.

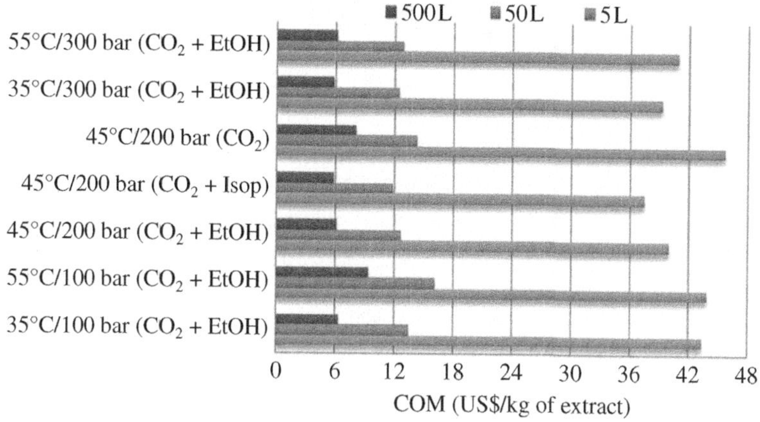

Figure 10.5 Cost of manufacture (COM) of extracts from *Anacardium occidentale* L. leaves using supercritical fluid extraction (SFE) under different extraction conditions and with different extractor capacities. *Source:* Leitão et al. (2013). Reproduced with permission of Elsevier.

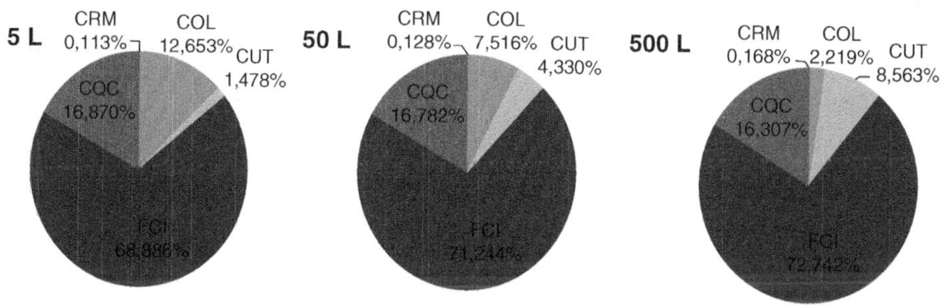

Figure 10.6 Share of costs in the cost of manufacture (COM) of extracts from *Anacardium occidentale* L. leaves using supercritical fluid extraction (SFE) with different extractor capacities. FCI, fixed capital investment; CRM, cost of raw materials; COL, cost of operational labour; CUT, cost of utilities; CQC, cost of quality control. *Source:* Leitão et al. (2013). Reproduced with permission of Elsevier.

oils, flavonoids, alkaloids and antioxidant compounds. In particular, the SFE process was found to be economically feasible for obtaining extracts from *A. occidentale* leaves for further application.

Vieira et al. (2013) investigated the chemical compositions and evaluated the economic feasibility of using natural antioxidant extracts obtained by UAE and ABE from jussara pulp (*Euterpe edulis*). The economic parameters used for the COM estimation of the crude extracts are described in Table 10.10. Table 10.11 indicates that the extraction time did not significantly influence the COM of the extracts, while the concentration of ethanol in water remarkably affected the manufacturing costs. Increasing temperature appeared to slightly increase the COM, while the increase in the ratio of solvent volume to feed mass (S/F) caused an oscillatory variation in the COM. The COL had the most effect on the final COM, followed by the CRM, FCI and CUT, in all operational conditions evaluated, except for the ratio of solvent volume to feed mass, in which CRM had a more significant impact than COL. The featureless effects of the FCI and CUT on the COM may be due to the low costs of the equipment, electricity and

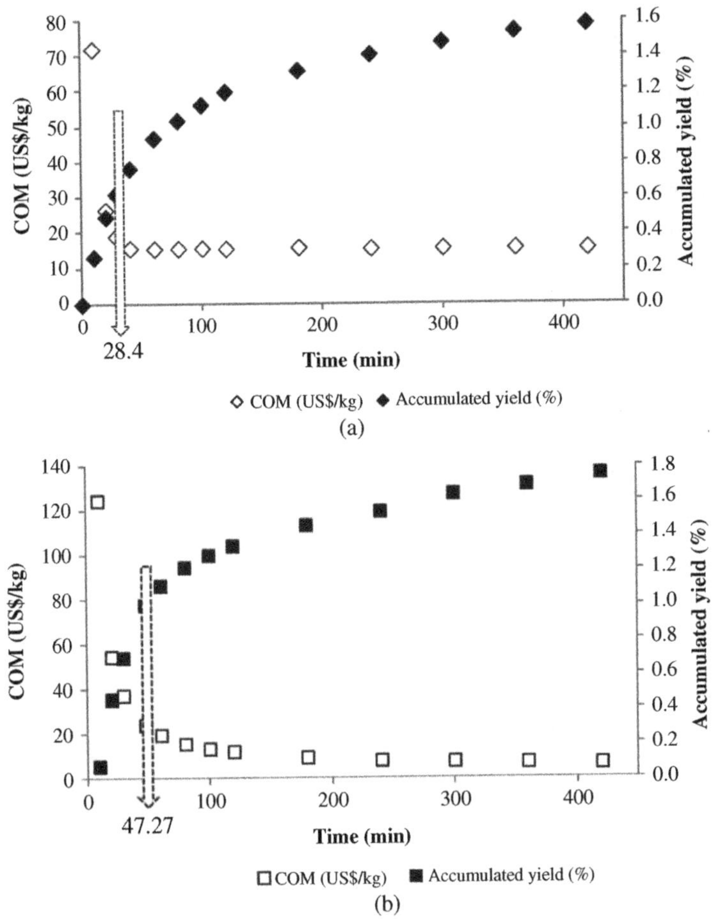

Figure 10.7 Cost of manufacture (COM) of extracts from *Anacardium occidentale* L. leaves using supercritical fluid extraction (SFE) under different experiment conditions: (a) 35 °C/300 bar and (b) 45 °C/200 bar. *Source:* Leitão et al. (2013). Reproduced with permission of Elsevier.

heat-exchange agents. Conversely, the high values of the COL and CRM indicate that the labour charge and raw materials costs were key factors in defining the COM in all situations evaluated. Jussara pulp is extremely costly (US$1.80/kg pulp), so the COM of the crude extracts obtained from jussara pulp by UAE and ABE were ranged from US$90.57 to US$165.34/kg and from US$87.32 to US$167.48 /kg, respectively, with the extracts produced by UAE having slightly higher COM than those obtained by ABE.

Osorio-Tobón et al. (2014) determined the process integration and economic feasibility for the extraction of curcuminoids from deflavoured turmeric (*Curcuma longa* L.) using pressurised liquids. The economic parameters used for the COM estimation for curcuminoid-enriched extracts from Jussara pulp (*Euterpe edulis*) are shown in Table 10.12. The outcomes showed that the extraction of curcuminoids was affected primarily by extraction temperature, ratio of solvent to feed (S/F) and pressure, where increasing the extraction temperature and S/F improved the extraction efficiency, whereas rising the pressure negatively affected the extraction process. Curcuminoid

Table 10.10 Economic parameters used to estimate the cost of manufacture (COM) of crude extracts from jussara pulp (*Euterpe edulis*) using ultrasound-assisted extraction (UAE) and agitated-bed extraction (ABE). *Source:* Vieira et al. (2013). Reproduced with permission of Elsevier.

	UAE	ABE		UAE	ABE
FCI (fixed capital investment)					
3 extraction tanks (US$)	84 400.00	18 600.00	Pulp of jussara (US$/kg)	1.80	1.80
1 Reservoir solvent tank (US$)	5000.00	5000.00	Ethanol (US$/kg)	0.85	0.85
3 Centrifuge pumps (US$)	30 000.00	30 000.00	Water (US$/kg)	0.04	0.04
1 Falling film evaporator (US$)	99 000.00	99 000.00	Citric acid (US$/kg)	0.55	0.55
1 Spray drier (US$)	167 000.00	167 000.00	*CUT (cost of utility)*		
Depreciation rate (%/year)	10	10	Electricity (US$/kWh)	0.138	0.138
COL (cost of operational labour)			Chilled water (US$/tonne)	0.34	0.34
Basic rate (US$/hr)	11.65	11.65	Steam (US$/tonne)	4.20	4.20
Number of operators	3	3	*Profitability*		
CRM (cost of raw materials)			Selling price (US$/kg)	25.00	25.00

yield was also dependent on the extractor capacity (Figure 10.8a). The optimum conditions for curcuminoid extraction were temperature 333 K and pressure 10 MPa. The curcuminoid yields obtained by PLE, Soxhlet extraction and LPSE were similar, but the PLE extraction time was three and six times less than that of Soxhlet extraction and LPSE, respectively, and the COM of PLE was two-and-a-half and two times lower. CRM played a very important role in COM, which increased with rising CRM. For example, when the CRM was US$0.85/kg and a two-0.05 m^3 vessel system was operated for 30 minutes, the COM was US$17.86/kg extract, which was approximately 1.7 and 5.3 times lower than when the CRM was US$7.91 and US$4.4/kg, respectively (Figure 10.8b). Therefore, PLE is an attractive and economically feasible technique for obtaining curcuminoid-rich extracts from deflavoured turmericrhizomes.

Veggi et al. (2014) investigated the economic evaluation of the production of phenolic-rich extracts from Brazilian plants (flame vine (*Pyrostegia venusta*) (FV), ice-cream bean (*Inga edulis*) (IC), dog's knot (*Heteropterys aphrodisiaca*) (DK) and common bean (*Phaseolus vulgaris* L.) (BE)) using SFE and subcritical fluid extraction. The economic parameters used to estimate the COM of phenolic-rich extracts are shown in Table 10.13, the COMs of crude extracts (COM_{EY}) and of the phenolic-rich fraction (COM_{TPC}) are given in Tables 10.14 and the parts of the COM of crude extracts using SFE for different solvent systems and extractor capacities are shown in Table 10.15. A lower COM_{EY} was obtained with CO_2 + EtOH and CO_2 + H_2O working

Table 10.11 Share of costs in the cost of manufacture (COM) of crude extracts from jussara pulp (*Euterpe edulis*) using ultrasound-assisted extraction (UAE) and agitated-bed extraction (ABE). *Source:* Vieira et al. (2013). Reproduced with permission of Elsevier.

Process variables	Level	UAE					ABE				
		COM (US$/kg of crude extract)	COM composition (%)				COM (US$/kg of crude extract)	COM composition (%)			
			CRM	COL	FCI	CUT		CRM	COL	FCI	CUT
Time (minutes)	5	110.39 ± 2.55	41.97	42.26	8.22	7.54	98.74 ± 2.42	42.9	43.19	6.27	7.63
	20	97.70 ± 3.19	41.97	42.26	8.22	7.54	94.27 ± 0.76	42.9	43.19	6.27	7.63
	40	90.57 ± 0.40	41.97	42.26	8.22	7.54	89.68 ± 0.34	42.9	43.19	6.27	7.63
	60	96.56 ± 0.42	41.97	42.26	8.22	7.54	88.02 ± 0.99	42.9	43.19	6.27	7.63
	80	95.55 ± 1.90	41.97	42.26	8.22	7.54	92.56 ± 0.55	42.9	43.19	6.27	7.63
	100	92.07 ± 0.10	41.97	42.26	8.22	7.54	92.13 ± 2.76	42.9	43.19	6.27	7.63
	120	93.76 ± 2.82	41.97	42.26	8.22	7.54	88.14 ± 2.18	42.9	43.19	6.27	7.63
	140	92.32 ± 2.66	41.97	42.26	8.22	7.54	87.32 ± 0.32	42.9	43.19	6.27	7.63
	180	96.98 ± 3.92	41.97	42.26	8.22	7.54	87.44 ± 0.16	42.9	43.19	6.27	7.63
Ethanol concentration in water (%, v/v)	0	103.95 ± 2.70	42.01	47.01	5.94	5.04	107.50 ± 3.10	42.04	47.05	5.94	4.97
	30	125.91 ± 0.67	42.87	45.00	5.68	6.45	118.34 ± 1.51	42.91	45.03	5.69	6.38
	50	134.13 ± 1.27	43.32	43.62	5.51	7.55	127.27 ± 2.74	43.35	43.65	5.51	7.48
	70	146.01 ± 5.87	43.28	41.86	6.08	8.78	134.11 ± 0.79	43.31	41.89	6.09	8.71
	90	165.34 ± 5.94	42.23	40.84	5.93	11.00	167.48 ± 2.80	42.26	40.87	5.94	10.93

(Continued)

Table 10.11 (Continued)

Process variables	Level	UAE					ABE				
		COM (US$/kg of crude extract)	COM composition (%)				COM (US$/kg of crude extract)	COM composition (%)			
			CRM	COL	FCI	CUT		CRM	COL	FCI	CUT
Temperature (°C)	25	122.55 ± 3.09	42.43	42.72	7.23	7.62	119.51 ± 4.31	43.25	43.56	5.50	7.69
	35	143.24 ± 4.01	35.72	46.91	10.99	6.38	128.98 ± 1.94	36.77	48.29	8.44	6.51
	45	136.80 ± 3.93	35.61	46.76	10.95	6.68	124.92 ± 1.64	36.65	48.13	8.41	6.81
	55	139.99 ± 2.66	35.50	46.62	10.92	6.97	129.61 ± 0.92	36.53	47.98	8.39	7.10
Solvent/feed ratio (ml/g)	5	98.57 ± 0.47	70.49	21.85	4.25	3.40	110.14 ± 0.29	71.29	22.1	3.21	3.40
	10	112.52 ± 1.45	57.34	31.88	5.39	5.39	102.84 ± 0.01	58.17	32.34	4.08	5.42
	15	109.26 ± 8.90	48.54	38.29	6.48	6.69	113.04 ± 1.36	49.39	38.96	4.92	6.74
	20	120.97 ± 3.27	42.43	42.72	7.23	7.62	124.78 ± 13.25	43.25	43.56	5.5	7.69
	25	98.14 ± 1.21	37.72	46.20	7.81	8.28	123.74 ± 11.63	38.51	47.16	5.95	8.37
	30	130.67 ± 2.19	34.22	48.71	8.24	8.82	125.80 ± 5.39	34.98	49.81	6.29	8.93

CRM, cost of raw materials; COL, cost of operational labour; FCI, fixed capital investment; CUT, cost of utilities.

Table 10.12 Economic parameters used to estimate the cost of manufacture (COM) of curcuminoids from deflavoured turmeric (*Curcuma longa* L.) using pressurised liquids. *Source:* Osorio-Tobón et al. (2014). Reproduced with permission of Elsevier.

	Industrial units[a]		
	$2 \times 0.05\ m^3$	$2 \times 0.3 m^3$	$2 \times 0.5 m^3$
Fixed capital investment (FCI)			
Low-pressure solvent extraction (LPSE)	US$60 000	–	–
Soxhlet unit extraction	US$115 000	–	–
pressurised liquid extraction (PLE)	US$117 000	US$317 000	US$417 000
Distiller	US$20 000	US$53 000	US$76 000
Depreciation rate	10%/year	10%/year	10%/year
Cost of operational labour (COL)	US$6.9/h	US$6.9h^{-1}	US$6.9/h
Number of workers	1	2	3
Cost of raw materials (CRM)			
Deflavoured turmeric	US$7.91/kg	US$7.91/kg	US$7.91/kg
	US$4.40/kg	US$4.40/kg	US$4.40/kg
	US$0.85/kg	US$0.85/kg	US$0.85/kg
	US$0.00/kg	US$0.00/kg	US$0.00/kg
Ethanol	US$0.85/kg	US$0.85/kg	US$0.85/kg
Water	US$0.04/kg	US$0.04/kg	US$0.04/kg
Cost of utilities (CUT)			
Electricity	US$0.092/kWh	US$0.092/kWh	US$0.092/kWh
Water steam	US$4.2/t	US$4.2/t	US$4.2/t

a) The first value indicates the number of extractor vessels, and the second value is their capacity in cubic metres.

at higher extraction capacity, mainly due to the increase in the extraction yield when modifier was added. A higher recovery of extracts using CO_2 + H_2O (except for IC) resulted in lower costs even with a longer process time (Table 10.14).

Table 10.15 shows that the price of raw material had a great influence on the final COM_{EY}. A smaller COM_{EY} was estimated for the process using $CO_2 + H_2O$, particularly for FV and DK extracts, the COMs of which were US$208.7 and US$353.3/kg extract, respectively, for the $0.5\ m^3$ extractor. The increase in the extraction yield and phenolic content had a negative effect on the COM, decreasing COM_{EY} and COM_{TPC}. An increase in the size of the extraction vessels resulted in a significant decrease in COM_{EY} and COM_{TPC}.

Increasing the batch size or number of batches per year increased the annual plant throughput and consequently led to a more economical process. The CRM and CUT increased with rising extractor size, while the FCI decreased, which suggested the feasibility of the processes at larger sizes. The COL, CUT and cost of quality control (CQC) represented a small fraction of the COM_{EY} when compared to CRM and FCI, in which the FCI and CRM for the three solvent systems were significantly different. For IC, DK

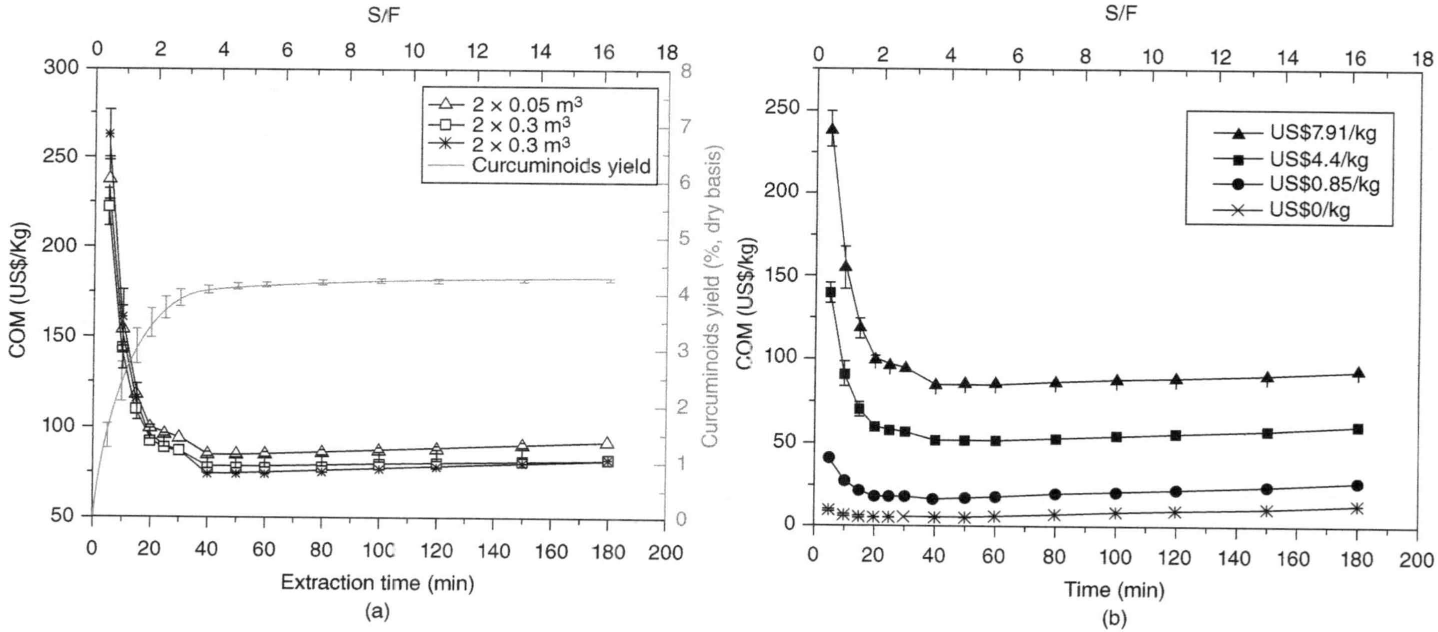

Figure 10.8 (a) Cost of manufacture (COM) of curcuminoid-rich extracts as a function of processing time for different extractor vessel capacities, with the raw material price at US$7.91/kg. (b) Influence of the cost of raw materials on the COM of curcuminoid-rich extracts at an extractor capacity of 0.05 m^3.
Source: Osorio-Tobón et al. (2014). Reproduced with permission of Elsevier.

Table 10.13 Economic parameters used to estimate the cost of manufacture (COM) of phenolic-rich extracts from Brazilian plants using supercritical and subcritical fluid extraction. *Source:* Veggi et al. (2014). Reproduced with permission of Elsevier.

Industrial SFE units	Chinese market	Raw materials	
2 extractors of 5 l	US$95 000.00	Flame vine (FV)	US$15.50/kg
2 extractors of 50 l	US$380 000.00	Ice-cream bean (IC)	US$50.00/kg
2 extractors of 500 l	US$1 500 000.00	Bean (BE)	US$2.20/kg
Depreciation rate	10%/year	Dog's knot (DK)	US$26.00/kg
Industrial freeze-dryer units		Pre-processing	US$40.00/ton
Drying area (m^2)	Chinese market	CO_2 (2% loss)	US$2.30/kg
20	US$38 000.00	Ethanol	US$0.716/l
200	US$150 000.00	Water	US$0.04/kg
300	US$192 000.00		
1500	US$500 000.00	*Utilities*	
2000	US$590 000.00	Electricity	US$0.092/kWh
3000	US$750 000.00	Cooling water	US$5.15/ton
		Steam	US$20.00/ton
Labour	US$4.00/h		
2 extractors of 5 l	1 operator		
2 extractors of 50 l	2 operators		
2 extractors of 500 l	3 operators		

and FV extracts using CO_2 and CO_2 + EtOH, the CRM fraction was higher than the FCI. In contrast, the extracts obtained using CO_2 + H_2O had a higher FCI than CRM, even for BE extracts. These data clearly demonstrate that CRM had a significant influence on the expense of the production of extracts. Although the cost of equipment had a great effect on the manufacturing cost, extraction yield was a more important factor in the economic evaluation of the processes studied. In this instance, the lowest COMs were achieved for FV using water as co-solvent, followed by extracts of DK and Bee using water and ethanol, respectively. The aqueous DK extract presented itself as the most viable commercial extract because of its high phenolic content and low cost.

10.3 Market Analysis of Recovered Bioactive Compounds

Market analysis of recovered bioactive compounds from agricultural wastes is undertaken in order to determine past and present consumption demand and so to predict consumption trends in the future. Consumer interest in healthy products is increasing for a number of reasons, including growing awareness of the link between diet and health, the greater convenience of meeting nutritional demands and the desire to prevent chronic diseases such as diabetes, hepatitis, cardiovascular diseases and cancer. Interest in healthy products among policy makers is also increasing, due to rising public health care costs, the increasing incidence of chronic and sedentary lifestyle-related

Table 10.14 Cost of manufacture of crude extracts (COM_{EY}) and of the phenolic-rich fraction (COM_{TPC}) using supercritical fluid extraction (SFE) with different solvent systems and with extractor capacities of 0.005, 0.05 and 0.5 m^3. *Source:* Veggi et al. (2014). Reproduced with permission of Elsevier.

Solvent system	Sample	$COM_{EY}^{0.005}$ (US$/kg extract)	$COM_{EY}^{0.05}$ (US$/kg extract)	$COM_{EY}^{0.5}$ (US$/kg extract)	$COM_{TPC}^{0.005}$ (US$/g phenolic compound)	$COM_{TPC}^{0.05}$ (US$/g phenolic compound)	$COM_{TPC}^{0.5}$ (US$/g phenolic compound)
CO_2	DK	$4120 \pm 112^{b.A}$	$3853 \pm 105^{a.A}$	$3599 \pm 98^{a.A}$	$1714 \pm 47^{a.A}$	$1603 \pm 44^{ab.B}$	$1497 \pm 41^{ab.B}$
	BE	$1304 \pm 348^{b.B}$	$920 \pm 246^{c.B}$	$833 \pm 234^{c.AB}$	$808 \pm 216^{c.A}$	$570 \pm 153^{b.B}$	$516 \pm 145^{c.A}$
	IC	$3856 \pm 759^{b.AB}$	$3570 \pm 703^{ab.B}$	$3252 \pm 319^{ab.B}$	$2000 \pm 394^{a.B}$	$1852 \pm 365^{a.B}$	$1687 \pm 166^{a.B}$
	FV	$2324 \pm 153^{a.A}$	$2132 \pm 133^{bc.A}$	$1928 \pm 127^{bc.A}$	$1001 \pm 66^{b.A}$	$918 \pm 58^{b.A}$	$830 \pm 55^{bc.A}$
CO_2 + EtOH	DKe	$1586 \pm 68^{a.A}$	$1265 \pm 54^{b.C}$	$1189 \pm 51^{b.B}$	$1795 \pm 77^{b.B}$	$1431 \pm 61^{b.B}$	$1345 \pm 57^{b.B}$
	BEe	$715 \pm 38^{a.B}$	$424 \pm 23^{c.B}$	$356 \pm 19^{d.B}$	$774 \pm 42^{d.A}$	$456 \pm 25^{c.B}$	$385 \pm 21^{c.A}$
	ICe	$2296 \pm 6^{a.A}$	$1984 \pm 5^{a.B}$	$1911 \pm 5^{a.B}$	$2316 \pm 6^{a.C}$	$2001 \pm 5^{a.B}$	$1927 \pm 5^{a.B}$
	FVe	$1451 \pm 13^{a.A}$	$1092 \pm 10^{b.B}$	$998 \pm 18^{c.B}$	$530 \pm 5^{c.B}$	$399 \pm 4^{d.B}$	$364 \pm 7^{c.B}$
$CO_2 + H_2O$	DKw	$2664 \pm 194^{b.A}$	$614 \pm 45^{c.B}$	$353 \pm 26^{c.C}$	$16917 \pm 1229^{c.C}$	$3901 \pm 284^{b.A}$	$2243 \pm 163^{b.A}$
	BEw	$6403 \pm 340^{a.A}$	$1232 \pm 66^{b.A}$	$551 \pm 30^{b.A}$	$1893 \pm 101^{b.A}$	$364.2 \pm 20^{c.A}$	$163 \pm 9^{c.A}$
	ICw	27617 ± 446^{bB}	$7581 \pm 123^{a.A}$	$5027 \pm 82^{a.A}$	$249956 \pm 404^{a.A}$	$6861 \pm 111^{a.A}$	$4550 \pm 74^{a.A}$
	FVw	$1948 \pm 102^{b.B}$	$413 \pm 22^{c.A}$	$209 \pm 11^{c.C}$	$703 \pm 37^{c.C}$	$149.0 \pm 8^{c.B}$	$75 \pm 4^{c.C}$

e, CO_2 + EtOH; *w*, CO_2 + H2O; DK, dog's knot; BE, common bean; IC, ice-cream bean; FV, flame vine.
Lowercase letters represent significant difference at 5% level of significance for samples evaluated from different raw materials using the same solvent. Capital letters represent significant difference at 5% level of significance for samples evaluated from the same raw material using different solvents.

Table 10.15 Shares in the cost of manufacture (COM) of Brazilian plant extracts using supercritical fluid extraction (SFE) with different solvent systems and with extractor capacities of 0.005, 0.05 and 0.5 m^3. *Source:* Veggi et al. (2014). Reproduced with permission of Elsevier.

Solvent system	Economic parameter	DK			BE			IC			FV		
		0.005 m^3	0.05 m^3	0.5 m^3	0.005 m^3	0.05 m^3	0.5 m^3	0.005 m^3	0.05 m^3	0.5 m^3	0.005 m^3	0.05 m^3	0.5 m^3
CO_2	CRM	63.6	77.7	82.9	22.6	36.8	46.7	74.0	83.8	87.0	57.2	74.2	80.6
	COL	7.0	1.6	0.3	15.6	5.1	1.0	4.0	0.9	0.2	8.7	2.1	0.4
	CUT	4.0	4.5	5.1	15.1	25.0	31.6	3.9	2.6	2.7	4.3	5.2	6.0
	FCI	21.6	14.5	11.1	38.2	28.0	18.3	16.0	11.8	9.8	25.0	16.4	12.0
	CQC	3.8	1.7	0.6	8.5	5.1	2.4	2.1	0.9	0.4	4.8	2.1	1.0
	COP ($\times 10^3$)	822	7692	71 844	290	2046	18 300	1536	14 228	140 114	629	5759	52 198
	Production (kg/year)	198	1984	19 841	230	2306	23 066	406	4063	40 637	270	2702	27 020
CO_2 + EtOH	CRM	45.4	66.7	76.4	11.2	22.9	32.0	58.3	76.0	82.7	37.7	60.8	72.7
	COL	11.0	3.2	0.6	17.9	7.3	1.5	8.1	2.1	0.3	13.1	4.2	0.7
	CUT	5.2	7.6	8.8	14.7	30.0	42.0	3.4	4.4	5.0	5.3	8.4	10.1
	FCI	31.0	19.2	13.0	44.5	32.6	21.2	24.8	15.3	11.2	35.1	22.4	14.7
	CQC	7.4	3.3	1.2	11.7	7.2	3.3	5.4	2.2	0.8	8.8	4.2	1.8
	COP ($\times 10^3$)	449	3580	34 098	226	1341	11 243	677	5848	56 306	356	2679	24 620
	Production (kg/year)	282	2827	28 653	316	3165	31 656	294	2941	29 420	243	2468	24 681
$CO_2 + H_2O$	CRM	5.3	19.0	38.1	0.9	3.1	7.3	9.0	29.0	51.4	3.8	14.3	30.6
	COL	26.4	10.0	2.0	27.5	10.9	2.7	25.4	8.6	1.6	27.0	10.7	2.3
	CUT	21.6	12.7	12.6	23.4	18.4	27.0	20.6	10.5	8.8	21.7	12.8	12.4
	FCI	35.7	47.2	39.3	36.8	54.5	51.7	34.6	42.1	32.1	36.3	50.2	45.1
	CQC	11.0	11.1	8.0	11.4	13.1	11.3	10.4	9.8	6.1	11.2	12.0	9.6
	COP ($\times 10^3$)	374	863	4963	352	678	3033	397	1089	7219	363	757	3908
	Production (kg/year)	143	1434	14, 342	55	551	5510	14	143	1430	188	1854	18 926

DK, dog's knot; BE, common bean; IC, ice-cream bean; FV, flame vine; CRM, cost of raw materials (%); COL, cost of labour (%); CUT, cost of utilities (%); FCI, fixed capital investment (%); CQC, cost of quality control; COP, operational cost (US$/year).

diseases, ageing populations, rising growth opportunities in related industries, the promotion of new research and development (R&D) applications and the increasing overall income in certain countries. Increasing health care costs have also led governments, health professionals and researchers to examine measures that promote well-being and reduce the risk of disease. Such rapid growth in any sector is accompanied by the need for appropriate policies and regulations and for a thorough examination of the issues and challenges related to consumers, firms/markets and government policies and regulations (Malla et al., 2013).

Recently, bioactive compounds have been applied in various industries, including nutraceuticals, medicines, pharmaceuticals and cosmetics (Figure 10.9). With regard to the role of bioactive compounds in the nutraceuticals and functional foods industries, nutraceuticals are dietary supplements with a high concentration of bioactive compounds found in foods, but they present in a non-food matrix and can be used to enhance health in dosages that could not be obtained from normal foods. They can be obtained from food extracts in which their physiological properties have been directly or indirectly evaluated. Functional foods are foods or ingredients that contain different amounts and different types of bioactive compounds. When these products are consumed, they produce a specific health benefit beyond their basic nutritional properties. When a bioactive compound is included in a food formulation for a specific purpose, the new product can be considered a functional food. Marketing studies have shown increasing demand for health-promoting food products (Gil-Chávez et al., 2013).

Malla et al. (2013) mentioned that a source of confusion in the industry is the lack of universally accepted terminology and definitions for healthy products such as functional foods and natural health products (NHPs). Definitions of 'functional food', although they differ across countries, do tend to have a common basis, but the global terminology applicable to NHPs is more diverse. To further compound the complexity, the terminology is constantly evolving with advances in scientific knowledge and the growth of international markets. For instance, in Canada, what used to be called nutraceuticals are now classified as NHPs, including amino acids, vitamins and minerals, herbal remedies, traditional medicines and essential fatty acids. Overlapping terminology is also common in Korea, Taiwan and Russia, where the definition of functional food is equivalent to that of a nutraceutical. The need to adopt common terminology in the industry is apparent, in order to increase credibility and facilitate trade.

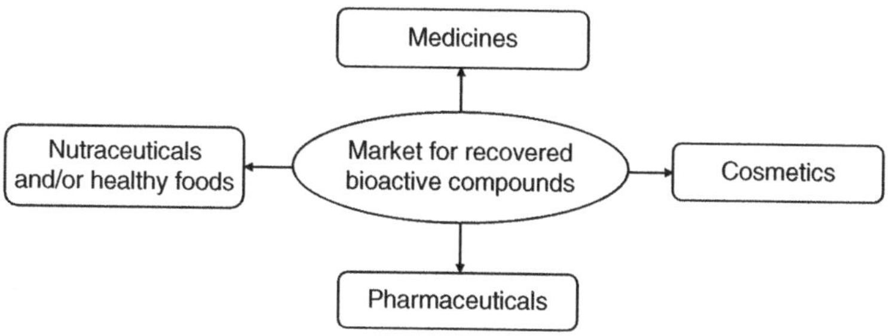

Figure 10.9 Potential markets for recovered bioactive compounds from agricultural wastes.

The global market for functional foods and NHPs is estimated to represent approximately 3% of the total food market, with the United States, European Union and Japan being the largest markets (Kotilainen et al., 2006). Malla et al. (2013) reported that the US functional food and beverage market alone had an estimated retail value of US$59 billion in 2007, approximately 8.6% of the total US food and beverage market, with an annual growth forecasted at 6.1%, while the EU market was estimated to be approximately US$8 billion in 2006, with a total 5-year forecast of 4.9% growth. Japan's health food industry was expected to grow from US$16.4 billion to about US$21.8 billion between 2007 and 2012, and the Canadian functional food market has been roughly 1% of the global market since 1997 (Table 10.16). The functional food and NHPs market has until now been dominated by developed countries, but emerging markets such as Brazil, China and India are considered potential future areas of growth. As such, it is apparent that trade is an important aspect of the health food industry as most countries are involved in the trade of both final and intermediate (ingredients) products. For example, most firms in the Canadian health food sector export their products, whereas the US market for functional foods is mostly domestically driven, while the European Union is one of the leading importers of crude medicinal plants and Japan is a global trading hub for functional foods and NHPs, importing intermediate ingredients and exporting finished products.

According to Prepared Foods (2011), the global market for food colours in 2009 was estimated at approximately $1.45 billion. By the middle of the next decade, this is expected to reach $1.6 billion, up 10% over its present levels. Between 2005 and 2009, the global market for natural colours increased by almost 35% in value, with much future growth expected to come from natural colours and colouring foodstuffs. Foods account for the majority of this market (~67%), followed by soft drinks (28%) and alcoholic beverages (5%). Europe accounts for 36% of the market, the United States 28%, Japan 10% and China 8%, with the remaining 18% coming from developed economies such as Canada and Australia and emerging markets like India and Brazil.

Phenolic compounds, particularly anthocyanins like cyanidin-3-glucoside and delphinidin-3-glucoside, are of great interest to the health food industry because of

Table 10.16 Estimated global market size for functional foods. *Source:* Adapted from Malla et al. (2013).

	1997		2000		2005	
	US$ billions	%	US$ billions	%	US$ billions	%
Canada	0.4	1	0.6	1	1.0	1.1
United States	13.6	35	17.4	34	25.7	28.7
Japan	9.6	24.7	12.8	25	22.6	25.2
European Union	12.3	31.6	16.4	32	34.4	38.4
Rest of world	3.0	7.7	4.1	8	5.9	6.6
Total	38.9	100	51.3	100	89.6	100

their antioxidant properties and great potential as natural food colourants (Melo et al., 2009; Santos et al., 2012; Vieira et al., 2013). Turmeric is an important commercial crop, grown for its aromatic rhizomes, with a global production of approximately 1.1 million tonnes per year, that is used primarily as a natural colouring agent to replace synthetic dyes in chutneys, pickles, mustard, butter and cheese, among other products (Osorio-Tobón et al., 2014). The global market for carotenoids as high value-added products reached US$1.2 billion in 2010, with an expected rise to US$1.4 billion by 2018. β-carotene is the most prominent carotenoid, with a projected rise in consumption from US$261 million in 2010 to US$334 million by 2018 (Cardenas-Toro et al., 2015).

In consideration of the market potential of bioactive compounds in the medicines and pharmaceuticals industries, Aitken & Kleinrock (2015) reported that the global volume and cost of medicines will reach 4.5 trillion doses and $1.4 trillion by 2020, and both will increase significantly from 2015. The largest countries using pharmaceuticals will be 'pharmerging' markets: those that have absolute spending growth more than $1 billion and GDP per capita of less than $30 000 at purchasing power parity from 2014 to 2018, including: Tier 1 – China; Tier 2 – Brazil, India and Russia; and Tier 3 – Algeria, Argentina, Bangladesh, Chile, Colombia, Egypt, Indonesia, Kazakhstan, Mexico, Nigeria, Pakistan, Philippines, Poland, South Africa, Saudi Arabia, Turkey and Vietnam. Two-thirds of global medicine volumes – mostly generic medicines – will be utilised in these countries, with dramatic increases due to broad-based health system expansions, while developed markets such as the United States, Japan, the top 5 European countries (Germany, France, Italy, Spain and the United Kingdom), Canada and South Korea, will continue to account for the majority of medicine spending, due to both higher prices per unit and the presence of newer medicines that bring meaningful clinical benefit to patients facing a wide range of diseases (Table 10.17). Spending will concentrate in developed markets, with more than half on original brands and noncommunicable diseases. Specialty therapies will be more significant in developed markets than in pharmerging markets, and different traditional medicines will be used in developed markets than in pharmerging ones. Spending growth will be promoted by brands in developed markets and by increased usage in pharmerging markets. Brand spending in developed markets will rise by $298 billion in the 5 years to 2020, driven by new products and price increases, primarily in the United States, but will be offset by about $90 billion in net price reductions. In particular, 943 new active substances introduced in the prior 25 years will be used in medicines by 2020, and new medicines in recent years will be weighted to specialty and biologics, leading to greater access for patients to breakthrough therapies and clusters of innovation around hepatitis C, a range of cancers, autoimmune diseases, heart disease and an array of other rare diseases (Figure 10.10). The use of novel technology will be key to the advancement of health care, especially in emerging markets where the expense of large-scale infrastructure projects would delay progress, and the outlook to 2020 includes higher levels of medicine spending and therefore higher revenues for manufacturers compared to the last five years.

In regards to the application of recovered bioactive compounds from agricultural wastes in the cosmetics industry, Research and Markets (2015) reported that the global cosmetic market (e.g. skin care, hair care, perfumes, oral cosmetics, etc.) reached

Table 10.17 Region and leading country spending on pharmaceuticals. *Source:* Adapted from Aitken & Kleinrock (2015).

US$ billions	2015	2010–15 CAGR	2020	2016–20 CAGR
Global	*1068.8*	*6.2%*	*1400–1430*	*4–7%*
Developed	*684.3*	*4.8%*	*870–900*	*3–6%*
United States	430.0	6.1%	560–590	5–8%
EU5	144.0	2.9%	170–200	1–4%
Germany	41.2	3.8%	52–62	2–5%
France	31.3	0.1%	30–38	−3 to 0%
Italy	25.1	3.1%	30–40	2–5%
United Kingdom	27.7	6.9%	28–38	3–6%
Spain	18.6	0.7%	20–28	1–4%
Japan	78.3	2.6%	79–89	0–3%
Canada	19.3	2.0%	23–33	3–6%
South Korea	12.7	2.3%	13–20	2–5%
Pharmerging	*249.2*	*11.9%*	*345–375*	*7–10%*
China	115.2	14.2%	150–180	6–9%
Tier 2	56.8	12.9%	85–95	9–12%
Brazil	28.1	13.8%	34–44	9–12%
Russia	16.6	13.0%	29–39	11–14%
India	12.1	10.9%	13–19	5–8%
Tier 3	77.2	8.2%	100–120	6–9%
Rest of world	*135.2*	*5.2%*	*150–180*	*1–4%*

US$460 billion in 2014 and is predicted to reach US$675 billion by 2020, at a growth rate of 6.4%. Among cosmetic products, skin care has the highest market share, while oral cosmetics represent the fastest-growing market during the forecasted period. The Asia Pacific is the largest cosmetic market, with a global market share of 35% in 2014, followed by Western Europe and North America.

The acceptance of bioactive compounds derived from agricultural wastes depends mainly on their biological properties and selling price, as well as socioeconomic factors such as cultural characteristics, consumption habits, income and geographic location. The applicability, transportability and storability of such products also play a role. Malla et al. (2013) indicated that a variety of socioeconomic factors have a significant effect on consumer acceptance, including age, gender, income and geographic location. Tangible product qualities such as functional properties, taste, side effects and selling price and intangible characteristics such as convenience, method of production and the credibility of information relating to health claims are also relevant. Given all these factors, a key strategy for selling recovered bioactive compounds from agricultural wastes is the assessment of market trends and evidence of consumer reactions to these products in order to satisfy consumer demands, meet the requirements of related industries and optimise profits for producers and traders.

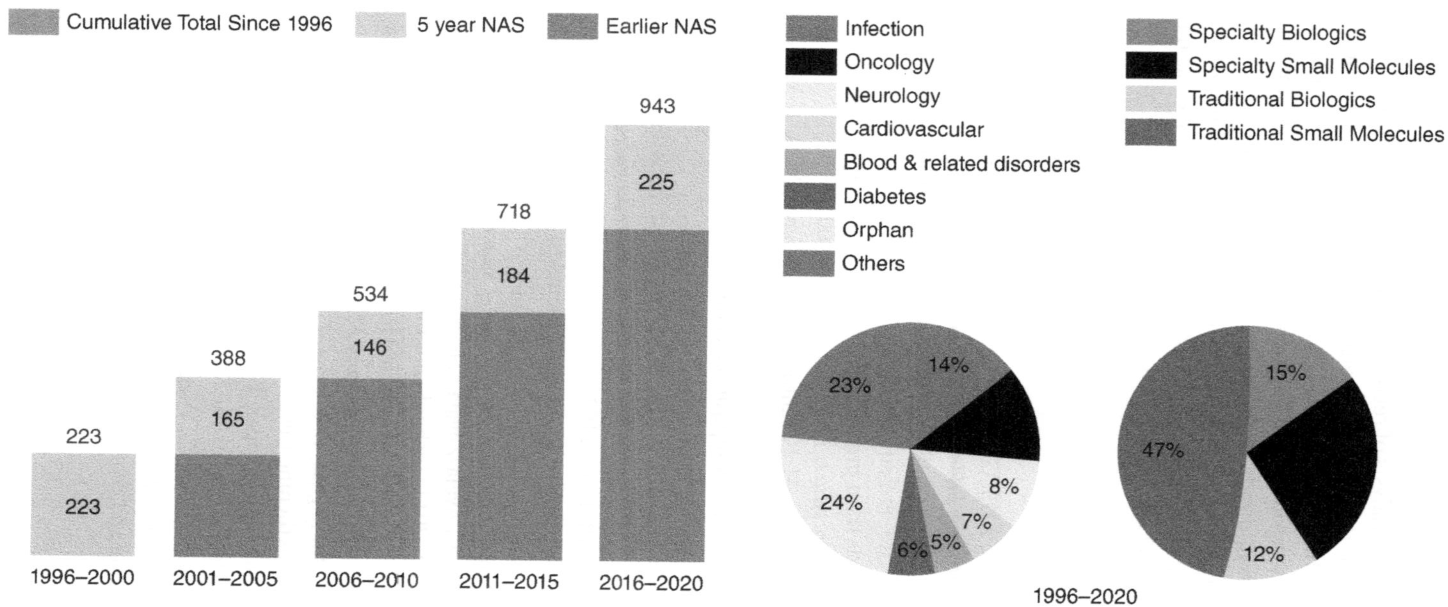

Figure 10.10 Global new active substances (NAS) available since 1996. *Source:* Adapted from Aitken & Kleinrock (2015).

10.4 Conclusion

The economic and market analysis of recovered bioactive compounds from agricultural wastes is important work that not only helps in determining the economic feasibility of new products and technologies but also guides the selection of potential markets for the consumption of these products. As such, this work is considered the final step in the recovery of bioactive compounds from agricultural wastes, with great meaning is promoting sustainable development in agricultural production and in providing valuable compounds for the long-term development of related industries such as nutraceuticals, medicines, pharmaceuticals and cosmetics.

References

Aitken, M. & Kleinrock, M. 2015. Global medicines use in 2020: outlook and implications. *IMS Institute for Healthcare Informatics*, 1–43.

Cardenas-Toro, F. P., Alcázar-Alay, S. C., Coutinho, J. P., Godoy, H. T., Forster-Carneiro, T. & Meireles, M. A. A. 2015. Pressurized liquid extraction and low-pressure solvent extraction of carotenoids from pressed palm fiber: experimental and economical evaluation. *Food and Bioproducts Processing*, **94**, 90–100.

Cavalcanti, R. N., Navarro-Díaz, H. J., Santos, D. T., Rostagno, M. A. & Meireles, M. A. A. 2012. Supercritical carbon dioxide extraction of polyphenols from pomegranate (*Punica granatum* L.) leaves: chemical composition, economic evaluation and chemometric approach. *Journal of Food Research*, **1**, 282–294.

Deenu, A., Naruenartwongsakul, S. & Kim, S. M. 2013. Optimization and economic evaluation of ultrasound extraction of lutein from *Chlorella vulgaris*. *Biotechnology and Bioprocess Engineering*, **18**, 1151–1162.

Farías-Campomanes, A. M., Rostagno, M. A. & Meireles, M. A. A. 2013. Production of polyphenol extracts from grape bagasse using supercritical fluids: yield, extract composition and economic evaluation. *Journal of Supercritical Fluids*, **77**, 70–78.

Gil-Chávez, G. J., Villa, J. A., Ayala-Zavala, J. F., Heredia, J. B., Sepulveda, D., Yahia, E. M. & González-Aguilar, G. A. 2013. Technologies for extraction and production of bioactive compounds to be used as nutraceuticals and food ingredients: an overview. *Comprehensive Reviews in Food Science and Food Safety*, **12**, 1–19.

Kotilainen, L., Rajalahti, R., Ragasa, C. & Pehu, E. 2006. Health Enhancing Foods: Opportunities for Strengthening the Sector in Developing Coutries. Agriculture and Rural Development Discussion Paper, 30. Washington, DC: International Bank for Reconstruction and Development: The World Bank.

Leitão, N. C. M. C. S., Prado, G. H. C., Veggi, P. C., Meireles, M. A. A. & Pereira, C. G. 2013. *Anacardium occidentale* L. leaves extraction via SFE: Global yields, extraction kinetics, mathematical modeling and economic evaluation. *The Journal of Supercritical Fluids*, **78**, 114–123.

Malla, S., Hobbs, J. E., Sogah, E. K. & Yeung, M. T. 2013. Assessing the functional foods and natural health products industry: a comparative overview and literature review. *Canadian Agricultural Innovation and Regulation (CAIRN) Network*, 1–151.

Melo, M. J., Pina, F. & Andary, C. 2009. Anthocyanins: nature's glamorous palette. In: Bechtold, T. & Musssak, R. (eds), *Handbook of Natural Colorants*. Chichester: John Wiley & Sons, 134–150.

Osorio-Tobón, J. F., Carvalho, P. I. N., Rostagno, M. A., Petenate, A. J. & Meireles, M. A. A. 2014. Extraction of curcuminoids from deflavored turmeric (*Curcuma longa* L.) using pressurized liquids: process integration and economic evaluation. *Journal of Supercritical Fluids*, **95**, 167–174.

Prado, J. M., Dalmolin, I., Carareto, N. D. D., Basso, R. C., Meirelles, A. J. A., Oliveira, J. V., Batista, E. A. C. & MEIRELES, M. A. A. 2012. Supercritical fluid extraction of grape seed: process scale-up, extract chemical composition and economic evaluation. *Journal of Food Engineering*, **109**, 249–257.

PREPARED FOODS. 2011. Trends in global colorings. Available from: http://www.preparedfoods.com/articles/109323-trends-in-global-colorings (last accessed 20 March 2017).

Research and Markets. 2015. Global cosmetics market – by product type, ingredient, geography, and vendors – market size, demand forecasts, industry trends and updates, supplier market shares 2014–2020. Available from: http://www.researchandmarkets.com/research/f2lvdg/global_cosmetics (last accessed 20 March 2017).

Santos, D. T., Veggi, P. C. & Meireles, M. A. A. 2010. Extraction of antioxidant compounds from jabuticaba (*Myrciaria cauliflora*) skins: yield, composition and economical evaluation. *Journal of Food Engineering*, **101**, 23–31.

Santos, D. T., Veggi, P. C. & Meireles, M. A. A. 2012. Optimization and economic evaluation of pressurized liquid extraction of phenolic compounds from jabuticaba skins. *Journal of Food Engineering*, **108**, 444–452.

Veggi, P. C., Cavalcanti, R. N. & Meireles, M. A. A. 2014. Production of phenolic-rich extracts from Brazilian plants using supercritical and subcritical fluid extraction: experimental data and economic evaluation. *Journal of Food Engineering*, **131**, 96–109.

Vieira, G. S., Cavalcanti, R. N., Meireles, M. A. A. & Hubinger, M. D. 2013. Chemical and economic evaluation of natural antioxidant extracts obtained by ultrasound-assisted and agitated bed extraction from jussara pulp (*Euterpe edulis*). *Journal of Food Engineering*, **119**, 196–204.

Index

Recovering Bioactive Compounds from Agricultural Wastes, First Edition. Edited by Van Tang Nguyen.

q

r

Printed and bound by CPI Group (UK) Ltd, Croydon, CR0 4YY
17/09/2026
14964302-0001